Essential Cosmology

Essential Cosmology

Edited by **Terry Santiago**

Callisto Reference

New York

Published by Callisto Reference,
106 Park Avenue, Suite 200,
New York, NY 10016, USA
www.callistoreference.com

Essential Cosmology
Edited by Terry Santiago

International Standard Book Number: 978-1-63239-318-0 (Hardback)

Printed in the United States of America.

Contents

Preface

In this book, extensive information regarding the field of cosmology has been provided. During the past few years, it has been witnessed that the field of cosmology has experienced a metamorphosis. From primarily being a search for three essential numbers - the cosmological constant, the expansion rate and deceleration parameter, it has now transformed into a precision science. This scientific domain is expected to unveil even the minutest details of fundamental processes that took place during initial stages of the Universe and also of the mechanisms driving the cosmic expansion and the progress of structures at the largest scales. For the achievement of these aims, we not only need formulation of novel observational and experimental strategies but also an in-depth comprehension of the underlying theoretical frameworks. Veteran researchers and scientists from across the globe have contributed significant information in this book.

Significant researches are present in this book. Intensive efforts have been employed by authors to make this book an outstanding discourse. This book contains the enlightening chapters which have been written on the basis of significant researches done by the experts.

Finally, I would also like to thank all the members involved in this book for being a team and meeting all the deadlines for the submission of their respective works. I would also like to thank my friends and family for being supportive in my efforts.

Editor

Chapter 1

Cosmological Constant and Dark Energy: Historical Insights

Emilio Elizalde

Additional information is available at the end of the chapter

1. Introduction

In this Chapter we are going to discuss about the large scale structure of the Universe. In particular, about the laws of Physics which allow us to describe and try to understand the present Universe behavior as a whole, as a global structure. These physical laws, when they are brought to their most extreme consequences---to their limits in their respective domains of applicability---are able to give us a plausible idea of how the origin of our Universe could happen to occur and also of how, expectedly, its future evolution and its end will finally take place.

The vision we have now of the so-called global or large-scale Universe (what astrophysicists term the extragalactic Universe) began to get shape during the second and third decades of the past Century. We should start by saying that, at that time, everybody thought that the Universe was reduced to just our own galaxy, the Milky Way. It is indeed true that a very large number of nebulae had been observed by then, but there was no clear proof that these objects were not within the domains of our own galaxy. Actually, the first nebulae had been already identified many centuries ago by Ptolemy who, in his celebrated work Almagest [1], reported five in AD 150. Later, Persian, Arabic and Chinese astronomers, among others, discovered some more nebulae, along several centuries of the History of Mankind. Concerning scientific publications, Edmond Halley [2] was the first to report six nebulae in the year 1715, Charles Messier [3] catalogued 103 of them in 1781 (now called Messier objects), while confessing his interest was "detecting comets, and nebulae could just be mistaken for them, thus wasting time." William Herschel and his sister Caroline published three full catalogues of nebulae, one after the other [4], between 1786 and 1802, where a total of 2510 nebulae where identified. However, in all these cases the dominant belief was that these objects were merely unresolved clusters of stars, in need of more powerful telescopes. On 26 April 1920, in the Baird auditorium of the Smithsonian Museum of Natural History a debate took place (called now by astronomers, in retrospective, the Great Debate), on the basis of two works by Harlow Shapley and Heber Curtis, later published in the Bulletin of the National Research Council. During the day, the two scientists presented independent technical results on "The Scale of the Universe" and then took part in a joint discussion in the evening. Shapley defended the Milky Way to be the entirety of the Universe and believed that objects as Andromeda and the Spiral Nebulae were just part of it. Curtis, on the contrary, affirmed that Andromeda and other nebulae were separate galaxies, or "island universes" (a term invented by the philosopher Immanuel Kant, who also believed that the spiral nebulae were extragalactic). Curtis showed that there were more novae in Andromeda than in the Milky Way and argued that it would be very unlikely within the same galaxy to have so many more novae in one small section of the galaxy than in the other sections. This led him to support Andromeda as a separate galaxy with its own signature age and rate of novae occurrences. He also mentioned Doppler redshifts found in many nebulae. Following this debate, by 1922 it had become clear that many nebulae were most probably other galaxies, far away from our own.

Figure 1. Claudius Ptolemaeus, c. AD 90 – c. AD 168

Figure 2. Edmond Halley, 1656 – 1742. Charles Messier, 1730 – 1817.

Figure 3. Sir Frederick William Herschel, 1738 – 1822. Caroline Lucretia Herschel, 1750 – 1848.

2. An expanding Universe

But it was Edwin Hubble [5] who, between 1922 and 1924, presented a definite proof that one of this nebulae, Andromeda, was at a distance of some 800.000 light years from us and, therefore, far beyond the limits of our own galaxy, the Milky Way. In this way, he definitely changed the until then predominant vision of our Universe, and opened to human knowledge the much more complex extragalactic Universe, whose theoretical study is one of the main goals of this Chapter [6].

Another very important fact, this one from the theoretical perspective, is that when Albert Einstein constructed, at the beginning of the second decade of last century and starting from very basic physical postulates–as the principles of covariance and equivalence of the laws of Physics–his theory of General Relativity (GR), scientists (himself included) where firmly convinced that our Universe was stationary. Static, in the more appropriate terminology, albeit rather counterintuitive, since this does not mean that celestial bodies do not move, but that stars and their clusters, in their wandering and distribution, would always have remained from the utmost far past, and would continue to do so into the utmost far future, as we see them at present, with no essential changes. No beginning or end of the Universe was foreseeable, nor needed or called for. But, to his extreme disappointment, Einstein realized that a Universe of this sort was not compatible with his equations, that is, the static universe is not a solution of Einstein's field equations for GR. The reason (not difficult to see) is that a universe of this kind cannot be stable: it will ultimately collapse with time owing to the attraction of the gravity force, against which there is no available protection. This led Einstein astray, until he came up with a solution. While keeping all physical principles that led him to construct his equations (there are ten of them, in scalar language, six of which are independent, but only one in tensorial representation), there was still the remaining freedom to introduce an extra term, a constant (with either sign) multiplied by the metric tensor. This is the now famous cosmological constant, but the problem was that it had no physical interpretation, of any sort. However, endowed with the right sign, it did produce a repulsive pressure to exactly counter the gravitational attraction and keep the universe solution static. Einstein was happy with this arrangement for some years (later it was proven that this solution was not stable, but this is considered nowadays to be just a technical detail that played no major role in the scientific discussion of the time).

Figure 4. Albert Einstein, 1879 – 1955. Edwin Hubble, 1889 – 1953.

The best known, by far, of the equations Einstein discovered (and probably the most famous equation ever written) is: $E = m c^2$ and corresponds to his Special Relativity theory (SR). It has a very deep physical meaning, since it establishes the equivalence between mass and energy, as two forms of one and the same physical quantity, thus susceptible to be transformed one into the other, and vice versa. The conversion factor is enormous (the velocity of light squared), meaning that a very small quantity of mass will give rise to an enormous amount of energy –as nuclear power plants prove every day (and very destructive bombs did in the past, to the shame of the Humankind). In any case, here we are not referring to this Einstein's equation (which will not be discussed any further), but to the so-called Einstein's field equations [7], actually only one in tensorial language, namely

$$R_{\mu\nu} - \frac{1}{2} g_{\mu\nu} R + g_{\mu\nu} \Lambda = \frac{8\pi G}{c^4} T_{\mu\nu}, \tag{1}$$

which he published in 1915. This is an extraordinary formula: it connects, in a very precise way, Mathematics with Physics, by establishing that the curvature, R, of space-time (a pure mathematical concept, the reference, coordinate system, so to say) is proportional to (namely, it will be affected or even determined by) the stress-energy tensor, T, which contains the whole of the mass-energy momentum (already unified by SR, as we just said) of the Universe. The proportionality factors are the universal Newton constant, G, the speed of light, c, to the fourth inverse power, and the numbers 8 and π, while Λ is the already mentioned cosmological constant, which multiplies g, the metric of space-time itself. This last term is the one that was absent in Einstein's initial formulation of GR.

Figure 5. Karl Schwarzschild, 1873 – 1916. Alexander Alexandrovich Friedmann, 1988 – 1925.

Soon Karl Schwarzschild (letter to Einstein from December 1915) found a solution to Einstein's equations (the original ones, without the cosmological constant), which corresponds to what is now know as a black hole (see below). Einstein was very surprised to see this so beautiful solution and wrote back to Schwarzschild congratulating him and admitting he had never thought that such a simple and elegant solution to his so complicated equations could exist.

$$ds^2=\left(1-\frac{2Gm}{c^2r}\right)^{-1}dr^2+r^2(d\theta^2+\sin^2\theta d\phi^2)-c^2\left(1-\frac{2Gm}{c^2r}\right)dt^2 \tag{2}$$

There is now evidence that Einstein himself had been working hard to find such solution but failed, probably because he was looking for a more general one. Schwarzschild's insight was namely to look for the simplest, with spherical symmetry. And Alexander Friedmann, in 1922, obtained another solution, which is derived by solving the now called Friedmann equations:

$$\begin{aligned}\left(\frac{\dot{a}}{a}\right)^2+\frac{kc^2}{a^2}-\frac{\Lambda c^2}{3}&=\frac{8\pi G}{3}\rho,\\ 2\frac{\ddot{a}}{a}+\left(\frac{\dot{a}}{a}\right)^2+\frac{kc^2}{a^2}-\Lambda c^2&=-\frac{8\pi G}{c^2}p.\end{aligned} \tag{3}$$

These are nowadays more commonly written in terms of the Hubble parameter, H,

$$\begin{aligned}H^2=\left(\frac{\dot{a}}{a}\right)^2&=\frac{8\pi G}{3}\rho-\frac{kc^2}{a^2},\\ \dot{H}+H^2=\frac{\ddot{a}}{a}&=-\frac{4\pi G}{3}\left(\rho+\frac{3p}{c^2}\right),\end{aligned} \tag{4}$$

and they are even much more interesting for cosmology than Schwarzschild's solution, because they correspond to the whole Universe. Friedmann's early death in 1925, at the age of 37, from typhoid fever, prevented him from realizing that, indeed, his solution would describe an expanding universe. This honor was reserved to the Belgian priest, astronomer and physicist Monsignor Georges Lemaître who, being not aware of Friedmann's important finding, went to re-discover essentially the same solution while he was working at the Massachusetts Institute of Technology on his second PhD Thesis, which he submitted in 1925. Before, Lemaître had already obtained a doctorate from Leuven university in 1920 and had been ordained a priest three years later, just before going to Cambridge University, to start working in cosmology under Arthur Eddington. In Cambridge, Massachusetts, he worked with Harlow Shapley, already quite famous (as mentioned above) for his work on nebulae.

The case is that, around the same time, Willem de Sitter had also been working on a universe solution (now called de Sitter space), which is the maximally symmetric vacuum solution of Einstein's field equations with a positive (therefore repulsive) cosmological constant

Λ, which corresponds to positive vacuum energy density and negative pressure. As a submanifold, de Sitter space is in essence the one sheeted hyperboloid

$$-x_0^2+\sum_{i=1}^{n} x_i^2=\alpha^2, \tag{5}$$

being α some positive constant which has dimensions of length. Topologically, de Sitter space is R × Sn^{-1}. A de Sitter universe has no ordinary matter content, but just a positive cosmological constant which yields the Hubble expansion rate, H, as

$$H\alpha\sqrt{\Lambda}. \tag{6}$$

Figure 6. Willem de Sitter, 1872 – 1934. Monsignor Georges Henri Joseph Édouard Lemaître, 1894 – 1966.

It is then immediate to obtain the scale factor as

$$a(t)=e^{Ht}, \tag{7}$$

where H is Hubble's constant and t is time. This was a very simple solution of Einstein's equations that undoubtedly corresponded to an expanding universe. In fact, in 1917 de Sitter had theorized, for the first time, that the Universe might be expanding. The big problem with his solution was, however, that it only could describe a universe devoid of matter, just a vacuum, and this seemed to be at that time not very useful or physically meaningful. Nowadays, on the contrary, this solution has gained extreme importance, as an asymptotic case to describe with good approximation the most probable final stages of the evolution of our Universe (if it will go on expanding forever) and also, as we shall see latter in more detail (even more in other Chapters), the initial stages, as the inflationary epoch: the fact that the de Sitter expansion is exactly exponential is very helpful in the construction of inflation ary models.

But let us continue with Lemaître. During his two-year stay in Cambridge, MA, he visited Vesto Slipher, at the Lowell Observatory in Arizona, and also Edwin Hubble, at Mount Wilson, in California, who had already accumulated at that time important evidence on the spectral displacements towards longer light wavelengths (redshift) of a large number of far distant nebulae. Actually, the most consistent earlier evidence of the redshift of distant nebulae had been gathered by Slipher who, already in 1912, had published his first results on the surprisingly large recessional velocity of the Andromeda nebula and, in 1914, at the American Astronomical Society's meeting at Evanston, Illinois, had announced radial velocities for fifteen spirals, reporting that "in the great majority of cases the nebula is receding; the largest velocities are all positive and the striking preponderance of the positive sign indicates a general fleeing from us or the Milky Way." Slipher was seeing the nebulae recede at up to 1,100 kilometers per second, the greatest celestial velocities that had ever been observed. He was so clear and convincing that chronicles say that when Slipher described his equipment and techniques along with his results, he received an unprecedented standing ovation. But the interpretation of the redshifts as true movements of the galaxies was not generally accepted then. De Sitter, for one, posited that the nebulae might only appear to be moving, the light waves themselves getting longer and longer as the light traveled towards Earth because of some interstellar processes.

Figure 7. Vesto Melvin Slipher, 1875 – 1969. Henrietta Swan Leavitt, 1868 – 1921.

When Hubble arrived at Mount Wilson, California, in 1919, the prevailing view of the cosmos was that the universe consisted entirely of the Milky Way Galaxy. Using the new Hooker telescope at Mt. Wilson, Hubble identified Cepheid variable stars in several spiral nebulae, including the Andromeda and the Triangulum nebulae. His observations, made in 1922–1923, proved conclusively that these nebulae were much too distant to be part of the Milky Way and should be considered as separate galaxies. This was the first clear evidence of the "island universe" theory. Hubble, who was then 35, found opposition to his results in the astronomy establishment and his finding was first published in the New York Times, on November 23, 1924, before being formally presented in the 1925 meeting of the American Astronomical Society. As said, most important in this discovery was the identification of the Cepheid variable stars in those nebulae, and this brings us to Henrietta S. Leavitt, who, in 1912, discovered the very important period-luminosity relation: a straight line relationship between the luminosity and the logarithm of the period of variability of these brilliant stars. Leavitt was a distinguished member of the so-called "women human computers" brought in at Harvard College by Edward C. Pickering to measure and catalog the brightness of stars in the observatory's photographic plate collection. In particular, her results came from the study, during several years, of 1,777 variable stars. Hubble did publicly recognize the importance of Leavitt's discovery for his own (saying even that she deserved the Nobel Prize). It is interesting to describe the now common explanation for the pulsation of Cepheid variables, which have been for many decades the "standard candles" for measuring distances at galaxy scales and were crucial, e.g., for the precise determination of Hubble's law. It is the so-called Eddington valve mechanism, based on the fact that doubly ionized helium is more opaque than singly ionized one. At the dimmest part of a Cepheid's cycle, the ionized gas in the outer layers of the star is more opaque. The gas is then heated by the star's radiation, temperature increases and it begins to expand. As it expands, it cools, and becomes single ionized and thus more transparent, allowing the radiation to escape. Thus the expansion stops, and gas falls back to the star due to gravitational attraction, and the process starts again.

In 1929 Hubble derived his important velocity-distance relationship for nebulae using, as he later wrote to Slipher, "your velocities and my distances." Hubble acknowledged Slipher's seminal contribution to his own work by declaring that "the first steps in a new field are the most difficult and the most significant. Once the barrier is forced, further development is relatively simple." Before that, however, we should go back again to Lemaître, who had visited in 1924-25 both Slipher and Hubble to learn about their results first hand. He also attended the meeting of the American Astronomical Society in Washington DC, in 1925, where Hubble announced his discovery that certain spiral nebulae, previously thought to be gaseous clouds within the Milky Way, were actually separate galaxies. Lemaître realized that the new galaxies could be used to test certain predictions of the general relativity equations and, soon after the meeting, he started to work on his own cosmological model. He realized the uniformity of the recession speed of the galaxies (yet nebulae), in different directions, and the fact that the redshift seemed to be proportional to the known distances to them, and concluded that the recession speed of these celestial objects could be better understood not as proper displacements of the galaxies, but much more naturally as a stretching of space itself, a true expansion of the fabric of our Universe! And this was not as crazy as it could seem at first sight, since his solution to Einstein's equations (recall, the same as Friedmann's) could be actually interpreted as corresponding to an expanding Universe. Theory and observations incredibly matched!

3. The Big Bang

Lemaître was still half-way to these conclusions when he submitted his PhD thesis at MIT in 1925, but he completed his work two years later and published it in an obscure Belgian journal in 1927. The retreat of distant nebulae, he wrote in his paper, is "a cosmical effect of the expansion of the universe." He even estimated a rate of expansion close to the figure that Hubble eventually calculated and published two years later. And at a scientific meeting in Brussels, in 1927, the young priest cornered Einstein and tried to persuade him. We do not know his exact words, but presumably they must have been something like: "Sehen Sie, Herr Einstein, your static model for the Universe, with a cosmological constant, does not stand, since it is not stable in the far past. But, on the other hand, you do not need a cosmological constant, your original equations are all right! In fact, I have found a solution to these equations which can be interpreted as describing an expanding Universe. And, on the other hand, the redshifts of distant galaxies, as found by astronomers, as Slipher and Hubble, most naturally account for an expansion of the space containing the galaxies, and not for arbitrary displacements of the galaxies themselves, since exactly the same pattern is seen in any direction!" But, as quoted later by Lemaître, Einstein's reply was utterly disappointing to him. He answered: "Monsieur Lemaître, I can find no mistake in your calculations, but your physical insight is abominable." Einstein, the great genius, the master of space and time, was not ready to imagine a universe in which this space-time was stretching! It took him more than two years to accept this. And now, when we are teaching the expanding universe to high-school student, or even to popular audiences, we pretend they should get this concept on the spot! Lemaître's paper was finally noticed by Eddington and with his help it was reprinted in 1931 in the Monthly Notices of the Royal Astronomical Society; it explained clearly, using Lemaître's (Friedmann's) solution, why Hubble saw the velocities of the galaxies steadily increase with distance. The same year, in the much prestigious journal Nature, Lemaître suggested that all the mass-energy of the universe was once packed within a "unique quantum," which he later called the "primeval atom." This was the logical conclusion of his looking back in time in the Universe evolution: an immediate consequence of his model was that long time ago the Universe was much smaller and that, going even more backwards, that it had had an origin. In 1933 he resumed his theory of the expanding Universe and published a more detailed version in the Annals of the Scientific Society of Brussels, finally achieving his greatest popularity (his name is now, however, rather forgotten by the younger generations of astronomers and physicists). From Lemaître's scenario arose the current vision of the Big Bang (albeit not this name, as we will soon see), a model that has shaped there since the thought of cosmologists as strongly as the idea of crystalline spheres, popularized by Ptolemy (who was already mentioned at the beginning), influenced natural philosophers through the Middle Ages. It took Einstein over two years to understand that Lemaître's model was right and, then, he abhorred of the cosmological constant by pronouncing his very famous sentence: "Away with the cosmological constant. This was the biggest blunder in my life" (in German: "Weg mit der kosmologischen Konstante. Dass war die grösste Eselei meines Lebens"). He clearly realized that, had he truly believed in his field equations, he could have predicted that the Universe was actually expanding (and not static), much before anybody else. (Quite in the same way, say, as Dirac predicted the existence of the positive electron, the positron, because it was a second solution of his quantum equation for the electron, impossible to get rid of by natural arguments.)

Figure 8. Albert Einstein and Georges Lemaître.

It is easy to understand that the Church, which had been so disappointed with the findings of Galileo several centuries ago and had condemned him for his defense of a sun-centered universe, was extremely happy with Lemaître's scenario. He was lauded and raised to the rank of monsignor and was made a fellow, and later president, of the Pontifical Academy of Sciences. But Lemaître always recoiled from any suggestion that his primeval atom had been inspired by the biblical story of Genesis. He insisted, throughout his life, that his theory about the origin of space and time sprang solely from the equations before him. However, the name Bib Bang, after which this theory is known today, was actually an occurrence of a rival scientist, Fred Hoyle, who in a BBC radio program, broadcasted on March 28th, 1949, pronounced these magic two words for the first time. Just the year before, Hoyle, Thomas Gold and Hermann Bondi had issued a theory, that was to became quite famous under the name of the Steady State theory, which involved a creation field (called the C-field), which created matter and energy constantly in wider regions of the Universe in a rather smooth manner. These researchers had realized the impossibility of the whole matter-energy of our Universe having been all packed once within a unique quantum or primeval atom. This could have no sense and a creation process needed to be involved. They were very clever to solve the question how to create matter-energy from 'nothing', constantly and at 'zero-cost,' since they realized that any positive amount of ordinary matter and energy would be compensated by the same amount of negative energy which corresponds to the associated gravitational potential (which in GR does also have negative energy content!). This observation was extremely important, since it anticipated the physical principles involved in inflationary theories; indeed, it has been widely recognized that the steady state theory anticipated inflation. Actually, the possibility that the negative energy of gravity could supply the positive energy for the matter of the universe was suggested by Richard Tolman already in 1932, although a viable mechanism for the energy transfer was not indicated. In any case, just translating this physics to Lemaître's scenario would mean that an unbelievably enormous amount of matter and energy should be created instantly, at the very moment of the origin of our Universe. After these considerations the reader should be prepared to understand the words that Fred Hoyle uttered on that occasion. In the BBC program Hoyle tried to push up his theory, as being much more reasonable, in contraposition to Lemaître's one. At a point, he refuted, in a very disrespectful manner, that: "The whole of the matter in the universe was created in one Big Bang in a particular time in the remote past." Hoyle could never imagine that these two words, pronounced with the purpose to absolutely discredit the rival theory, would serve from that moment on to identify what is nowadays the most accepted theory of the Universe, a name that any school child knows. This was clearly not Hoyle's intention. Before going on, a last word on Lemaître's primeval atom scenario. It may seem incredible that this wrong, physically unsustainable idea (again, the whole energy of the universe could never in the past have been concentrated in a nutshell) can be still found nowadays in popular books on cosmology that are being issued by scientific writers having no idea about the physical principles underlying inflation, quantum gravity, or even the more primitive steady state theory. The creation of matter and energy from a void, de Sitter state is key to inflationary models and, as already said, the steady state theory gave a first clear hint to how this could be done while respecting all basic physical principles including energy conservation.

In the year 1963, Arno Allan Penzias and Robert Woodrow Wilson started a project, at the Bell Labs in New Jersey, on the recalibration of a 20-foot horn-reflector, that had been previously employed for a number of years for satellite work, and which they wanted to prepare for use in radioastronomy. Even if, at that time, there were at several other places much more powerful radio-telescopes available, this seven meter, very modest horn reflector had some special features they wanted to exploit for high-precision measurements in the 21 cm band, a wavelength at which the galactic halo would be bright enough in order to be detected, and at which the line corresponding to neutral hydrogen atoms could be observed. They wanted, in particular, to detect the presence of hydrogen in clusters of galaxies (this development is very nicely described, and in much more detail, in the Nobel Lecture by Wilson [8]). After having carried out a number of measurements during several months, Penzias and Wilson did not manage to get rid of a very light but persistent noise, which translated into temperature was an excess of some 2 to 4 K, and which it was exactly the same in all directions, day and night. Indeed, the antenna temperature should have been only the sum of the atmospheric contribution, so-called temperature of the sky (due to microwave absorption by the terrestrial atmosphere), of 2.3 K, and the radiation from the walls of the antenna and ground, of 1 K. Unless Penzias and Wilson could understand what they first called "antenna problem" their 21 cm galactic halo experiment would not be feasible. So they went through a number of possible reasons for the temperature excess and tested for them. They considered the possibility of some terrestrial source and pointed their antenna towards different directions, in particular to New York City, but the variation was always insignificant. They also took into account the possible influence of the radiation from our Galaxy, but they checked this could not contribute decisively, either. They also ruled out discrete extraterrestrial radio sources as the source of the excess radiation as they had a spectrum similar to that of the Galaxy. For some time they lived with the antenna temperature problem and concentrated on measurements in which it was not critical. One day they discovered that a pair of pigeons was roosting up in the horn and had covered part of it with (in their own words) 'what all city dwellers know well.' They cleaned the mess, and later, in the spring of 1965, they thoroughly cleaned out the horn-reflector and put aluminum tape over the riveted joints, but only a small reduction in antenna temperature was obtained. In this way, a whole year passed.

Figure 9. Arno Allan Penzias (born April 26, 1933). Robert Woodrow Wilson (born January 10, 1936).

Figure 10. Big Bang detection: Penzias and Wilson's 20-foot horn-reflector.

At the same time, in Princeton, only 60 km away, R.H. Dicke, P.J.E. Peebles and D.T.

Figure 11. Robert H. Dicke, 1916 – 1997. Philip J. E. Peebles, born April 25, 1935. David T. Wilkinson, 1935 – 2002.

Wilkinson where working on a paper where they tried to guess the characteristics that a microwave radiation that would come from a very dense universe, in its origin (possibly pulsating), should have; that is to say, under the conditions that they thought could correspond to those of the Big Bang. The sequence of events which led to the unraveling of the mystery began one day when Penzias was talking to Bernard Burke of MIT about other matters and mentioned the unexplained noise. Burke recalled hearing about the work of the theoretical group in Princeton on radiation in the universe. In the preprint, Peebles, following Dicke's suggestion calculated that the universe should be filled with a relic blackbody radiation at a minimum temperature of 10 K. Shortly after sending the preprint, Dicke and his coworkers visited Penzias and Wilson and were quickly convinced of the accuracy of their measurements. They agreed to a side-by-side publication of two letters in the Astrophysical Journal - a letter on the theory from Princeton [9] and one on the measured excess temperature from Bell Laboratories [10]. Penzias and Wilson were careful to exclude any discussion of the cosmological theory of the origin of background radiation from their letter, because they had not been involved in any of that work and thought, that their measurement was independent of the theory and might outlive it. After the meeting, an experimental group was set up in Princeton to complete their own measurement with the expectation that the background temperature would be about 3 K. There was the great expectation that what Penzias and Wilson had detected could be in fact the Big Bang itself! However, the final confirmation of this extraordinarily important cosmological discovery took several years yet.

And the first additional evidence did not actually come from the experimental group at Princeton, but from a totally different, indirect measurement. Indeed, it came out from rescuing from oblivion a measurement that had been made thirty years earlier by W.S. Adams and T. Dunhan Jr., who had discovered several faint optical interstellar absorption lines which were later identified with the molecules CH, CH+, and CN. In the case of CN, in addition to the ground state, absorption was seen from the first rotationally excited state. This was reanalyzed in 1965-66, and it was realized that the CN is in equilibrium with the background radiation, since there is no other significant source of excitation where these molecules are located. In December 1965, P.G. Roll and D.T. Wilkinson [11] completed their measurement of 3.0 ± 0.5 K at 3.2 cm, the first confirming microwave measurement, which was followed shortly by T.F. Howell and J.R. Shakeshaft's value of 2.8 ± 0.6 K at 20.7 cm [12] and then by Penzias and Wilson's one of 3.2 K ± 1 K at 21.1 cm [13]. By mid 1966 the intensity of the microwave background radiation had been shown to be close to 3 K between 21 cm and 2.6 mm, almost two orders of magnitude in wavelength. This was already very close to the present, highly accurate value of 2,725 K.

In the same way that the first experimental evidence for the cosmic microwave background radiation was obtained (but unrecognized) long before 1965, it soon was realized that the theoretical prediction had been made, at least sixteen years before Penzias and Wilson's detection, by George Gamow (a former student of Friedmann) in 1948, and improved by R.A. Alpher and R.C. Herman, in 1949 [14]. Those authors are now recognized as the first who theoretically predicted the cosmic radiation associated to the Big Bang, for which they calculated a value of 5 K, approximately (a very nice figure that they later spoiled, bringing it to 28 K). We will finish this section with the well known fact that Arno Penzias and Robert Wilson were laureated with the 1978 Nobel Prize in Physics by their very important discovery, which can be considered as one the milestone findings in Human History. The Universe had indeed an origin, the fabric of space was stretching and, as clearly understood by Lemaître, Friedmann's solution to Einstein's equations was a unique, real description of our Universe. The stationary universe, also under its more modern form of the steady state theory, was dead.

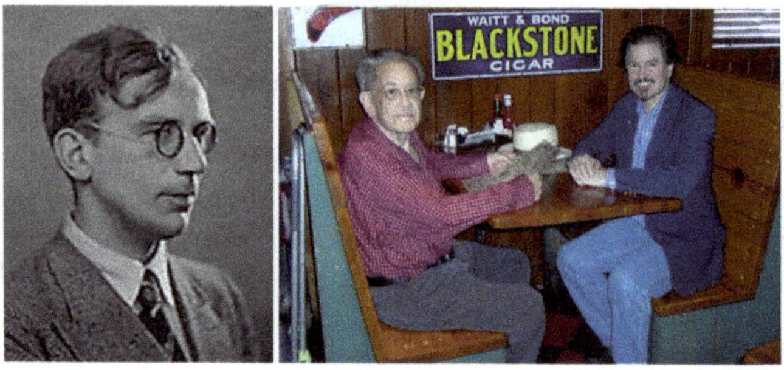

Figure 12. George Gamow, 1904 – 1968. Ralph Asher Alpher, 1921 – 2007 (with Victor S. Alpher, in Tampa, Florida).

4. The Big Bang modified: Inflation

However, the original Big Bang theory had to be modified, what occurred at the beginning of the eighties, in order to solve several very serious discrepancies it had accumulated when comparing it with the most accurate astronomical observations of the cosmos, specifically, concerning what happened during the very first second in the history of the Universe. It was realized that the expansion during this first second could by no means be an ordinary one, understanding by this the one that has taken place later in its evolution, say, kind of a linear one. A very special stage had to be devised to account for what occurred in this initial instant of time (well, in fact one second is a very, very long time at this scale). This stage is generically called inflation, and its formulation is mainly due to Allan Guth, Katsuhiko Sato, Andrei Linde, Andreas Albrecht, Paul Steinhardt, Alexei Starobinsky, Slava Mukhanov, G.V. Chibisov, and a large list of other scientist (the number and classes of models are actually still growing, nowadays). The name inflation comes from the fact that the Universe expansion had to be enormous, incredibly big during an extremely small instant of time (of the order of 10^{-33} seconds). In this infinitesimal fraction of a second the Universe expanded from the size of a peanut to that of the present Milky Way (in volume, an increase of at least 75 orders of magnitude). Actually, in the inflationary theory the Universe begins incredibly small, some 10^{-24} cm, a hundred billion times smaller than a proton. And, at the same time, during inflation it cools down abruptly (supercooling) by 5 orders of magnitude, from some 10^{27} K to 10^{22} K. This relatively low temperature is maintained during the inflationary phase. When inflation ends the temperature returns to the pre-inflationary temperature; this is called reheating or thermalization because the large potential energy of the inflaton field decays into particles and electromagnetic radiation, which fills the universe, starting in this way the radiation dominated phase of the Universe. Because the very nature of inflation is not known, this process is still poorly understood. As explained before, energy conservation is consistent with physics during the whole process: this lies in the subtle behavior of gravity, already present in Newtonian physics, where we know that the energy of the gravitational potential is always negative, a fact which is maintained in GR. The development and shaping of the concept of inflation constitutes, for different reasons, another brilliant page in the history leading to our present knowledge of the cosmos.

The first to come up to this very revolutionary idea was Allan Guth, born in 1947 and who studied (both graduation and PhD) at MIT, from 1964 to 1971. During the following nine years he was, successively, a PostDoc at Princeton, Columbia, Cornell and Stanford (SLAC), all of them top class Universities. But Guth did not manage to jump over this level and get a real contract. In 1978 he was at Cornell while his career was up in the air and he badly needed to find a permanent job to support his wife and son. Someday, a fellow PostDoc called Henry Tye (now a professor at Cornell) proposed him to study jointly the problem of monopole production in the very early Universe. Guth got interested in this subject so that when Robert Dicke (whom we have already mentioned before) came to give a seminar, he attended it with much interest. Guth was very intrigued by Dicke's conclusion that the traditional Big Bang theory had severe problems and that it was leaving out something important. There was the problem of flatness (also called Dicke's coincidence): the fact that the matter density of the Universe was so close to the critical mass corresponding to a flat (Euclidean) Universe. Also the horizon problem, namely the fact that the Universe is so perfectly homogeneous and isotropic at large enough scales, which is in absolutely good agreement with the cosmological principle. And to these problems Guth and Tye added, as a results of their specific study, the problem of absence of magnetic monopoles, which should actually be very abundant in the present Universe, but it is the case that (with the only exception of Blas Cabrera, who reported finding one in 1982) nobody has ever seen any of them. One should note, however, that John Preskill, at Harvard at that time, had published a result in the same direction before Guth and Tye. Anyway, all these problems and the sudden interest of Guth on cosmology kept him busy for two years. His personal description of how, in a sleepless night, he suddenly thought of a mechanism in order to solve these severe problems, all of them at once, is better than any science fiction story. Looking back at the situation, now we can say that it was very risky on his side, being just a PostDoc fellow without a tenured job, to propose such a revolutionary mechanism as the inflationary model, what he did in 1980. Guth first made public his ideas in a seminar at SLAC on January 23, 1980. In August, he submitted his paper, entitled "The Inflationary Universe: A Possible Solution to the Horizon and Flatness Problems," to the Physical Review, and was published in January 1981 [15]. Soon after, he captured the interest of several universities and got several offers which he rejected, until he had the possibility to come back to MIT, as an associate visiting professor in 1980. His scientific career has been growing since then. Not without some problems at the beginning, however: it was discovered that his initial model had an important flaw, which was corrected by Andrei Linde (now at Stanford) and, independently, by Paul Steinhardt (Princeton) and Andreas Albrecht (Davies). The modified theory was given the name of "new inflation." The works of Katsuhiko Sato, who about the same time as Guth proposed a much related theory and of Alexei Starobinsky, who argued at about the same time that quantum corrections to gravity would replace the initial singularity of the universe with an exponentially expanding de Sitter phase, must be mentioned, as well. In particular the last one is getting more and more popular recently, in a unifying context to be explained later. The Spanish researcher Jaume Garriga, at Barcelona University, has published influential papers in this area, too.

Figure 13. Allan Guth Andrei Linde Fritz Zwicky, 1898 – 1974

Nowadays, under the name of inflation there are over fifty different theories which have evolved from Guth's original idea. Borrowing of energy from the gravitational field is the basic principle of the inflationary paradigm, completely different from the classical Big Bang theory, where all matter-energy in the universe was assumed to be there from the beginning (as explained above). In Guth's words: "Inflation provides a mechanism by which the entire universe can develop from just a few ounces of primordial matter." As a final consequence of all these developments, the so called standard cosmological model, or FLRW (Friedmann-Lemaître-Robertson-Walker) model emerged. The two last names appear here because, between 1935 and 1937, the mathematicians Howard P. Robertson and Arthur G. Walker rigorously proved that the FLRW metric is the only one possible, on a spacetime that is spatially homogeneous and isotropic. In other words, they showed that the solution to Einstein's equations found by Friedmann and later by Lemaître was unique in describing the Universe we live in. Let us pause to ponder, for a second, the extraordinary beauty of this cosmological model as a description of the Universe: to the uniqueness of Einstein's field equation (the only freedom being the cosmological constant) we add up the fact that the solution is also single. We have arrived to just one possible mathematical description of our Universe, and the inflation paradigm opens a possible way to understand how it could be created, without violating the basic conservation principles of Physics. This last point will however require further elaboration.

5. Dark matter

Before that, however, we need go back in time and explain about another very important problem in cosmology which appeared for the first time, in a compelling, clear way, in 1933 when the Swiss astrophysicist Fritz Zwicky, at CALTECH, unveiled it from his detailed observations of the most exterior galaxies of the Coma cluster. It should be mentioned, however, that two years before Zwicky, Einstein and de Sitter had already published a paper where they considered a most probable theoretical existence of enormous amounts of matter in the Universe which did not emit light. It had also been postulated by Jan Oort, one year before Zwicky, to account for the orbital velocities of some stars in the Milky Way. But Zwicky's calculations, based on the use of the virial theorem, where much more convincing. According to them, the gravity of the visible galaxies in the Coma cluster was too small in order to possibly account for the large speeds of the more exterior galaxies. A big amount of mass was missing! This was called the missing mass problem and Zwicky referred to this unseen matter as dunkle Materie (dark matter). Since those years, more and more different observations indicated the presence of dark matter in the universe, such as the anomalous rotational speeds of galaxies, gravitational lensing by galaxy clusters, such as the Bullet Cluster, the temperature distribution of hot gas in galaxies and clusters, and other.

Very famous astronomers in this context now are Vera Rubin and Kent Ford for their seminal papers published around 1975. It so happened that during some forty years after Zwicky's discovery no other corroborating observations appeared and the problem was almost forgotten. But in the early 70s, Vera Rubin, a young astronomer at the Carnegie Institution of Washington, presented findings based on a new, very sensitive spectrograph that could measure the velocity curve of edge-on spiral galaxies to a great degree of accuracy. In 1975, in a meeting of the American Astronomical Society, Rubin and Ford announced their important discovery that most stars in spiral galaxies orbit at roughly the same speed, implying that their mass densities were uniform well beyond the locations of most of the visible stars in the galaxy. In 1980 they published a paper [16] which has had enormous influence in modern cosmology, where they summarized the results of over a decade of work on this subject. Their results have shaken the very grounds of Newton's universal law of gravity since they undoubtedly indicate either that Newton's results are not applicable to the Universe at large distances (the error obtained is certainly enormous) or that a very important part of the mass of spiral galaxies must be located in the galactic halo region, which is extremely dark in relation with the central part.

At the beginning and for some time these results met very strong skepticism by the community of astronomers. But Rubin, a brave and stubborn scientist, never changed her conviction that her results were correct. They have been subsequently checked to enormous precision and there is now no more doubt that an important problem to be explained is facing us. The most accepted conclusion is the existence of dark matter, that is, ordinary matter made up of particles that we cannot see for some reason. There are many candidates for dark matter but, while this is the most generally accepted conclusion, there still remains open the other mentioned possibility, namely that Newton's laws need to be modified at large distances (modified gravities, MOG, MOND, and other theories). Actually, Rubin herself is a convinced supporter of this second possibility. The debate continues and it is very lively nowadays.

To finish with this point let us summarize that, talking in terms of dark matter, for what we now know it must constitute an enormous amount of ordinary (that is, gravitating) matter, ten times as abundant as visible galaxies. And we infer its existence not just by the clear gravitational effects we have mentioned, as the observed anomalies in the rotation curves of spiral galaxies (just described), and which account for the rotational speeds of the exterior stars of the galaxy as a function of the distance of the start to the galactic center, but also from the rotation of the so-called satellite galaxies of our Milky Way (and of Andromeda, too), some of which can already be measured with enough accuracy as they turn around our own galaxy (resp. Andromeda), in a way very similar to how planets describe orbits around the Sun. The extraordinary regularities found in the trajectories of such satellite galaxies constitute a really thrilling, very active research field at present. A different way to trace the presence of dark matter is through gravitational lensing (both macro and micro lensing). Its effects are very apparent there, as a notorious amplification of the power of gravitational lenses, compared with the case that the effect would be just due to the visible stellar objects. In clusters as, for instance, Abell 1689, the observed, very strong effects cannot by any means be explained as being produced by its visible mass only. And in the case of the Bullet cluster one clearly detects an enormous mass acting as a gravitational lens and which is completely separated from the barionic, visible mass which emits X rays.

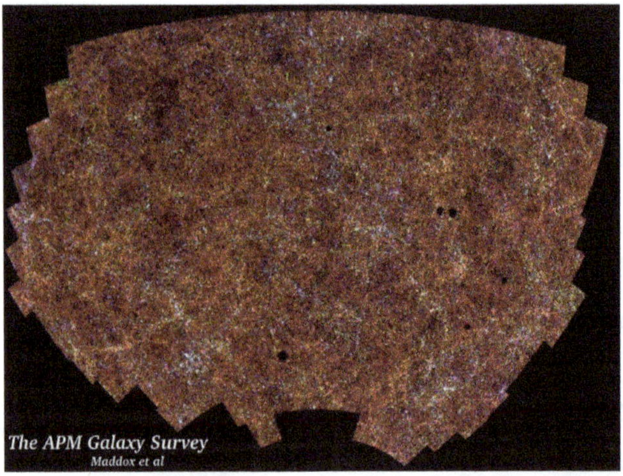

Figure 14. Two Million Galaxies: S. Maddox (Nottingham U.) et al., APM Survey, Astrophys. Dept. Oxford U.

We certainly do not know yet what dark matter is made of, neither why we cannot see it. But we do know that the discovered neutrino mass (neutrinos being indeed invisible!) is not enough to account for it; and also that adding up the masses of big, Jupiter like planets (so called MACHOs, which are also very difficult to see) is again not enough in order to explain the missing amount of mass. But astroparticle physicists got indeed a good number of other possible candidates, as axions, neutralinos and other (they come from the breakdown of certain fundamental symmetries in particle and quantum field theories). What we know is that they must be elusive particles, very weekly coupled with any of the known physical fields since, on the contrary, its presence would have been detected already. It is for this reason that the generic name WIMPs (weakly interacting massive particles) has been proposed to generically name any particle of this sort as a dark matter candidate.

6. The Universe in depth

Another very important landmark in the knowledge of the cosmos at large scale was the publication, in 1986 of the first map of the Universe in three dimensions. In fact, it was only a very thin slice of an angular sector of the same but it was extremely important and completely changed the vision astronomers and other scientists had of it. Up to then, the only representations of the cosmos were in form of two dimensional projections on the celestial sphere, as still is (and serves as a very good example) the APM Galaxy Survey, which contains two million galaxies. Even if, in comparison, the Harvard CfA strip of Valérie de Lapparent, Margaret Geller and John Huchra [17], contained only a total of 1,100 galaxies, what was most important was that for 584 of them their distance from us could be determined (through the observation of their cosmological redshift). And this allowed, for the first time in History, to see a part of our Universe in the elusive third dimension: the distance from us. Actually the plot looks again two-dimensional, since the slice is represented as flat but, again, the spatial structures created by the disposition of the galaxies and clusters, away from us, had never been seen before.

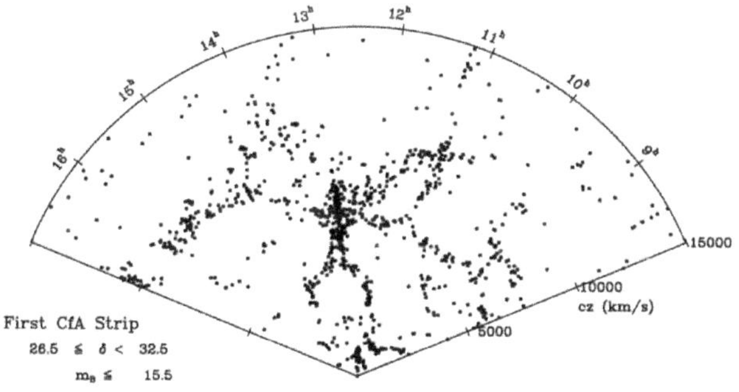

Figure 15. The first slice of the CfA Survey, by Valerie de Lapparent, Margaret Geller and John Huchra, published in 1986.

The impact of this work was spectacular, also due in part to the shapes of these point structures, showing that the distribution of galaxies in space was anything but random, with galaxies actually appearing to be distributed on surfaces, almost bubble like, surrounding large empty regions, or "voids." Anyone could easily identify what looked like a human being (the man), another shape looked like a thumb imprint (God's thumb) pointing towards us, and so on. But the most intriguing fact, for scientists, was the presence in the whole picture of such very large regions devoid of any galaxy (voids), while they concentrated on the verge of these voids, and forming filaments and large walls (as the so-called Great Wall)

Many astronomers, but also a good number of prestigious theoretical physicists and even mathematicians who had never before dealt with cosmological issues started to work on this point distribution, trying to find some fundamental model that could possibly generate such peculiar pattern in the Universe evolution. Astronomers, on their side, tried to find new observational confirmation of this large-scale behavior of galaxies and clusters. Collaborations of pure theoreticians and astronomers flourished, as was the case of Edward Witten with Jeremiah Ostriker. That same year, in Spain, in the historical Peñíscola Castle, we had a five-day workshop of GIFT (Interuniversity Group of Theoretical Physics) where Ricky Kolb and Mike Turner were invited to present such recent and astonishing developments. This author was there and felt immediately captivated by such map. Coming back from the workshop he handed a problem to Enrique Gaztañaga (who was, by the way, in search of a subject for his PhD Thesis): to provide an effective mathematical characterization of the point distribution, more simple than the usual higher-order point correlation statistics, and to try to generate such point distribution from a phenomenological model by taking into account, essentially, the gravitational attraction. This was the origin of our large-scale cosmology group in our Institute ICE-CSIC and IEEC, in Barcelona. When more and more precise surveys, of millions of galaxies with redshifts, as the 2d Field, where carried out, all these spectacular forms have smoothly disappeared:

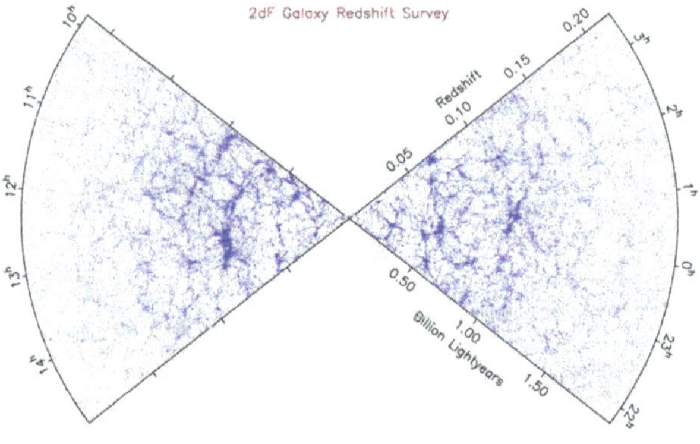

Figure 16. The 2dF Galaxy Redshift Survey (2dFGRS), by M.M. Colless, et al, 2001.

almost all were generated by errors in the computation of distances, due to the fact that the redshift produced by the Universe expansion gets mixed with the redshift coming from the proper movements of the galaxy with respect to other celestial bodies in its neighborhood and from the movement of the observer, which are sometimes not easy to disentangle. Some of the big structures remain, however, as is the case of the Great Wall, and of the voids surrounded by galaxies on their surfaces. Moreover, on top of slices we have now true 3-dimensional representations of the observed data, together with computer simulations depicting a very rich and marvelous web structure. However, the problem to obtain this large scale pattern starting from a fundamental theory remains, to large extent, open.

Summarizing a lot, cosmologists know now that our Universe is not static nor in a steady state. Quite on the contrary, it had a very spectacular origin some 13,730 million years ago, what we know with an error of less than 1%, according to the most recent (7th year) data from the WMAP (Wilkinson Microwave Anisotropy Probe) satellite, and to the first data coming from the PLANCK mission, and from different terrestrial observatories. All astronomical tests that have been carried out until now have confirmed, without the slightest doubt and with increasing accuracy, the new Big Bang theory, that is, the one which includes inflation (although, concerning this last, a too-large number of different, competing models still remains). But this is by no means the last word.

7. The expansion accelerates

Until the end of last Century, cosmologists were convinced that the expansion of the fabric of the Universe, originated in the Big Bang, was uniform. Up to then the main challenge of cosmology at large scale was to determine if the mass-energy density, ϱ, of our cosmos was large enough (above critical) so that it would be able to completely stop this expansion at some point in the future—an instant after which the Universe would begin to contract, to finally finish in a so-called Big Crunch—or if, quite on the contrary, this energy density ϱ was smaller, subcritical, and thus unable to stop the Universe expansion completely, ever in the future. In this case, expansion would continue forever, even if, of course, there was no doubt that the action of gravity would certainly decelerate the expansion rate, this was crystal clear. The most precise observations carried out until then indicated that the actual value of ϱ was indeed very close to the critical value, ϱ_c, being in fact quite difficult to determine if it was, in fact, above or below such value.

This situation radically changed just before the end of the Century, because of two different analyses of very precise observations carried out —with the big Hubble Space Telescope— on type Ia supernovae by two teams, each comprising some thirty scientists. The two groups wanted to measure with high precision the deceleration, caused by gravity attraction, on the expansion rate of the Universe, by calibrating the variation in this expansion rate with distance. To their enormous surprise, the values obtained by both teams were completely unexpected, and matched with each other. The first to issue results, in 1998, was the High-z Supernova Search Team an Australian-American project, led by Brian Schmidt and Adam Riess, while the other group, with the name Supernova Cosmology Project and led by Saul Perlmutter at Lawrence Berkeley National Laboratory, published independent results the year after, 1999. The author cannot help mentioning that one of the members in this last collaboration is the Spanish astronomer Pilar Ruiz Lapuente, from Barcelona University. The common and very clear conclusion of the two observations was that the expansion of the Universe is nowadays accelerating and not decelerating, and that it has been accelerating for a long period of time in the past. This was one of these moments in History where you have something in front of your eyes that you really do not believe. You cannot explain it with the scientific tools at your hand. The impact of this discovery on our knowledge of the Universe was extraordinary and the three researchers who led the teams have been awarded the 2011 Nobel Prize in Physics. The first conclusion seems quite clear: in order that this acceleration can occur a force must be present, as we already know since Galileo, XVI C, and Newton, XVII C, but in this case the force must be acting constantly at the level of the whole cosmos! The question is now, what kind of force can have this property in order to produce the desired acceleration?

Figure 17. Saul Perlmutter Brian Schmidt Adam Riess

Thinking for a while, it is not difficult to explain the problem even to a non-specialist. An expanding Universe, as in the case of the Bing Bang theory, does not need any force to expand forever, just an initial impulse, for a short interval of time, as when we throw a stone in the air. In this case, owing to the enormous mass of the Earth we are sure the stone will stop flying and come back; but if the Earth was the size of a mountain this same stone would never return. As already explained, at cosmological level everything just depends on the mass density of the whole Universe being larger or smaller than the critical value, ϱ_c, which marks the difference between the situation when the Universe would continue expanding forever and the one in which it would stop expanding, to start contracting back. But now, in order that the stone can accelerate, a force must act on it all the time, as with an accelerating car. As in the case of dark matter, nobody knows yet what produces this acceleration of the Universe expansion, and this missing energy is generically called dark energy. In fact a (too large) number of possible explanations have arisen, which can be roughly classified into three types.

The first one is the most natural and immediate, but in no way the simplest to match. Coming back where we started, with Einstein's equation, the only possibility to provide a repulsive force there is by introducing again the cosmological constant, Λ, with the appropriate sign. There is no other freedom but, fortunately, we still have this one! Regretfully, however, as with Einstein's, there is a big question mark behind it, namely, what is the physical nature of Λ? Where does it come from? This brings us to explain about another crucial revolution which took place in Physics during the first thirty years of the past Century: Quantum Mechanics. This is probably the most radical change in our conception of the world that has ever happened. In spite of Richard Feynman saying that "nobody can understand QM," the fact that it works to enormous precision for the description of nature is witnessed by the unchallenged 14 to 15 digit matching in the results of some particle physics experiments. Already Wolfgang Pauli in the 20's, and then Yakov Zel'dovich in the 60's, among others, clearly realized that if the fluctuations of the quantum vacuum—which are always there owing to W. Heisenberg's uncertainty principle and have a magnitude of the order of Planck's constant— are taken into account in Einstein's equation (as a valid form of energy satisfying the equivalence principle), then their contribution at cosmological scale (which happens to go together with Λ) would be enormously big. In principle, infinitely so, albeit we know that through a regularization and renormalization process the number is rendered finite. But even then it is still enormous: some 60 to 120 orders of magnitude larger than needed in order to explain the observed Universe acceleration. This is the famous cosmological constant problem, which was around since the first attempts to reconcile General Relativity and Quantum Physics appeared (although, at first, the problem was just to explain why vacuum fluctuations yielded a zero contribution, not a very small one, as now). Some very important physicists, as the Nobel Prize laureate Steven Weinberg [18], have been working for years on this problem, without real success. The reader must be adverted that, in these discussions, the concepts of cosmological constant and of quantum vacuum fluctuations are taken as one and the same thing, the reason being that there is no other possible contribution to Λ which is known up to now.

Another possible explanation is that there might exist some peculiar energy fluid filling the Universe (of course not of ordinary nature, as in the case of dark matter). There are many different models, with fancy names, for this fluid: quintessence, k-essence, Chaplygin gas, Galileons, and many more. The third possible explanation is the most radical of all, from the theoretical viewpoint and as seen from the whole description of the History of the Universe as summarized in the present article: maybe something is in error when trying to apply Einstein's General Relativity to cosmological scales, so that this marvelous theory may need be modified at these scales (as also Newton's equations might have to be modified too, in order to account for the missing dark matter). The reader will find full details of these thrilling issues in the other chapters of the Book. Let me here just note that modifications of Einstein's equation usually proceed by way of introducing additional terms with higher order powers on the Ricci curvature, R, a general function f(R), and/or higher order derivatives. In fact some of these terms are difficult to avoid when one considers quantum corrections to Einstein's equation, as Alexey Starobinsky and collaborators did already at the beginning of the eighties, finding in this way a model that would, to start, produce inflation and which can be modified to possibly account for its present acceleration, too. Scientists do not know yet the right answer, nor if the Universe acceleration is actually constant. To this end, the derivative of the acceleration should be obtained, what is still impossible with the quality of present data.

We should add that, for some time, the interpretation of the Ia Supernovae results as implying a Universal acceleration were controverted, some possible explanations involving a non-

Copernican view of the position of our local galaxy group in the cosmos were published (we could be in one of these enormous voids surrounded by very massive structures), and even very recently some alternative interpretation has appeared. However, Type Ia Supernovae are very good standard candles for the redshift range where the observations were carried out, since they have a very strong and consistent brightness along considerable cosmological distances; moreover, since 1990 several other independent proofs have been added to check the results. Among them, the impact of acceleration on the fluctuations of the cosmic microwave background, where measures have been carried out on the imprint of the acceleration on the gravitational potential wells which contribute to the integrated Sachs-Wolfe effect (and translate into colder and hotter spots in the CMB map). Also, the effect of acceleration on the gravitational lenses, and the one that it has on the large scale structures of the Universe, on the basis of the phenomenon known as acoustic baryon oscillations (BAO). All these observations are absolutely independent from each other and this contributes to the fact that there remains little doubt today that the Universe expansion accelerates. The Dark Energy Survey project (DES) is being set to provide new measurements, integrating all these different techniques, with participation of a group of our Institute ICE-CSIC and IEEC, led by Enrique Gaztañaga.

Figure 18. Jaume Garriga, UB Enrique Gaztañaga, CSIC Sergei D. Odintsov, ICREA Emilio Elizalde, CSIC

As already mentioned, a promising possibility to explain the acceleration consists in modifying Einstein's equation, that is GR itself, at least at large scales, entering the so-called f(R) or scalar-tensor theories, in their different variants. In our group of the Institute for Space Science (ICE-CSIC) and of the Catalan Institute of Space Research (IEEC), led by Sergei D. Odintsov and the author of this Chapter, we are presently working on this kind of models with a long list of international collaborators. As clearly stated at the beginning, all present day cosmology is based on Einstein's equation, thus, in making this step we are entering a new age in our knowledge of the cosmos. Yet to be seen is if it will finally be a successful one. As advanced, there are different ways to depart from GR, one of the most popular is by extending the Hilbert-Einstein action by the addition of a function, f(R), in principle arbitrary, of the Ricci curvature, R. A theory of this kind was first proposed by Hagen Kleinert and Hans-Jürgen Schmidt, and independently by Salvatore Capozziello, in 2002. Already from the beginning this theory was related with quintessence, in which a scalar field with time evolution is incorporated to GR. The discussion about f(R) theories being in fact equivalent to scalar-tensor ones is still open today. At the classical level they are most probably equivalent, but at the quantum level the answer seems to be clearly negative. The recent and extremely important discovery of the Higgs field will surely give a spectacular thrust to this kind of models. In fact, in a paper of 2004 by Elizalde, Odintsov and Shin'ichi Nojiri (now at Nagoya University, Japan)[19] there was an independent proposal of the so-called quintom dark energy: one phantom plus one quintessence scalar which could have a relation with the discovered Higgs.

And with this we have reached the very final stage of our general description of our knowledge of the cosmos at large scales. There are still no observations to confirm or disprove these last theories. A lot more about them and all the most recent developments is to be found in the other chapters of this Book. Some very promising results seem to indicate that, within f(R) theories, there is the possibility to build, with blocks of a really fundamental theory, as string or M theory, a fully-fledged model which could describe all the stages of the evolution of the cosmos, from the Big Bang through inflation, reheating and recombination, to the present accelerated expansion and on towards the end of the Universe in a de Sitter asymptotic phase, which is the most plausible one (although some compelling models with future singularities, as the Big Rip, or either pulsating universes, cannot be excluded with present data). Adding up our knowledge of the Universe, we must shamefully confess that over 95% of it is, as of today, 'terra ignota.' But this is actually good for Science, since it means that, in front of us, there is a lot to be discovered, hopefully soon!

8. The origin of the Universe

Even more uncertain is the explanation of the creation of the Universe, of the very instant when it came to being. We are more or less acquainted with the corresponding passage of the Bible. Looking now at the descriptions of scientists, Stephen Hawking and Roger Penrose did important work on the subject, which has been influential for several decades, with the conclusion (obtained again under very general and natural conditions) that such instant is (or it was until recently) a mathematical singularity and, therefore, beyond reach of any kind of physical interpretation. This result was quite disappointing but, fortunately, it just affects classical theories and does not take into account quantum corrections which generically soften the singularities, or even make them completely disappear. Making the story short, there are new models (Alex Vilenkin and also Andrei Linde have been working on them since over twenty years ago) in which one can sidestep the singularity problem: by combining inflation with quantum fluctuations of the vacuum state of a primordial system in which a spark or miniscule particle---a "twist in matter and space-time" so-called "Hawking-Turok instanton"---would be able, at zero-energy cost (as explained already), to ignite inflation which, on its turn, would amplify the negligibly small quantum fluctuations (of Planck's constant magnitude) of the vacuum, giving rise, in this way, to the cosmic fluctuations (of order 10^{-5}) which we clearly observe on the CMB plot below. This is the most ancient map of the Universe that we have been able to capture until now. It corresponds to when it was some 370,000 years old. Just before that, the Universe was like a very dense and hot soup of quarks, gluons and elementary particles. It was absolutely dark, light being unable to travel in it, since photons, even if continuously created, where destroyed immediately, through recombination with the neighboring particles at such high densities. But the Universe was expanding and the temperature went down until it reached a value below the ionization threshold of the lightest of all atoms: that of hydrogen. All of a sudden, hydrogen precipitated at cosmic scale and, in this way, for the very first time in History, the very first light of the first cosmic dawn started to fill out the entire Universe. And this light is still reaching us from the most remote corners of the cosmos, and we can see it in all its brightness with the very curious eyes of our satellites as COBE, WMAP and PLANCK, which have transformed it into images, each time more and more clear, of the most ancient map of the Universe we now have. Putting all pieces together the so-called standard cosmological model, or ΛCDM (Cold Dark Matter with a cosmological constant, Λ) could be constructed and remains unchallenged till now.

Figure 19. Stephen Hawking Roger Penrose Alex Vilenkin

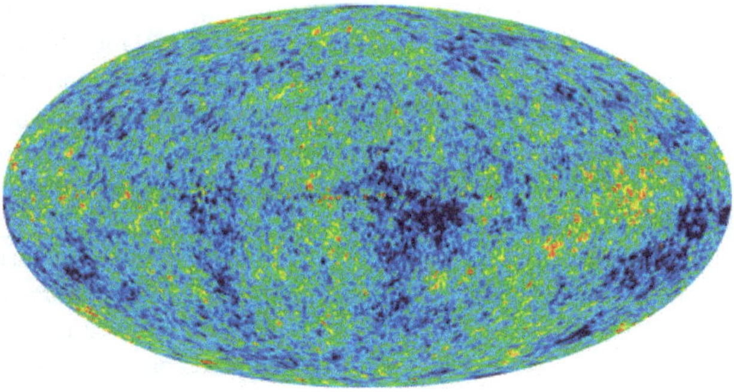

Figure 20. CMB Seven Year Microwave Sky, NASA/WMAP Science Team.

In order to proceed further into the observation of the origin --- eventually until the very origin of time --- we will need much better eyes. To start, those capable of processing the information hidden in the primordial gravitational waves, what we expect to be able to do in one to two decades from now (projects LISA, BBO, DECIGO, etc.). In that way we will obtain pictures of a much younger Universe and inflation could be eventually confirmed. But what is a real challenge for present day Physics, at least without involving any form of the anthropic principle (which in its strong version states that the properties of the Universe, the universal constants must be such that they need allow intelligent life to exist, that is, our presence as observers), is to develop a model for the origin and evolution of a single Universe like ours. The most advanced, and only feasible, theories will always produce a multiverse, that is, an uncountable collection of universes, of all possible kinds of sizes and properties, one of which, by mere chance, would be the one we happen to live in. But, until no observational proof of the existence of a multiverse is obtained, these theories will yet stay beyond the frontiers of Physics, and rather in the domain of science fiction. It must be acknowledged that these theories have been built up by very competent scientists and that they do not contravene any of the basic laws of nature. But in order to enter its realm, as in the case of the other theories discussed in this Chapter, compelling observational evidence must first be found.

As last word, among theorists there is still the much extended idea that the ultimate answer will be found, sooner or later, within string (or M) theory, the so-called "theory of everything." But a too common mistake at different moments in the History of Science has been the strong belief that one already had on its hands the final theory, that all what was left to do was just polish it a bit, fill up some small holes, and carry out more precise calculations. Errors in the past have been flagrant and were committed by some of the most brilliant scientists of each generation. The author of this Chapter defends the idea that a new theory will emerge, sooner or later, which will be very different from the ones we now have at disposal, and which will radically change our vision of the world, as much as General Relativity and Quantum Mechanics did one hundred years ago.

Author details

Emilio Elizalde*

Address all correspondence to: elizalde@ieec.uab.es

Consejo Superior de Investigaciones Científicas, ICE/CSIC and IEEC Campus UAB, Facultat de Ciències, Spain

References

[1] Ptolemy, Claudius. (1515). *University of Vienna: Almagest*, http://www.univie.ac.at/hwastro/rare/1515_ptolemae.htm.

[2] Halley, Edmund. (1714-16). An account of several nebulae or lucid spots like clouds, lately discovered among the fixt stars by help of the telescope. *Philosophical Transactions*, 39, 390-392.

[3] Messier, Charles. (1781). Catalogue des Nébuleuses & des amas d'Étoiles. *Connoissance des Temps*, 1784, 227-267, [Bibcode: 1781CdT.1784.227M].

[4] Herschel, William, & Herschel, Caroline. (1786). Catalogue of One Thousand New Nebulae and Clusters of Stars. *Philosophical transactions*, Vol.Royal Society GB.

[5] Hubble, Edwin P. (1937). *The Observational Approach to Cosmology (Oxford).*

[6] Bartusiak, Marcia. ,(2010). The Day We Found the Universe. (Random House Digital)

[7] Einstein, Albert. (1915). Die Feldgleichungen der Gravitation. *Sitzungsberichte der Preussischen Akademie der Wissenschaften zu Berlin*, 844-847, http://nausikaa2.mpiwg-berlin.mpg.de/cgi-bin/toc/toc.x.cgi?dir=6E3MAXK4&step=thumb.

[8] Wilson, Robert W. (1978). The Cosmic Microwave Background Radiation. *Nobel Lecture*, http://nobelprize.org/nobel_prizes/physics/laureates/1978/wilson-lecture.pdf.

[9] Dicke, R. H., Peebles, P. J. E., Roll, P. G., & Wilkinson, D. T. (1965). *Ap. J.*, 142, 4-14.

[10] Penzias, A. A., & Wilson, R. W. (1965). *Ap. J.*, 142, 420.

[11] Roll, P. G., & Wilkinson, D. T. (1966). *Physical Review Letters*, 16, 405.

[12] Howell, T. F., & Shakeshaft, J. R. (1966). *Nature*, 210, 138.

[13] Penzias, A. A., & Wilson, R. W. (1967). *Astron. J.*, 72, 315.

[14] Gamow, G. (1948). The Origin of Elements and the Separation of Galaxies. The evolution of the universe. On the Relative Abundance of the Elements. One, Two, Three...Infinity. *Physical Review*, 74, 505, *Nature*, 162, 680, AlpherR.A.HermanR., 1948, *Physical Review*, 74, 1577, GamowG, 1974, *Viking Press*, revised 1961; Dover P.

[15] Guth, Alan H. (1981). Inflationary universe: A possible solution to the horizon and flatness problems. *Phys. Rev. D*, 23, 347.

[16] Rubin, V., Thonnard, N., & Ford, W. K. , Jr. (1980). Rotational Properties of 21 Sc Galaxies with a Large Range of Luminosities and Radii from NGC 4605 (R=4kpc) to UGC 2885 (R=122kpc). *Astrophysical Journal*, 238, 471.

[17] de Lapparent, V., Geller, M. J., & Huchra, J. P. (1986). A slice of the universe. *Astrophysical Journal, Letters to the Editor*, 302, L1-L5, March 1.

[18] Weinberg, Steven. (1989). The cosmological constant problem. Dreams of a Final Theory: the search for the fundamental laws of nature. The First Three Minutes. *Rev. Mod. Phys.*, 61, 1-23, *Vintage Press*, 1993, *Basic Books, 1977; updated 1993.*

[19] Elizalde, E., Nojiri, S., & Odintsov, S. D. (2004). Late-time cosmology in (phantom) scalar-tensor theory: dark energy and the cosmic speed-up. *Phys. Rev.*, D70, 04359.

Chapter 2

Extragalactic Compact Sources in the *Planck* Sky and Their Cosmological Implications

Luigi Toffolatti, Carlo Burigana,
Francisco Argüeso and José M. Diego

Additional information is available at the end of the chapter

1. Introduction

As of mid August 2012, the *Planck* cosmic microwave background anisotropy probe[1] [1,2] – launched into space on 14 May 2009 at 13:12:02 UTC, by an Ariane 5 ECA launcher, from the Guiana Space Centre, Kourou, French Guiana – is still successfully operating. The spacecraft accumulated data with its two instruments, the High Frequency Instrument (HFI) [3], based on bolometers working between 100 and 857 GHz, and the Low Frequency Instrument (LFI) [4], based on radiometers working between 30 and 70 GHz, up to the consumption of the cryogenic liquids on January 2012, achieving $\simeq$ 29.5 months of integration, corresponding to about five complete sky surveys. A further 12 months extension is on-going for observations with LFI only, cooled down with the cryogenic system provided by HFI. Moreover, *Planck* is sensitive to linear polarization up to 353 GHz.

Thanks to its great sensitivity and resolution on the whole sky and to its wide frequency coverage that allows a substantial progress in foreground modeling and removal, *Planck* will open a new era in our understanding of the Universe and of its astrophysical structures (see [5] for a full description of the *Planck* Scientific programme). *Planck* will improve the accuracy of current measures of a wide set of cosmological parameters by a factor from $\sim$ 3 to $\sim$ 10 and will characterize the geometry of the Universe with unprecedented accuracy. *Planck* will shed light on many of the open issues in the connection between the early stages of the Universe and the evolution of the cosmic structures, from the characterization of primordial conditions and perturbations, to the late phases of cosmological reionization.

---

* This paper is based largely on the *Planck* Early Release Compact Source Catalogue, a product of ESA and the *Planck* Collaboration. Any material presented in this review that is not already described in *Planck* Collaboration papers represents the views of the authors and not necessarily those of the *Planck* Collaboration.

[1] *Planck* (http://www.esa.int/Planck) is a project of the European Space Agency - ESA - with instruments provided by two scientific Consortia funded by ESA member states (in particular the lead countries: France and Italy) with contributions from NASA (USA), and telescope reflectors provided in a collaboration between ESA and a scientific Consortium led and funded by Denmark.

The *Planck* perspectives on some crucial selected topics linking cosmology to fundamental physics (the neutrino masses and effective number of species, the primordial helium abundance, various physics fundamental constants, the parity property of CMB maps and its connection with CPT symmetry with emphasis to the Cosmic Birefringence, the detection of the stochastic field of gravitational waves) will also show how *Planck* represents an extremely powerful *fundamental and particle physics laboratory*. Some of these analyses will be carried out mainly through a precise measure of CMB anisotropy angular power spectrum (APS) in temperature, polarization and in their correlations, whereas others, in particular those related to the geometry of the Universe and to the research of non-Gaussianity signatures, are based on the exploitation of the anisotropy pattern. The most ambitious goal is the possible detection of the so-called B-mode APS.

The first scientific results[2], the so-called *Planck* Early Papers [3] have been released in January 2011 and published by Astronomy and Astrophysics (EDP sciences), in the dedicated Volume 536 (December 2011). A further set of astrophysical results has been presented on the occasion of the Conference *Astrophysics from radio to sub-millimeter wavelengths: the Planck view and other experiments*[4] held in Bologna on 13-17 February 2012. Several articles have been already submitted in 2012 and others are in preparation, constituting the set of so-called *Planck* Intermediate Papers.

The outline of this Chapter is as follows: in Section 2 we briefly sketch the main characteristics and the capabilities of the ESA *Planck* mission; in Section 3 we discuss the most recent detection methods for compact source detection; in Section 4 the SZ effect, detected by *Planck* in many cluster of galaxies and its importance for cosmological studies are analyzed; Section 5 is dedicated to summarize current results obtained by *Planck* data on the properties of EPS; finally, Section 6, discusses the very important results up to now achieved by the analysis of CIB anisotropies detected by *Planck*.

2. The ESA *Planck* mission: Overview

CMB experimental data are affected by uncertainties due to instrumental noise (crucial at high multipoles, ℓ, i.e. small angular scales), cosmic and sampling variance (crucial at low ℓ, i.e. large angular scales) and from systematic effects. The uncertainty on the angular power spectrum is given by the combination of three components, cosmic and sampling variance, and instrumental noise, and it is approximately given by [9]:

$$\frac{\delta C_\ell}{C_\ell} = \sqrt{\frac{2}{f_{sky}(2\ell+1)}}\left(1+\frac{A\sigma^2}{NC_\ell W_\ell}\right). \tag{1}$$

Here f_{sky} is the sky coverage, A is the surveyed area, σ is the instrumental rms noise per pixel, N is the total pixel number, W_ℓ is the beam window function that, in the case of

[2] http://www.sciops.esa.int/index.php?project=PLANCK&page=Planck_Published_Papers

[3] The *Planck* Early papers describe the instrument performance in flight including thermal behaviour (papers I–IV), the LFI and HFI data analysis pipelines (papers V–VI), and the main astrophysical results (papers VII-XXVI). These papers have complemented by a subsequent work, published in 2012, based on a combination of high energy and *Planck* observations (see [8]).

[4] http://www.iasfbo.inaf.it/events/planck-2012/

LFI			
Frequency (GHz)	30	44	70
FWHM	33.34	26.81	13.03
N of R (or feeds)	4 (2)	6 (3)	12 (6)
EB	6	8.8	14
NET	159	197	158
δT/T [μK/K] (in T)	2.04	3.14	5.17
δT/T [μK/K] (in P)	2.88	4.44	7.32

HFI		
Frequency (GHz)	100	143
FWHM in T (P)	9.6 (9.6)	7.1 (6.9)
N of B in T (P)	– (8)	4 (8)
EB in T (P)	33 (33)	43 (46)
NET in T (P)	100 (100)	62 (82)
δT/T [μK/K] (in T)	2.04	1.56
δT/T [μK/K] (in P)	3.31	2.83

HFI		
Frequency (GHz)	217	353
FWHM in T (P)	4.6 (4.6)	4.7 (4.6)
N of B in T (P)	4 (8)	4 (8)
EB in T (P)	72 (63)	99 (102)
NET in T (P)	91 (132)	277 (404)
δT/T [μK/K] in T (P)	3.31 (6.24)	13.7 (26.2)

HFI		
Frequency (GHz)	545	857
FWHM in T	4.7	4.3
N of B in T	4	4
EB in T	169	257
NET in T	2000	91000
δT/T [μK/K] in T	103	4134

Table 1. *Planck* performance. The average sensitivity, $\delta T/T$, per (FWHM)^2 resolution element (FWHM: Full Width at Half Maximum of the beam response function, is indicated in arcmin) is given in CMB temperature units (i.e., equivalent thermodynamic temperature) for 29.5 (plus 12 for LFI) months of integration. The white noise (per frequency channel for LFI and per detector for HFI) in 1 sec of integration (NET, in $\mu K \cdot \sqrt{s}$) is also given in CMB temperature units. The other acronyms here used are: N of R (or B) = number of radiometers (or bolometers), EB = effective bandwidth (in GHz). Adapted from [6, 7] and consistent with [3, 4]. Note that at 100 GHz all bolometers are polarized and the equivalent temperature value is obtained by combining polarization measurements.

a Gaussian symmetric beam, is $W_\ell = \exp(-\ell(\ell+1)\sigma_B^2)$, with $\sigma_B = FWHM/\sqrt{8 \ln 2}$ the beamwidth which defines the angular resolution of the experiment. For $f_{sky} = 1$ the first term in parenthesis defines the "cosmic variance", an intrinsic limit on the accuracy at which the APS of a certain cosmological model defined by a suitable set of parameters can be derived with CMB anisotropy measurements[5]. It typically dominates the uncertainty on the APS at low ℓ because of the small, $2\ell+1$, number of modes m for each ℓ. The second term in parenthesis characterizes the instrumental noise, that never vanishes in the case of real experiments. Note also the coupling between experiment sensitivity and resolution, the former defining the low ℓ experimental uncertainty, namely for W_ℓ close to unit, the latter determining the exponential loss in sensitivity at angular scales comparable with the beamwidth. We computed an overall sensitivity value, weighted over the channels, defined by $1/\sigma_j^2 = \sum_i 1/\sigma_{j,i}^2$, where $j = T$ and i indicates the sensitivity of each frequency channel, listed in Table 1. FWHM values of 13 and 33 arcmin are used in Fig.1 to define the overall combination of *Planck* sensitivity and resolution, i.e. the computation of the effective beam window function[6], relevant for the sensitivity at high ℓ. Finally, to improve the signal to noise ratio in the APS sensitivity, especially at high multipoles, a multipole binning is usually applied. Of course, the real sensitivity of the whole mission will have to also include the potential residuals of systematic effects. The *Planck* mission has been designed to suppress

[5] Note that the cosmic and sampling variance (74% sky coverage excluding the sky regions mostly affected by Galactic emission) implies a dependence of the overall sensitivity on r at low multipoles, relevant to the parameter estimation; instrumental noise only determines the capability of detecting the B mode.

[6] In fact, it is possible to smooth maps acquired at higher frequencies with smaller beamwidths to the lowest resolution corresponding to a given experiment. We adopt here FWHM values of 33 and 13 arcmin, which correspond to the lowest resolution of all the *Planck* instruments (i.e., 30 GHz channel) and to the lowest resolution of the so called cosmological channels (i.e., 70 GHz channel), respectively (see Table 1).

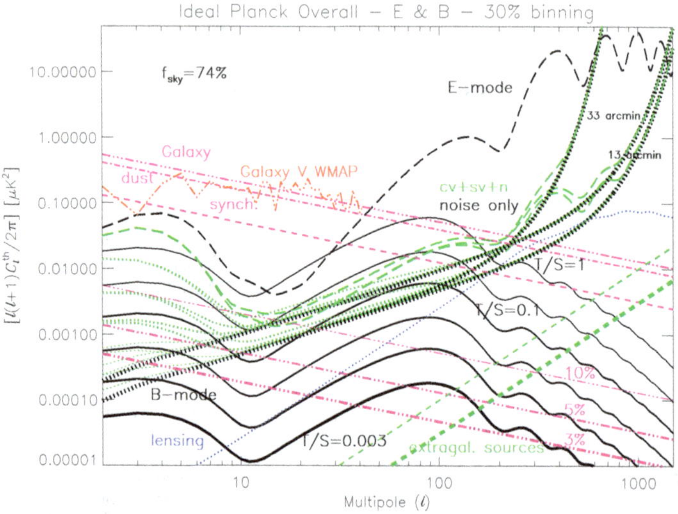

Figure 1. CMB E polarization modes (black long dashes) compatible with *WMAP* data and CMB B polarization modes (black solid lines) for different tensor-to-scalar ratios $T/S = r$ of primordial perturbations are compared to the *Planck* overall sensitivity to the APS assuming two different FWHM angular resolutions (33 and 13 arcmin) and the overall sensitivity corresponding to the whole mission duration (and also to two surveys only: upper curve in black thick dots, labeled 13 arcmin). The expected noise is assumed to be properly subtracted. The plots include cosmic and sampling variance plus instrumental noise (green dots for B modes, green long dashes for E modes, labeled with cv+sv+n; black thick dots, noise only) assuming a multipole binning of 30%.The B mode induced by lensing (blue dots) is also shown. Galactic synchrotron (purple dashes) and dust (purple dot-dashes) polarized emissions produce the overall Galactic foreground (purple three dots-dashes). *WMAP* 3-yr power-law fits for uncorrelated dust and synchrotron have been used. For comparison, *WMAP* 3-yr results (http://lambda.gsfc.nasa.gov/) derived from the foreground maps using HEALPix tools (http://healpix.jpl.nasa.gov/) [12] are shown (red three dots-dashes broken line): power-law fits provide (generous) upper limits to the power at low multipoles. Residual contamination levels by Galactic foregrounds (purple three dot-dashes) are shown for 10%, 5%, and 3% of the map level, at increasing thickness. We plot also as thick and thin green dashes realistic estimates of the residual contribution of un-subtracted extragalactic sources, $C_\ell^{\mathrm{res,PS}}$ and the corresponding uncertainty, $\delta C_\ell^{\mathrm{res,PS}}$.

potential systematic effects down to $\sim \mu K$ level or below. Fig.1 compares CMB polarization modes with the ideal sensitivity of *Planck* (including also a 15% level of HFI data loss because of cosmic rays; see [10]) and the signals coming from astrophysical foregrounds as discussed below.

CMB anisotropy maps are contaminated by a significant level of foreground emission of both Galactic and extragalactic origin. For polarization, the most critical Galactic foregrounds are certainly synchrotron and thermal dust emission, whereas free-free emission gives a negligible contribution. Other components, like spinning dust and "haze", are still poorly known, particularly in polarization. Synchrotron emission is the dominant Galactic foreground signal at low frequencies, up to ~60 GHz, where dust emission starts to dominate. External galaxies are critical only at high ℓ, and extragalactic radio sources are likely the most crucial in polarization up to frequencies ~200 GHz, the most suitable for CMB anisotropy experiments. We parameterize a potential residual from non perfect cleaning of CMB maps from Galactic foregrounds simply assuming that a certain fraction

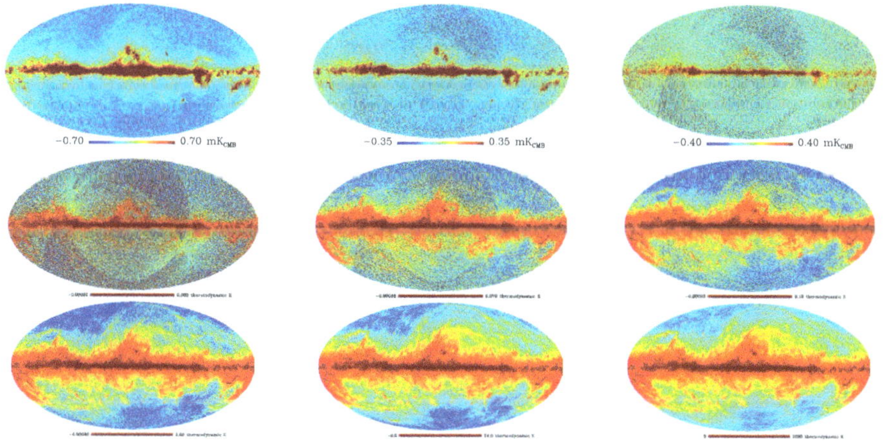

Figure 2. CMB removed *Planck* full-sky maps. From left to right and from top to bottom: 30, 44, 70; 100, 143, 217; 353, 545, and 857 GHz, respectively. Credits: Zacchei, et al., A&A, Vol. 536, A5, 2011; *Planck* HFI Core Team, A&A, Vol. 536, A6, 2011b, reproduced with permission © ESO.

of the foreground signal *at map level* (or, equivalently, its square at power spectrum level) contaminates CMB maps. Of course, one can easily rescale the following results to any fraction of residual foreground contamination. The frequency of 70 GHz, i.e. the *Planck* channel where Galactic foregrounds are expected to be at their minimum level, at least at angular scales above $\sim$ one degree, is adopted as reference.

For what concerns CMB temperature fluctuations produced by undetected EPS [13], we adopt the recent (conservative) estimate of their Poisson contribution to the (polarized) APS [14] at 100 GHz[7] by assuming a detection threshold of $\simeq$ 0.1 Jy. We also assume a potential residual coming from an uncertainty in the subtraction of this contribution computed by assuming a relative uncertainty of $\simeq$ 10% in the knowledge of their degree of polarization and in the determination of the source detection threshold, implying a reduction to $\simeq$ 30% of the original level. Except at very high multipoles, their residual is likely significantly below that coming from Galactic foregrounds.

The first publications of the main cosmological (i.e. properly based on *Planck* CMB maps) implications are expected in early 2013, together with the delivery of a first set of *Planck* maps and cosmological products coming from the first 15 months of data. They will be mainly based on temperature data. Waiting for these products, a first multifrequency view of the *Planck* astrophysical sky has been presented in the Early Papers: Fig. 2 reports the first LFI and HFI frequency (CMB subtracted) maps. These maps are the basis of the construction of the *Planck* Early Release Compact Source Catalog (ERCSC) (see [15] and *The Explanatory Supplement to the Planck Early Release Compact Source Catalogue*), the first *Planck* product delivered to the scientific community.

[7] We adopt here a frequency slightly larger than that considered for Galactic foregrounds (70 GHz) because at small angular scales, where point sources are more critical, the minimum of foreground contamination is likely shifted to higher frequencies.

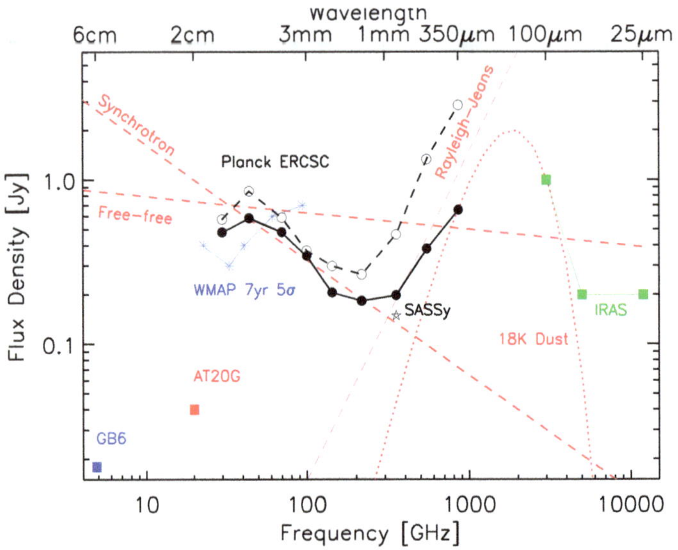

Figure 3. The *Planck* ERCSC flux density limit quantified as the faintest ERCSC sources at $|b| < 10°$ (dashed black line) and at $|b| > 30°$ (solid black line) is shown relative to other wide area surveys. See Fig. 5 of [15] for more details. Credit: *Planck* Collaboration, A&A, vol. 536, A7, 2011, reproduced with permission © ESO.

Fig. 3 compares the sensitivity of *Planck* ERCSC with those of other surveys from radio to far-infrared wavelengths. Of course, by accumulating sky surveys and refining data analysis, the *Planck* sensitivity to point sources will significantly improve in the coming years. The forthcoming *Planck* Legacy Catalog of Compact Sources (PCCS), to be released in early 2013 and to be updated in subsequent years, will represent one of the major *Planck* products relevant for multi-frequency studies of compact or point–like sources.

2.1. Extragalactic point sources vs. non-Gaussianity

The cosmological evolution of extragalactic sources and its implications for the CMB and the CIB will be discussed in the following Sections, 5 and 6. On the other hand, statistical analyses of the extragalactic source distribution in the sky can be applied to test cosmological models. In this context, the possibility of probing the Gaussianity of primordial perturbations appears particularly promising. Primordial perturbations at the origin of the large scale structure (LSS) may leave their imprint in the form of small deviations from a Gaussian distribution, [16, 17] that can appear in different kinds of configurations, such as the so-called local type, equilateral, enfolded, orthogonal. For example, the local type of deviation from Gaussianity is parameterized by a constant dimensionless parameter f_{NL} (see, e.g., [18–20]) $\Phi = \phi + f_{NL}(\phi^2 - \langle\phi^2\rangle)$, where Φ denotes Bardeen's gauge-invariant potential (evaluated deep in the matter era in the CMB convention) and ϕ is a Gaussian random field. Extragalactic radio sources are particularly interesting as tracers of the LSS since they span large volumes extending out to substantial redshifts. The radio sources from the

NRAO VLA Sky Survey (NVSS), the quasar catalogue of Sloan Digital Sky Survey Release Six (SDSS DR6 QSOs) and the MegaZ-LRG (DR7), the final SDSS II Luminous Red Galaxy (LRG) photometric redshift survey, have been recently analysed by [21] (see this work and references therein for a thorough analysis on the subject). Through a global analysis of the constraints on the amplitude of primordial non-Gaussianity by the angular power spectra obtained from extragalactic radio sources (mapped by these surveys) and, moreover, from their cross-correlation power spectra with the WMAP CMB temperature map, [21] set limits on $f_{NL} = 48 \pm 20$, $f_{NL} = 50 \pm 265$ and $f_{NL} = 183 \pm 95$ at 68% confidence level for local, equilateral and enfolded templates, respectively. These results have been found to be stable with respect to potential systematic errors: the source number density and the contamination by Galactic emissions, for NVSS sources; the use of different CMB temperature fluctuation templates and the contamination of stars in the SDSS and LRG samples. Such tests of non–Gaussianity would have profound implications for inflationary mechanisms - such as single-field slow roll, multifields, curvaton (local type) - and for models which have effects on the halo clustering can be described by the equilateral template (related to higher-order derivative type non-Gaussianity) and by the enfolded template (related to modified initial state or higher-derivative interactions). Fundamental progress on this topic will be achieved by combining forthcoming LSS surveys with the CMB maps provided by *Planck*.

3. Methods for compact source detection in CMB maps

Compact sources, in CMB literature, are defined as spatially bounded sources which subtend very small angular scales in the images, such as galaxies and galaxy clusters. On the other hand, diffuse components, such as the CMB itself and Galactic foregrounds, do not show clear spatial boundaries and extend over large areas of the sky. Due to the fact that compact sources are spatially localized, the techniques for detecting them differ from those applied for the separation of the diffuse components. Most of the detection methods use scale diversity, i.e. different power at different angular scales, to enhance compact sources against diffuse components. Sources must be detected against a combination of CMB, instrumental noise and Galactic foregrounds. From the point of view of signal processing, the source is the signal and the other components are the noise.

Point sources are "compact" sources in the sense that their typical observed angular size is much smaller than the beam resolution of the experiment. Therefore, they appear as point-like objects convolved with the instrumental beam. Radio sources and far–IR sources are usually seen as point-like sources. Galaxy clusters, which are detected through the thermal SZ effect [22], have a shape that is obtained as the convolution of the instrumental beam with the cluster profile. In contrast to point sources, the cluster profile has to be taken into account for cluster detection. However, since the projected angular scale of clusters is generally small, techniques that are useful for point sources can be adapted for clusters, too.

The thermal SZ effect has a general dependence with frequency, that makes the use of multichannel images very convenient for cluster detection. On the contrary, the flux of each individual point source has its own frequency dependence. Despite this fact, the combination of several channels can also improve point source detection. We will review techniques applied to single-frequency channels in a first subsection and then we will discuss more recent methods, that use multichannel information.

3.1. Single channel detection

We now focus on techniques for detecting point–like sources. Galaxy clusters can be detected by similar methods, but taking into account the cluster profile (see, e.g., [23-25]). Since multichannel methods for cluster detection improve significantly the performance of single channel techniques, we leave a more detailed study of clusters for the next subsection.

As mentioned before, compact source detection techniques make use of scale diversity. For example, SEXTRACTOR [26] – where maps are pre–filtered by a Gaussian kernel the same size as the beam – approximates the image background by a low-order polynomial and then subtracts the background from the image. The object is detected after connecting the pixels above a given flux density threshold. SEXTRACTOR has been used for elaborating the *Planck* ERCSC [15] in the highest frequency channels, from 217 to 857 GHz. However, CMB emission and diffuse foregrounds are complex and cannot be modeled in a straightforward way. Thus, apart from this important application, SEXTRACTOR has had a limited use in CMB astronomy.

A standard method which has been used often for compact source detection is the common matched filter (MF) [27]. The MF is just a linear filter with suitable characteristics for amplifying the source against the background. The image $y(\vec{x})$ is convolved with a filter $\psi(\vec{x})$:

$$\omega(\vec{x}) = \int y(\vec{u})\psi(\vec{x} - \vec{u})\, d\vec{u} \tag{2}$$

The MF is defined as the linear filter that is an unbiased estimator of the source flux and minimizes the variance of the filtered map. In order to satisfy these mathematical constraints, if we assume that the beam is circularly symmetric and the background a homogeneous and isotropic random field, the MF must be defined in Fourier space as

$$\psi(q) = k\frac{\tau(q)}{P(q)} \tag{3}$$

where $\tau(q)$ is the Fourier transform of the beam, $P(q)$ the background power spectrum and k the normalization constant. With this definition, the MF gives the maximum amplification of the compact source with respect to the background. Once the source has been amplified, point sources are detected in the filtered map as peaks above a given threshold, typically5σ, with σ the r.m.s deviation of the filtered map. The MF has been used both with simulations [28] and real data [29]. In this last paper, the WMAP team estimated the fluxes by fitting to a Gaussian profile plus a planar baseline. In [24] the MF was also applied to the detection of clusters. The use of wavelets for source detection is an interesting alternative to the MF. Wavelets are compensated linear filters, i.e their integral is zero, that help to remove the background contribution and yield a high source amplification. Since the beam is approximately Gaussian, the Mexican Hat Wavelet (MHW), constructed as the Laplacian of a Gaussian function, adapts itself very well to the detection problem. The MHW depends on the scale R, a parameter that determines the width of the wavelet:

$$\psi(q) \propto (qR)^2 \exp\big(- \frac{(qR)^2}{2}\big) \tag{4}$$

The MHW has been succesfully applied to simulated CMB data [30]. The scale R is fixed in order to obtain the maximum amplification and the determination of the power spectrum is not necessary. A family of wavelets that generalize the MHW, the Mexican Hat Wavelet Family (MHWF) was presented in [31]. The performance of this family was compared with the MF in [28] and it produced similar results when implemented on Planck simulations. The MHWF was also applied to point source detection in WMAP images [32] by using a non-blind method. This method yielded a larger number of detections than the MF used by the WMAP group. The general expression of the MHWF is:

$$\psi(q) \propto (qR)^{2n} \exp\Big(- \frac{(qR)^2}{2}\Big) \tag{5}$$

n being a natural number. Further improvements can be obtained if we admit any real exponent such as in the Bi-parametric Adaptive Filter (BAF) [33].

The MF and the diverse types of wavelets do not use any prior information about the average number of sources in the surveyed patch, the flux distribution of the sources or other properties. Therefore, useful information is being neglected by these methods. In contrast, Bayesian methods provide a natural way to incorporate information about the statistical properties of both the source and the noise. Several Bayesian methods have been proposed in the literature for the detection problem [24, 34, 35]. These methods construct a posterior probability $Pr(\theta|D,H)$ by using Bayes' theorem

$$Pr(\theta|D,H) = \frac{Pr(D|\theta,H)Pr(\theta|H)}{Pr(D|H)} \tag{6}$$

where θ are the relevant parameters (positions, fluxes, sizes, etc.), D the data, and H the underlying hypothesis. In Bayesian terminology $Pr(D|\theta,H)$ is the likelihood, $Pr(\theta|H)$ is the prior and $Pr(D|H)$ is the Bayesian evidence. Different Bayesian techniques can differ in the priors or in the way of exploring the complicated posterior probability. PowellSnakes [34] is an interesting method, which has been applied with success to the compilation of the ERCSC for *Planck* frequencies between 30 and 143 GHz [15]. It uses Powell's minimization and physically motivated priors. This method can be also applied to cluster detection just by introducing a suitable prior on the cluster size.

A simple Bayesian way of determining the position of the sources and estimating their number and flux densities has been presented in [35]. Whereas by the MF, or by wavelets, sources are detected above an arbitrary threshold, Bayesian methods select them in a more natural way, for instance by comparing the posterior probability of two hypothesis: presence or absence of the source. In the next subsection we will explore multichannel methods that help improve the detection performance

3.2. Multi-channel detection

The flux density distribution, f_ν, of extragalactic radio sources as a function of frequency, ν, is usually approximated by a power law, although this approximation is only valid in limited frequency intervals, i.e. $f_\nu \propto (\nu/\nu_0)^\alpha$, with ν_0 being some frequency of reference. Nevertheless, the so called "spectral index", α, changes from source to source and this formula is not reliable when the range of frequencies is wide enough. In [36] a scheme for channel combination was proposed that makes the spectral behavior irrelevant. This method is called matrix multifilters (MTXFs) and relies on the application of a set ($N \times N$ matrix) of linear filters which combine the information of the N channels in such a way that: 1) an unbiased estimator of the source flux at each channel is obtained; and 2) the variance of the estimator is minimum. Note that the method does not mix the images in a single map, but it produces N maps in which the sources are conveniently amplified. The method defaults to the MF when there is no cross-correlation among the channels. When there is a non negligible correlation among the channels, as is the case for microwave images taken at different frequencies where CMB and Galactic foregrounds are present in all the images, this method gives a clear increase of the amplification when compared with the MF.

Now, we discuss a method tailored for cluster detection through the thermal SZ effect. Matched Multifilters (MMF) [37] combine N channels in a single image, incorporating the information about the spectral behavior (thermal SZ effect) and with the N filters depending on a scale parameter S. The filters are constructed in the usual way, by imposing unbiasedness and minimum variance. The MMF is given (in matrix notation) by

$$\Upsilon(q) = \alpha \mathbf{P}^{-1} F, \ \alpha^{-1} = \int d\mathbf{q}\, F^t \mathbf{P}^{-1} F, \tag{7}$$

where F is the column vector $[f_\nu \tau_\nu]$, which incorporates the spectral behavior f_ν and the shape of the cluster at each frequency τ_ν and $\mathbf{P}$ is the cross-power spectrum. Since the cluster size is not known a priori, the images are convolved with a set of filters with different scales S_i, and it has been proven that the amplification is maximum when the chosen scale coincides with the cluster size. A common pressure profile is assumed for the clusters. The detection is performed by searching for the maxima of the filtered map above a given threshold. The estimated amplitude of the thermal SZ effect is given by the amplitude at the maxima.

MMF can also be adapted to detect the fainter kinematic SZ effect. In this case an orthogonality condition with respect to the spectral behavior of the thermal SZ effect is imposed. Together with the usual unbiasedness and minimum variance conditions, this last constraint helps cancel out the thermal SZ effect contamination [38]. A MMF can also be designed for point source detection, by incorporating the (unknown) spectral behavior of the sources as a set of parameters in the filter, it has also been proven that the amplification is maximum when these parameters coincide with the real source spectrum. By changing the parameters and selecting those which give the maximum amplification, in [39] the authors were able to increase the number of point source detections in the WMAP 7-year maps.

Finally, a multi-channel Bayesian method has been developed recently, Powell-Snakes II [40]. This method constructs a posterior distribution by combining the likelihood and the prior information of the different channels. It is an extension of Powell-Snakes I and uses prior information on the positions, number of sources, intensities, sizes and spectral parameters. The method is suitable both for point sources and for clusters. It is worth noting that maximizing the likelihood when the sources are well separated, i.e. in the absence of source blending, amounts to using the MMF presented above. Here we have briefly summarized the most important topics on the subject: a more detailed review can be found in [41].

4. Sunyaev-Zeldovich effect in clusters of galaxies

The Sunyaev-Zeldovich effect (SZ, [22]) accounts for the interaction between a hot plasma (in a cluster environment) and the photons of the CMB. When CMB photons cross a galaxy cluster, some of these photons interact with the free electrons in the hot plasma through inverse Compton scattering. The temperature change observed in a given direction, θ, and at the frequency ν, can be described as

$$\Delta T(\theta,\nu) = C_0 \int n_e(l)T(l)dl \tag{8}$$

where C_0 contains all the relevant constants including the frequency dependence ($g_x = x(e^x+1)/(e^x-1)-4$, with $x = h\nu/kT$), n_e is the electron density and T is the electron temperature. The integral is performed along the line of sight.

The same electrons that interact with the CMB photons emit X-rays through the bremstrahlung process:

$$S_x(\theta,\nu) = S_0 \frac{\int n_e^2 T^{1/2} dl}{D_\ell(z)^2} \tag{9}$$

where $D_\ell(z)$ is the luminosity distance. The quantity S_0 contains all the relevant constants and frequency dependence. Combining X-ray and SZ observations it is thus possible to reduce the degeneracy between different models due to their different dependency on T and especially with n_e.

Due to the nature of the microwave radiation, water vapour (and hence our atmosphere) presents a challenge for studying this radiation from the ground. Observations have to be carried out through several windows where the transmission of the microwave light is maximized. In recent years, ground–based experiments have benefited from important progress in the development of very sensitive bolometers. These bolometer arrays when combined with superb atmospheric conditions – found in places like the South Pole and the Atacama desert (with extremely low levels of water vapour) – have allowed, for the first time at all, galaxy clusters to be mapped in great detail through the SZ effect. The South Pole Telescope (or SPT; see, e.g., [42]) and the Atacama Cosmology Telescope (or ACT; see, e.g., [43]) are today the most important ground–based experiments carrying out these observations.

From space, the *Planck* satellite – even though it lacks the spatial resolution of ground–based experiments – complements them by applying a full–sky coverage, a wider frequency range and a better understanding of Galactic and extragalactic foregrounds. In particular, *Planck* is better suited than ground–based experiments to detect large angular scale SZ signals like nearby galaxy clusters or the diffuse SZ effect. In fact, ground–based experiments can have their large angular scales affected by atmospheric fluctuations that need to be removed, carefully. This removal process can distort the modes that include the large angular scales signal. On the contrary, *Planck* data does not suffer from these limitations and its relatively poor angular resolution (if compared to some ground experiments) can be used to its advantage. The wide frequency coverage and extremely high sensitivity of *Planck* allows for detailed foreground (and CMB) removal that could overwhelm the weak signal of the SZ effect. *Planck* data will help improve the understanding of the distribution and the characteristics of the plasma in clusters. The conclusions derived on the internal structure of clusters will ultimately have an impact on other works that focus on deriving cosmological parameters. In fact, cosmological studies cannot by themselves disentangle among the uncertainties in the physics inside galaxy clusters.

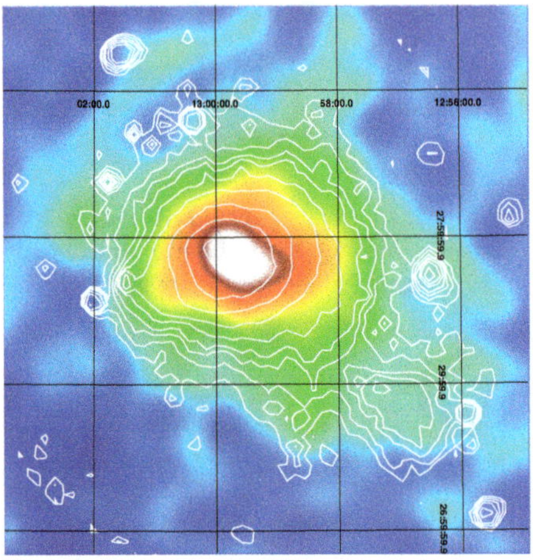

Figure 4. Fig. 1 from [44]. Coma cluster as seen by *Planck*. Contours show the the X-ray emission from Coma. Credit: *Planck* Collaboration, A&A, Vol. 536, A8, 2011, reproduced with permission © ESO.

The *Planck* satellite is currently detecting hundreds of clusters of galaxies through the thermal SZ effect. One of the peculiarities of the SZ effect is that the change in the CMB temperature in the direction of a cluster is independent of the distance to that cluster. This makes the SZ an ideal tool to explore the high redshift Universe. *Planck* is perfect for studying the most massive clusters in the Universe and is expected to see clusters beyond $z = 1$. Earlier results on galaxy clusters obtained by *Planck* have been presented in a subset of the *Planck* Early results papers and, more specifically, can be found in [44–49] and also in [50], where new results based on additional data are starting to be presented.

Among the first results published by the *Planck* collaboration on the SZ effect, the Coma cluster (see Fig. 4) constitutes one of the most spectacular ones. Coma is a nearby massive cluster that is well resolved by *Planck*. Fig. 4 shows the power of *Planck* to study the SZ effect with unprecedented quality. In the near future, studies based on *Planck* data alone or combined with X-ray data will reveal new details about the internal structure of this and other clusters.

Planck's earlier results include the detection of almost 200 clusters through their SZ signature ([44]). *Planck* is particularly sensitive to phenomena that increase the pressure, like mergers or superclusters. One such supercluster was detected by *Planck* [45]. Most of the clusters seen by *Planck* in the early analysis were known nearby objects but some of them were

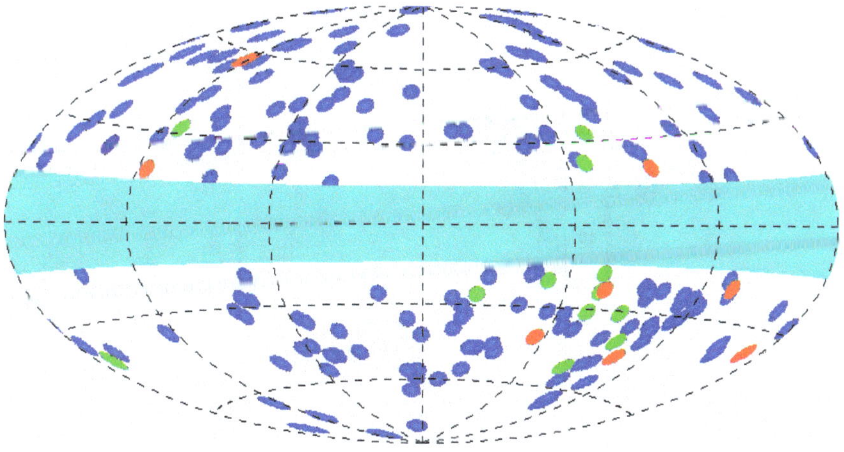

Figure 5. Fig. 3 from [44]. Distribution of Planck clusters that were already known (blue) and the new cluster candidates (green and red). Credit: *Planck* Collaboration, A&A, Vol. 536, A8, 2011, reproduced with permission © ESO.

new clusters, that have been later confirmed by X-ray and/or optical follows up. Fig. 5 shows the distribution in the sky of clusters of galaxies as seen by *Planck*. This includes the most massive clusters in the nearby and intermediate distance Universe. The redshift independence of the SZ effect can be appreciated in Fig. 6 (next page), which shows the relative flatness of the selection function of *Planck* as compared to cluster selections based on X-ray luminosity. New analysis based on better data will improve the selection function by reducing the limiting mass as a function of redshift. It is expected [5] that *Planck* will increase the number of known clusters in a significant way and, in particular, it will explore the high–redshift regime, detecting the most massive clusters at these high redshifts.

One of the most interesting conclusions derived by these earlier results comes from the combination of X-ray and SZ data. Earlier studies based on X-ray data were able to conclude that there exists a universal profile that accurately describes the gas pressure in galaxy clusters [51]. The newly discovered (by *Planck*) SZ clusters seem to follow well this profile but small deviations were observed when comparing the mean SZ profile with the average profile derived from X-ray observations. Fig. 7 summarizes one of the main results of the paper [45] where it can be appreciated how the average profile of the SZ observations (red thick line) flattens towards the cluster center ($R_{500} \ll 1$) when compared to the average of a sample of cool, core relaxed X-ray clusters (thick blue line). This fact suggests that the new clusters detected by *Planck* tend to be non–cool core, morphologically disturbed clusters. This would explain why these clusters where not found by previous X-ray surveys but *Planck*, that is sensitive to the total pressure rather than to the distribution of the gas, has no problem in finding them.

Many other relevant results can be found in the first series of papers from the *Planck* ERCSC, including studies of scaling relations between SZ quantities and optical or X-ray ones. More recently, a new analysis based on 2.5 full sky surveys has studied the relationship between the Compton parameter and weak lensing mass estimates [50]. As shown in Fig. 8, this work is very promising and could allow – in the near future – the use of the Compton parameter

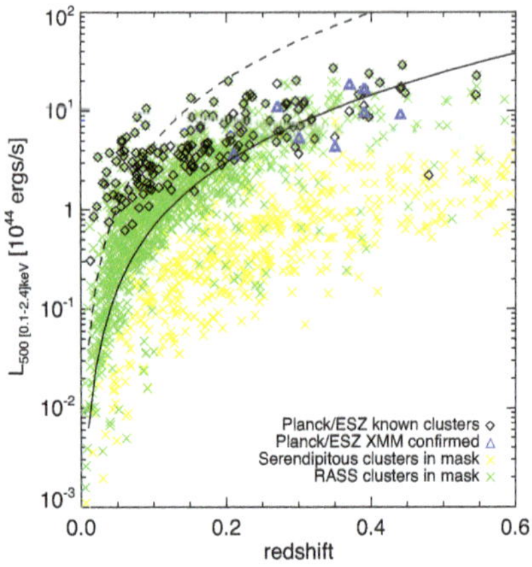

Figure 6. Fig. 21 from [44]. Redshift distribution as a function of luminosity for the 158 clusters from the Planck Early SZ sample (diamonds and triangles) identified with known X-ray clusters, compared with serendipitous and RASS clusters (crosses). See [44] for more details. Credit: *Planck* Collaboration, A&A, Vol. 536, A8, 2011, reproduced with permission © ESO.

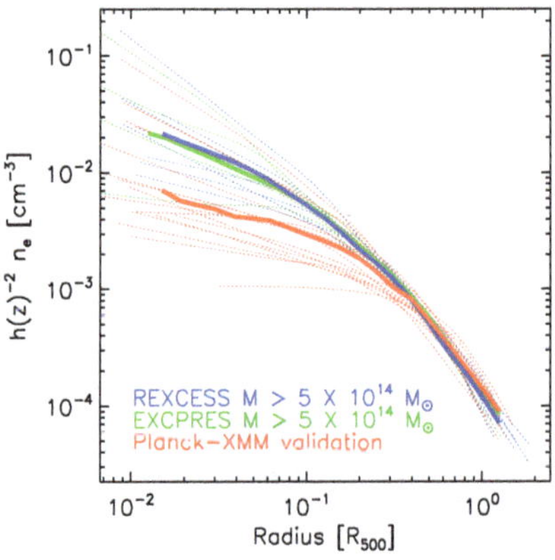

Figure 7. Fig. 10 from [45]. Scaled density profiles, derived from X-ray data, of the new *Planck* SZ clusters compared to those of similar mass systems from representative X–ray samples. Thick lines show the mean profile of each sample. Credit: *Planck* Collaboration, A&A, Vol. 536, A9, 2011, reproduced with permission © ESO.

as a mass proxy in cosmological studies, where a good mass estimator is crucial to derive accurate cosmological parameters from the analysis of a cluster sample.

After *Planck*'s data release, science based on the SZ effect will change dramatically. *Planck* is expected to release more than 2 full sky surveys of data early in 2013, thus opening the door for multiple studies to be carried out by the scientific community. Cluster science based on the SZ effect will motivate many of these studies. Of particular interest will be those works that combine SZ effect and X-ray data. The different dependence of the SZ effect and X-ray emission with electron density and temperature allows for deprojection techniques to reduce the uncertainties of the models. Also, the combination of X-ray and SZ data can be particularly powerful to study the clumpiness of the gas and deviations from spherical symmetry. An area where future *Planck* data will be used extensively will be the detection of new cluster candidates. The legacy *Planck* cluster catalogue will contain the most significant cluster signals. Hundreds of weaker SZ (and unknown) clusters will still be present in the public data but not in the legacy catalogue. Many groups will dig into the *Planck* data searching for these weaker signals. Among them there will be the most distant clusters at $z > 1$ that will be crucial for future cosmological studies.

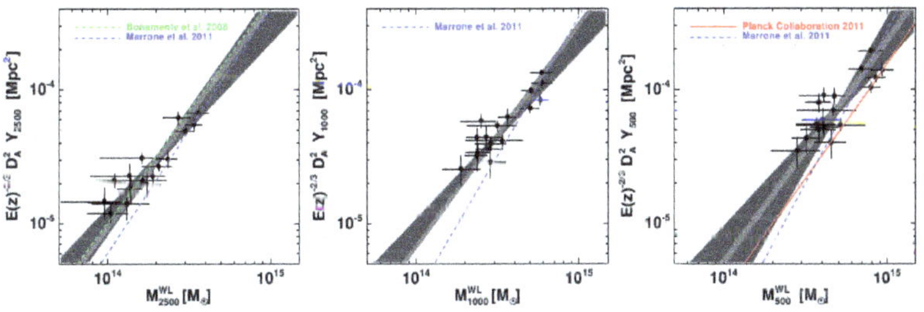

Figure 8. Fig. 1 from [50]. Correlation between the Compton parameter and the weak lensing mass. Credit: *Planck* Collaboration, A&A, submitted (ms AA/2012/19398), 2012, reproduced with permission © ESO.

Another area where *Planck* data might contribute significantly is in the study of energetic phenomena in galaxy clusters. The SZ effect is sensitive to the temperature of the Plasma (or more generally, to the speed distribution of the electrons). Hot clusters have an SZ spectrum that deviates from the standard shape. The shift (or relativistic correction) is more dramatic at higher frequencies ($\nu > 100$ GHz). Current X-ray missions like *Chandra* have some trouble determining the temperature of the plasma for clusters with high temperatures. On the contrary, the relativistic corrections to the SZ effect can dramatically boost the SZ signal in *Planck* at frequencies $\nu > 500$ GHz making, it easier to detect these clusters at these frequencies and also to derive constraints on the physical state of the plasma. A strong deviation in the spectrum could be an indication that very energetic phenomena (very high temperatures, shock waves, etc.) are operating on the cluster at large scales.

5. Extragalactic radio and far–IR sources

The *Planck* ERCSC [15] provides positions and flux densities of compact sources found in each of the nine *Planck* frequency maps. The flux densities are calculated using aperture photometry, with careful modeling of *Planck*'s elliptical beams[8]. These data on sources detected during the first 1.6 full-sky surveys offers, among other things, the opportunity of studying the statistical and emission properties of extragalactic radio and far–IR sources over a broad frequency range, never before fully explored by blind surveys.

[8] Flux densities taken from the ERCSC should be divided by the appropriate colour correction to give the correct flux values for an assumed narrow band measurement at the central frequency.

As shown by [15], their Table 1, the full-sky surveys of the *Planck* satellite are – and will be, for years to come – unique in the millimetre, at $\lambda \lesssim 3$ mm, and submillimetre domains. The lack of data in this frequency range represented the largest remaining gap in our knowledge of bright extragalactic sources (i.e., normal and star–forming galaxies and Active Galactic Nuclei, AGNs) across the electromagnetic spectrum. In the course of its planned surveys, *Planck* has been able to measure the integrated flux of many hundreds of "radio" sources – i.e., sources at intermediate to high–redshift, dominated by synchrotron emission due to hot electrons in the inner jets of the Active Galactic Nucleus (AGN) of the source – and of many thousands "'far–IR" sources – i.e., low–redshift dusty galaxies or sources with emission dominated by interstellar dust in thermal equilibrium with the radiation field – thus providing the *first complete full-sky catalogue* (ERCSC) of bright submillimetre sources. Thanks to this huge amount of new data it is thus possible to investigate the spectral energy distributions (SEDs) of EPS in a spectral domain very poorly explored before and, at the same time, their cosmological evolution, at least for some relevant source populations.

5.1. Synchrotron sources: "blazars"

The most recent estimates on source number counts of radio (synchrotron) sources up to $\sim 50 - 70$ GHz, and the optical identifications of the corresponding point sources (see, e.g., [52]), show that these counts are dominated by radio sources whose average spectral index is "flat", i.e., $\alpha \simeq 0.0$ (with the usual convention $S_\nu \propto \nu^\alpha$). This result confirms that the underlying source population is essentially made of Flat Spectrum Radio Quasars (FSRQ) and BL Lac objects, collectively called "blazars"[9], with minor contributions coming from other source populations [13, 54]. At frequencies > 100 GHz, however, there is now new information for sources with flux densities below about 1 Jy, coming from the South Pole Telescope (SPT) collaboration [55], with surveys over 87 deg^2 at 150 and 220 GHz, and from the Atacama Cosmology Telescope (ACT) survey over 455 deg^2 at 148 GHz [56].

The "flat" spectra of blazars are generally believed to result from the superposition of different components in the inner part of AGN relativistic jets, each with a different synchrotron self-absorption frequency [57]. At a given frequency, the observed flux density is thus dominated by the synchrotron-emitting component which becomes self-absorbed, and, in the equipartition regime, the resulting spectrum is approximately flat. Given their sensitivity and full sky coverage, *Planck* surveys are uniquely able to shed light on this transition from an almost "flat" to a "steep" regime in the spectra of blazar sources, which can be very informative on the ages of sources and on the inner jet processes which determine the acceleration of the electrons [58].

To study the spectral properties of the extragalactic radio sources in the *Planck* ERCSC [59] used a reference 30 GHz sample above an estimated completeness limit $S_{lim} \simeq 1.0$ Jy. Not all of these sources were detected at the $\geq 5\sigma$ level in each of the *Planck* frequency channels considered. Whenever a source was not detected in a given channel they replaced its (unknown) flux density by a 5σ upper limit, where for σ they used the average r.m.s.

---

[9] Blazars are jet-dominated extragalactic objects characterized by a strongly variable and polarized emission of the non-thermal radiation, from low radio energies up to high energy gamma rays [53].

error estimated at each *Planck* frequency. Finally, for estimating spectral index distributions, these upper limits have been redistributed among the flux density bins by using a Survival Analysis technique and, more specifically, by adopting the Kaplan-Meyer estimator[10][59].

In the sample analyzed by [59], the 30–100 GHz median spectral index is very close to the $\alpha \sim -0.39$ found by [60] between 20 and 95 GHz, for a sample with 20 GHz flux density $S >$ 150 mJy. Moreover, the 30–143 GHz median spectral index is in very good agreement with the one found by [56] for their bright ($S_\nu > 50$ mJy) 148 GHz-selected sample with complete cross-identifications from the Australia Telescope 20 GHz survey, i.e $\alpha_{20}^{148} = -0.39 \pm 0.04$. Fig.9 presents the contour levels of the distribution of α_{70}^{143} vs. α_{30}^{70} (obtained using Survival Analysis) in the form of a 2D probability field: the colour scale can be interpreted as the probability of a given pair of spectral indices and a bending down, i.e. $\alpha < -0.5$, at high frequencies is displayed. In the whole, the results of [59] show that in their sample selected at 30 GHz a moderate steepening of spectral indices of EPS at high radio frequencies, i.e. $\gtrsim 70 - 100$ GHz, is clearly apparent[11].

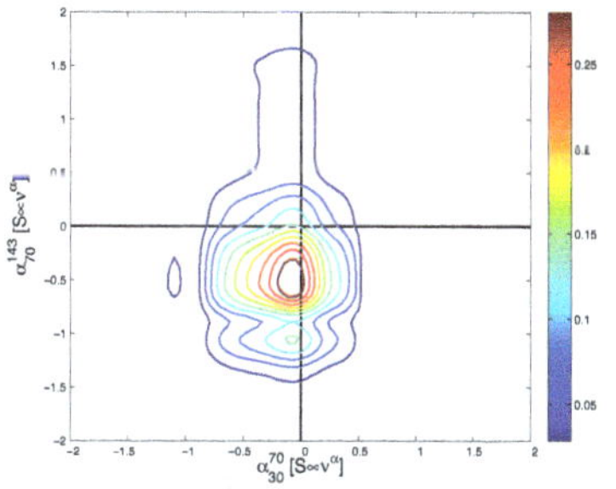

Figure 9. Fig. 7 from [59]. Contour levels of the distribution of α_{70}^{143} vs. α_{30}^{70} obtained by Survival Analysis, i.e., taking into account the upper limits to flux densities at each frequency. The colour scale can be interpreted as the probability of having any particular pair of values of the two spectral indices. The maximum probability corresponds to $\alpha_{30}^{70} \simeq -0.18$ and $\alpha_{70}^{143} \simeq -0.5$. Credit: Planck Collaboration, A&A, Vol. 536, A13, 2011, reproduced with permission © ESO.

As already noted, at high radio frequencies ($\nu > 30$ GHz) most of the bright extragalactic radio-sources are blazars. From the contour plot of Fig. 9 it is possible to see that the maximum probability of the spectral indices of blazars corresponds to $\alpha_{30}^{70} \simeq -0.18$ and $\alpha_{70}^{143} \simeq -0.5$. A secondary maximum can also be seen at $\alpha_{70}^{143} \simeq -1.2$. In a companion paper, i.e. [61], a detailed discussion on the modelling of the spectra of this source class is also presented. In this paper, spectral energy distributions (SEDs) and radio continuum spectra are presented for a northern sample of 104 extragalactic radio sources, based on the *Planck* ERCSC and simultaneous multifrequency data[12]. The nine *Planck* frequencies, from 30 to 857 GHz, are complemented by a set of quasi-simultaneous observations ranging from radio to gamma-rays. SED modelling methods are discussed, with an emphasis on proper, physical

[10] Since the fraction of upper limits is found to be always small (it reaches approximately 30% only in the less sensitive channel at 44GHz), the spectral index distributions are reliably reconstructed at each frequency.

[11] Some hints in this direction were previously found by other works on the subject. Additional evidence of spectral steepening is also presented in [61] by the analysis of a complete sample of blazars selected at 37 GHz.

[12] The great amount of data present in the *Planck* ERCSC complemented with quasi-simultaneous ground-based observations at mm wavelengths have also enabled the study of the very interesting spectral properties of the rare peculiar and/or extreme radio sources detected by the *Planck* surveys [62].

modelling of the synchrotron bump using multiple components, and a thorough discussion on the original accelerated electron energy spectrum in blazar jets is presented. The main conclusion is that, at least for a fraction of the observed mm/sub-mm blazar spectra, the energy spectrum could be much harder than commonly thought, with a power-law index ~ 1.5 and the implications of this hard value are discussed for the acceleration mechanisms effective in blazar shocks.

It has also been shown by [59] that differential number counts at 30, 44, and 70 GHz are in good agreement with those derived from *WMAP* [29] data at nearby frequencies. The model proposed by [54] is consistent with the present counts at frequencies up to 70 GHz, but over-predicts the counts at higher frequencies by a factor of about 2.0 at 143 GHz and about 2.6 at 217 GHz[13]. As shown above, the analysis of the spectral index distribution over different frequency intervals, within the uniquely broad range covered by *Planck* in the mm and sub-mm domain, has highlighted an average *steepening* of source spectra above about 70 GHz. This steepening accounts for the discrepancy between the model predictions of [54] and the observed differential number counts at HFI frequencies.

In the fall of 2011, a successful explanation of the change detected in the spectral behavior of extragalactic radio sources (ERS) at frequencies above 70-80 GHz has been proposed by [63]. This paper makes a first attempt at constraining the most relevant physical parameters that characterize the emission of blazar sources by using the number counts and the spectral properties of extragalactic radio sources estimated from high–frequency radio surveys[14]. As noted before, a relevant steepening in blazar spectra with emerging spectral indices in the interval between -0.5 and -1.2, is commonly observed at mm/sub-mm wavelengths. [63] interpreted this spectral behavior as caused, at least partially, by the transition from the optically–thick to the optically–thin regime in the observed synchrotron emission of AGN jets [64]. Indeed, a "break" in the synchrotron spectrum of blazars, above which the spectrum bends down, thus becoming "steep", is predicted by models of synchrotron emission from inhomogeneous unresolved relativistic jets [65, 66]. Based on these models, [63] estimated the value of the frequency ν_M (and of the corresponding radius r_M) at which the break occurs on the basis of the flux densities of ERS measured at 5 GHz and of the most typical values for the relevant physical parameters of AGNs.

As displayed in Fig. 10, high frequency ($\nu \geq 100$ GHz) data on source number counts are the most powerful for distinguishing among different cosmological evolution models (see [63] for more details on the models plotted in Fig. 10)[15]. As clearly shown, these most recent data on number counts require spectral "breaks" in blazars' spectra and clearly favor the

[13] This implies that the contamination of the CMB APS by radio sources below the 1 Jy detection limit is lower than previously estimated. No significant changes are found, however, if we consider fainter source detection limits, i.e., 100 mJy, given the convergence between predicted and observed number counts.

[14] The main goal of [63] was to present physically grounded models to extrapolate the number counts of ERS, observationally determined over very large flux density intervals at cm wavelengths down to mm wavelengths, where experiments aimed at accurately measuring CMB anisotropies are carried out.

[15] The two most relevant models of [63], i.e. **C2Co** and **C2Ex**, assume different distributions of r_M – i.e., the smallest radius in the AGN jet from which optically-thin synchrotron emission can be observed – for BL Lacs and FSRQs, with the former objects that generate, in general, the synchrotron emission from more compact regions, implying higher values of ν_M (above 100 GHz for bright objects). These two models differ only in the r_M distributions for FSRQs: in the **C2Co** model the emitting regions are more compact, implying values of ν_M partially overlapping with those for BL Lacs, whereas in the **C2Ex** model they are more extended, thus predicting very different values of ν_M for FSRQs and BL Lacs.

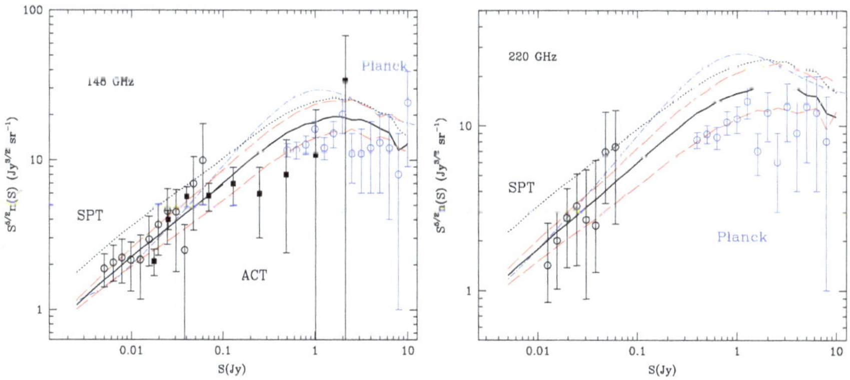

Figure 10. Comparison between predicted and observed differential number counts at 148 GHz (*left panel*) and at 220 GHz (*right panel*). Filled circles: ACT data; open black circles: SPT data; open blue circles: *Planck* ERCSC counts [59] at 143 GHz (left panel) and 217 GHz (right panel). The plotted lines indicate predictions of different models, as follows: **C0**(dotted lines), **C1** (thick continuous lines), **C2Ex** (lower red long–dashed lines) and **C2Co** (upper red long dashed lines) and the [54] model (blue dash–dotted line). Credit: Tucci M., et al., A&A, Vol. 533, A57, 2011, reproduced with permission © ESO.

model **C2Ex**. According to this, most of the FSRQs (which are the dominant population at low frequencies and at Jy flux densities), differently from BL Lacs, should bend their flat spectrum before or around 100 GHz. The **C2Ex** model also predicts a substantial increase of the BL Lac fraction at high frequencies and bright flux densities[16]. On the whole, the results of [63] imply that the parameter r_M should be of parsec–scales, at least for FSRQs, in agreement with the theoretical predictions of [67], whereas values of $r_M \ll 1\,\mathrm{pc}$ should be only typical of BL Lac objects or of rare, and compact, quasar sources.

5.2. Far–IR sources: Local dusty galaxies

The full-sky coverage of the *Planck* ERCSC provides an unsurpassed survey of galaxies at submillimetre (submm) wavelengths, representing a major improvement in the numbers of galaxies detected, as well as the range of far-IR/submm wavelengths over which they have been observed. The analysis done by [68] presented the first results on the properties of nearby galaxies using these data. They matched the ERCSC catalogue to IRAS-detected galaxies in the Imperial IRAS Faint Source Redshift Catalogue (IIFSCz) [69], so that they could measure the SEDs of these objects from 60 to 850 μm. This produced a list of 1717 galaxies with reliable associations between *Planck* and IRAS, from which they selected a subset of 468 for SED studies, namely those with strong detections in the three highest frequency Planck bands and no evidence of cirrus contamination. This selection has thus provided a first *Planck* sample of local, i.e. at redshift < 0.1, dusty galaxies, very important for determining their emission properties and, in particular, the presence of different dust components contributing to their submm SEDs. Moreover, the richness of data on

[16] This is indeed observed: a clear dichotomy between FSRQs and BL Lac objects has been found in the *Planck* ERCSC. Almost all radio sources show very flat spectral indices at LFI frequencies, i.e. $\alpha_{LFI} \geq -0.2$, whereas at HFI frequencies, BL Lacs keep flat spectra, i.e. $\alpha_{HFI} \geq -0.5$, with a high fraction of FSRQs showing steeper spectra, i.e. $\alpha_{HFI} \leq -0.5$.

extragalactic point sources gathered by *Planck* has allowed the measurement of the submm number density of bright ($S > 0.5 - 2$ Jy) dusty galaxies (and of synchrotron-dominated sources) for the first time.

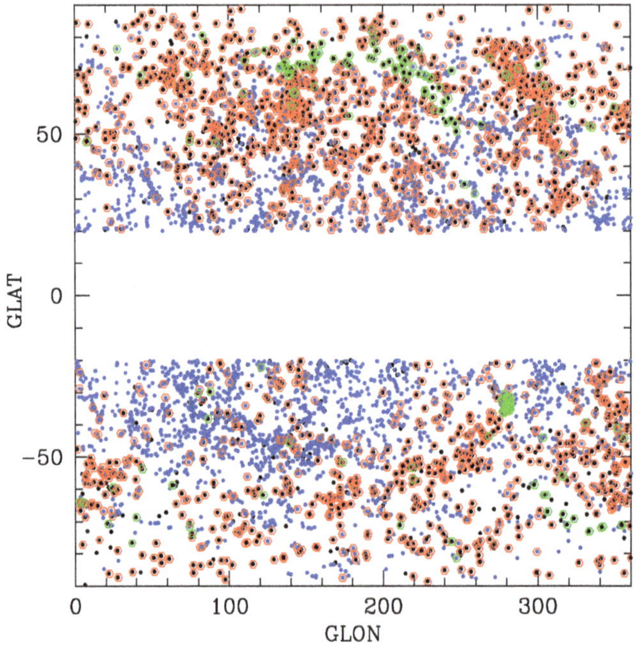

Figure 11. Fig. 2 from [68]. Sky plot of ERCSC sources in Galactic coordinates. ERCSC point-sources (black filled hexagons) and ERCSC sources flagged as extended (blue filled hexagons) are shown. Red hexagons are sources associated with IIFSCz IRAS FSC galaxies. Green hexagons are ERCSC sources not associated with IIFSCz, but associated with bright galaxies in NED (only for $|b| > 60°$ for extended sources). Credit: Planck Collaboration, A&A, Vol. 536, A16, 2011, reproduced with permission © ESO.

Fig. 11 shows the sky distribution of ERCSC sources at $|b| > 20°$, with sources flagged as extended in the ERCSC shown as blue filled hexagons, and point-sources shown in black. Associations with the IIFSCz are shown as red circles. The extended sources not associated with IIFSCz sources have a strikingly clustered distribution, which matches the areas of our Galaxy with strong cirrus emission, as evidenced by IRAS 100 μm maps. Therefore, the majority of these are cirrus sources and not extragalactic (see [68] for more details).

The studies of nearby galaxies detected by *Planck* [68] confirm the presence of cold dust in local galaxies and also largely in dwarf galaxies. The SEDs are fitted using parametric dust models to determine the range of dust temperatures and emissivities. They found evidence for colder dust than has previously been found in external galaxies, with temperatures $T <$ 20 K. Such cold temperatures are found by using both the standard single temperature dust model with variable emissivity β, or a two dust temperature model with β fixed at 2. In

[68] it is also found that some local galaxies are both luminous and cool, with properties similar to those of the distant submm galaxies uncovered in deep surveys. This suggests that previous studies of dust in local galaxies have been biased away from such luminous cool objects. In most galaxies the dust SEDs are found to be better described by parametric models containing two dust components, one warm and one cold, with the cold component reaching temperatures as low as 10 K[17]. The main conclusion of [68] is that cold ($T < 20$ K) dust is thus a significant and largely unexplored component of many nearby galaxies. Furthermore, a new population of cool submm galaxies is detected, with presence of very cold dust ($T = 10 - 13$ K) showing a more extended spatial distribution than generally assumed for the gas and dust in galaxies.

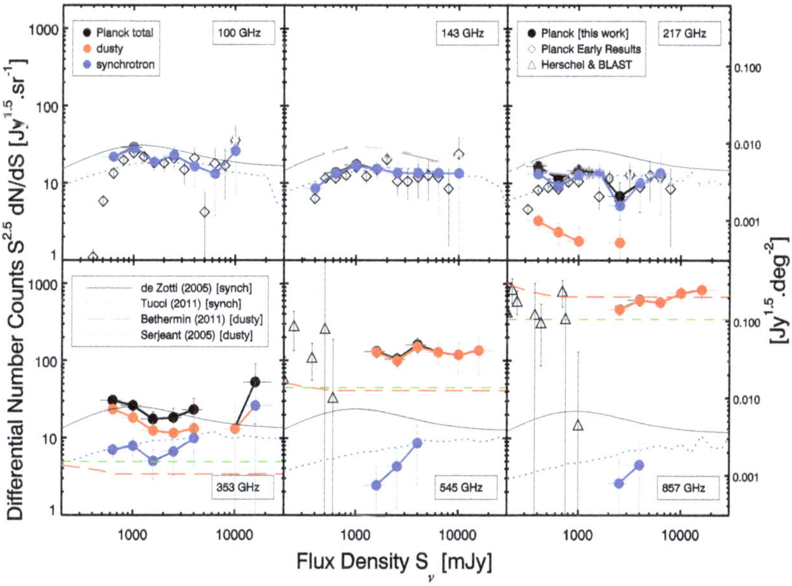

Figure 12. Fig. 9 from [70]. *Planck* differential number counts, normalized to the Euclidean value (i.e. $S^{2.5}dN/dS$), compared with models and other data sets. *Planck* counts: total (black filled circles); dusty (red circles); synchrotron (blue circles). Four models are also plotted: [54], dealing only with synchrotron sources – solid line; [63] dealing only with synchrotron sources – dots; [71] dealing only with dusty sources – long dashes; [72] dealing only with local dusty sources – short dashes. Other data sets: *Planck* early counts for 30 GHz-selected radio galaxies [59] at 100, 143 and 217 GHz (open diamonds); *Herschel* ATLAS and HerMES counts at 350 and 500μm from [73] and [74]; BLAST at the same two wavelengths, from [75], all shown as triangles. Left vertical axes are in units of $Jy^{1.5}$ sr^{-1}, and the right vertical axis in $Jy^{1.5}.deg^{-2}$. Credit: Planck Collaboration, A&A, submitted (ms AA/2012/20053), 2012, reproduced with permission © ESO.

Very recently, using EPS samples selected from the first *Planck* 1.6 full-sky surveys, i.e. from the *Planck* ERCSC, [70] have derived number counts of extragalactic sources from 100 to 857 GHz (3 mm to 350 μm). Three zones (deep, medium and shallow) of approximately

[17] Fits to SEDs of selected objects using more sophisticated templates derived from radiative transfer models confirm the presence of the colder dust found through parametric fitting.

homogeneous coverage are used to ensure a clean completeness correction[18]. For the first time, bright number counts of EPS at 353, 545 and 857 GHz (i.e., 850, 550 and 350 μm) have been calculated[19]. *Planck* number counts are found to be in the Euclidean regime in this frequency range, since the ERCSC comprises only bright sources ($S > 0.3$ Jy). The estimated number counts appear generally in agreement with other data sets, when available (see [70] for more details).

Using multi-frequency information to classify the sources as dusty- or synchrotron-dominated (and measure their spectral indices), the most striking result of [70] is the estimated contribution to the number counts by each population. These new estimates of number counts of synchrotron and of dust–dominated EPS (displayed in Fig. 12) have allowed new constraints to be placed on models which extend their predictions to bright flux densities, i.e. $S > 1$ Jy. A very relevant result is that the model **C2Ex** of [63] (see Section 5.1) is performing particularly well at reproducing the number counts of synchrotron-dominated sources up to 545 GHz. On the contrary, [70] highlights the failure of many models for number count predictions of dusty sources to reproduce all the high-frequency counts. The model of [71] agrees marginally at 857 GHz but is too low at 545 GHz and also at lower frequencies, whereas the model of [72] is marginally lower at 857 GHz, fits the data well at 545 GHz, but is too low at 353 GHz. The likely origin of the discrepancies is an inaccurate description of the galaxy SEDs used at low redshift in these models. Indeed a cold dust component, detected by [68], is rarely included in the models of galaxy SEDs at low redshift. On the whole, these results already obtained by the exploitation of the *Planck* ERCSC data are providing valuable information about the ubiquity of cold dust in the local Universe, at least in statistical terms, and are guiding to a better understanding of the cosmological evolution of EPS at mm/sub-mm wavelengths.

6. Cosmic Infrared Background anisotropies

The Cosmic Infrared Background (CIB) is the relic emission, at wavelengths larger than a few microns, of the formation and evolution of the galaxies of all types, including AGNs and star-forming systems [76–78][20]. The CIB accounts for roughly half of the total energy in the optical/infrared Extragalactic Background Light (EBL) [77], although with some uncertainty, and its SED peaks near 150 μm. Since local galaxies give rise to an integrated infrared output that amounts to only about a third of the optical one [79], there must have been a strong evolution of galaxy properties towards enhanced far–IR output in the past. Therefore, the CIB, made up by high density, faint and distant galaxies[21] is barely resolved into its constituents. Indeed, less than 10% of the CIB is resolved by the *Spitzer* satellite at 160 μm

[18] The sample, prior to the 80 % completeness cut, contains between 217 sources at 100 GHz and 1058 sources at 857 GHz over about 12,800 to 16,550 deg^2 (31 to 40 % of the sky). After the 80 % completeness cut, between 122 and 452 and sources remain, with flux densities above 0.3 and 1.9 Jy, at 100 and 857 GHz, respectively.

[19] More specifically, number counts have been provided of synchrotron-dominated sources at high frequency (353 to 857 GHz) and of dusty-dominated galaxies at lower frequencies (217 and 353 GHz).

[20] An important goal of studies of galaxy formation has thus been the characterization of the statistical behavior of galaxies responsible for the CIB - such as the number counts, redshift distribution, mean SED, luminosity function, clustering – and their physical properties, such as the roles of star-forming vs. accreting systems, the density of star formation, and the number density of very hot stars

[21] The CIB records much of the radiant energy released by processes of structure formation occurred since the last scattering epoch, four hundred thousand years after the Big Bang, when the CMB was produced.

[75], about 10% by *Herschel* at 350 μm [73]. Thus, in the absence of foreground (Galactic dust) and CMB emissions, and when the instrument noise is subdominant, maps of the diffuse emission at the angular resolution probed by the current surveys reveal a web of structures, characteristic of CIB anisotropies. With the advent of large area far-IR to millimeter surveys (*Herschel, Planck*, SPT, and ACT), CIB anisotropies thus constitute a new tool for structure formation and evolution studies.

CIB anisotropies are expected to trace large-scale structures and probe the clustering properties of galaxies, which in turn are linked to those of their hosting dark matter halos. Because the clustering of dark matter is well understood, observations of anisotropies in the CIB constrain the relationship between dusty, star-forming galaxies at high redshift, i.e. $z \geq 2$, and the underlying dark matter distribution. The angular power spectrum of CIB anisotropies has two contributions, a white-noise component caused by shot noise and an additional component caused by spatial correlations between the sources of the CIB. Correlated CIB anisotropies have already been measured by many space–borne as well as ground–based experiments (see [80] for more details). Depending on the frequency, the angular resolution and size of the survey, these measurements can probe two different clustering regimes. On small angular scales ($\ell \geq 2000$), they measure the clustering within a single dark matter halo and, accordingly, the physics governing how dusty, star–forming galaxies form within a halo. On large angular scales, i.e. $200 \leq \ell \leq 2000$, CIB anisotropies measure clustering between galaxies in different dark matter halos. These measurements primarily constrain the large-scale, linear bias, b, of dusty galaxies, which is usually assumed to be scale-independent over the relevant range.

Thanks to the exceptional quality of the *Planck* data, [80] were able to measure the clustering of dusty, star-forming galaxies at 217, 353, 545, and 857 GHz with unprecedented precision. The CIB maps were cleaned using templates: HI for Galactic cirrus; and the *Planck* 143 GHz maps for CMB. Having HI data is necessary to cleanly separate CIB and cirrus fluctuations. After careful cleaning, they obtained CIB anisotropy maps that reveal structures produced by the cumulative emission of high-redshift, dusty, star–forming galaxies. The maps are highly correlated at high *Planck* frequencies, whereas they decorrelate at lower *Planck* HFI frequencies. [80] then computed the power spectra of the maps and their associated errors using a dedicated pipeline and ended up with measurements of the APS of the CIB anisotropy, C_ℓ, at 217, 353, 545, and 857 GHz, with high signal-to-noise ratio over the range $200 < l < 2000$. These measurements compare very well with previous measurements at higher ℓ[22].

Moreover, from *Planck* data alone [80] could exclude a model where galaxies trace the (linear theory) matter power spectrum with a scale-independent bias: that model requires an *unrealistic* high level of shot noise to match the small-scale power they observed. Consequently, an alternative model that couples the dusty galaxy, parametric evolution model of [71] with a halo model approach has been developed (see [80], again, for more details). Characterized by only two parameters, this model provides an excellent fit to our measured anisotropy angular power spectrum for each frequency treated independently. In the near future, modelling and interpretation of the CIB anisotropy will be aided by the use

[22] The SED of CIB anisotropies is not different from the CIB mean SED, even at 217 GHz. This is expected from the model of [71] and reflects the fact that the CIB intensity and anisotropies are produced by the same population of sources.

of cross-power spectra between bands, and by the combination of the *Planck* and *Herschel* data at 857 and 545/600 GHz and *Planck* and SPT/ACT data at 220 GHz.

7. Acknowledgements

LT, FA and JMD acknowledge partial financial support from the Spanish Ministry of Science and Innovation (MICINN) under project AYA2010–21766–C03-01. CB acknowledges partial financial support by ASI through ASI/INAF Agreement I/072/09/0 for the Planck LFI Activity of Phase E2 and by MIUR through PRIN 2009 grant n. 2009XZ54H2. The analysis of the *Planck* full-sky maps has been performed by means of the HEALPix package [12]. We thank the Editorial Board of Astronomy and Astrophysics (European Southern Observatory; ESO) for having granted us the permission to reproduce many Figures originally published in (or submitted to) the same Journal. Credits are indicated in each one of the Figures we used. Many thanks are due to Douglas Scott, for carefully reading the original manuscript and for his very useful suggestions. We also warmly thank the *Planck* Collaboration and, in particular, all the members of the *Planck* Working Groups 2, 5 and 6 and of the LFI Core Team A09, with whom we shared the analysis and the interpretation of *Planck* data as for the subject here discussed, i.e. "extragalactic compact sources". Finally, we also thank the members of the *Planck* Science Team (ST) and of the *Planck* Editorial Board (EB) for granting us the permission of publishing this Chapter.

Author details

Luigi Toffolatti[1,2,*], Carlo Burigana[3,4],
Francisco Argüeso[2,5] and José M. Diego[2]

* Address all correspondence to: ltoffolatti@uniovi.es

1 Department of Physics, University of Oviedo, Oviedo, Spain
2 IFCA-CSIC, University of Cantabria, Santander, Spain
3 National Institute of Astrophysics - Institute of Space Astrophysics and Cosmic Physics (INAF-IASF), Bologna, Italy
4 Department of Physics, University of Ferrara, Ferrara, Italy
5 Department of Mathematics, University of Oviedo, Oviedo, Spain

References

[1] Tauber J.A., Mandolesi N., Puget J.-L., et al. (2010) Planck pre-launch status: the Planck mission. Astronomy and Astrophysics, 520, A1:1-22.

[2] Planck Collaboration I. (2011) Planck early results. I. The Planck Mission. Astronomy and Astrophysics, Vol. 536, A1:1-16.

[3] Planck HFI Core Team (2011a). Planck early results. IV. First assessment of the High Frequency Instrument in-flight performance. Astronomy and Astrophysics, Vol. 536, A4:1-20.

[4] Mennella A., Butler C., Curto A., et al. (2011). Planck early results. III. First assessment of the Low Frequency Instrument in-flight performance. Astronomy and Astrophysics, Vol. 536, A3:1-29.

[5] *Planck* Collaboration, 2005, ESA Science Report, ESA-SCI(2005)/01, arXiv:0604069.

[6] N. Mandolesi, et al., (2010) Planck pre-launch status: The Planck-LFI programme Astronomy and Astrophysics, Vol. 520, A3:1-24.

[7] J.-M. Lamarre, et al., (2010) Planck pre-launch status: The HFI instrument, from specification to actual performance. Astronomy and Astrophysics, Vol. 520, A9:1-20.

[8] P. Giommi, et al., (2012) Simultaneous Planck, Swift, and Fermi observations of X-ray and γ-ray selected blazars. Astronomy and Astrophysics, Vol. 541, A160:1-59.

[9] Knox, L. (1995) Determination of inflationary observables by cosmic microwave background anisotropy experiments. Physical Review D, Vol. 52:4307-4318.

[10] Planck HFI Core Team (2011b). Planck early results. VI. The High Frequency Instrument data processing. Astronomy and Astrophysics, Vol. 536, A6:1-47.

[11] Zacchei A., Maino D., Baccigalupi C., et al. (2011) Planck early results. V. The Low Frequency Instrument data processing. Astronomy and Astrophysics, Vol. 536, A5:1-19.

[12] K.M. Górski, et al. (2005) HEALPix: A Framework for High-Resolution Discretization and Fast Analysis of Data Distributed on the Sphere. The Astrophysical Journal, Vol. 622, 759-780.

[13] Toffolatti L., Argüeso F., De Zotti G., et al. (1998) Extragalactic source counts and contributions to the anisotropies of the cosmic microwave background: predictions for the Planck Surveyor mission. Monthly Notices of the Royal Astronomical Society, 297:117-127.

[14] Tucci M., and Toffolatti L. (2012) The impact of polarized extragalactic radio sources on the detection of CMB anisotropies in polarization. Advances in Astronomy, Special Issue: "Astrophysical Foregrounds in Microwave Surveys", Hindawi Publ. Co., Editor: C. Burigana, Vol. 2012, Article ID 624987:1-17 (doi:10.1155/2012/624987).

[15] Planck Collaboration VII. (2011) Planck Early results. VII. The Planck Early Release Compact Source Catalogue. Astronomy and Astrophysics, Vol. 536, A7:1-26.

[16] Komatsu E., et al. (2010) Non-Gaussianity as a Probe of the Physics of the Primordial Universe and the Astrophysics of the Low Redshift Universe. Astro2010: The Astronomy and Astrophysics Decadal Survey, Science White Papers, Vol. 158.

[17] Bartolo N., Komatsu E., Matarrese S., and Riotto A. (2010) Non-Gaussianity from inflation: Theory and observations. Physics Report, Vol. 402:103-114.

[18] Verde L., Wang L.M., Heavens A., and Kamionkowski M. (2000) Large-scale structure, the cosmic microwave background, and primordial non-gaussianity. Monthly Noticesof the Royal Astronomical Society, Vol. 313:L141-L145.

[19] Komatsu E., and Spergel D.N. (2001) Acoustic signatures in the primary microwave background bispectrum, Physical Rev. D, Vol. 63:063002

[20] Babich D., Creminelli P., and Zaldarriaga M. (2004) The shape of non-Gaussianities, Journal of Cosmology and Astroparticle Physics, Vol. 08:009-018.

[21] Xia J.Q., Baccigalupi C., Matarrese S., Verde L., Viel M. (2011) Constraints on primordial non-Gaussianity from large scale structure probes. Journal of Cosmology and Astroparticle Physics, Vol. 08:033 (JCAP08(2011)033).

[22] Sunyaev, R., Zeldovich, Ya.B. (1972) The Observations of Relic Radiation as a Test of the Nature of X-Ray Radiation from the Clusters of Galaxies. Comments on Astrophysics and Space Physics 4:173-179.

[23] Herranz D., Sanz J.L., Barreiro R., Martínez-González E. (2002) Scale-adaptive Filters for the Detection/Separation of Compact Sources. The Astrophysical Journal 580:610-625.

[24] Schulz A. E., White M. (2003) Survey of Galaxy Clusters with the Sunyaev-Zeldovich Effect. The Astrophysical Journal 586:723-730.

[25] Hobson M., McLachlan C. (2003) A Bayesian Approach to Discrete Object Detection in Astronomical Data Sets. Monthly Notices of The Royal Astronomical Society 338:765-784.

[26] Bertin E., Arnouts S. (1996) Sextractor:Software for Source Extraction. Astronomy and Astrophysics Supplement 117:393-404.

[27] Tegmark M., de Oliveira-Costa A. (1998) Removing Point Sources from Cosmic Microwave Background Maps. The Astrophysical Journal Letters 500:L83-L86.

[28] López-Caniego et al. (2006) Comparison of Filters for the Detection of Point Sources in Planck Simulations. Monthly Notices of The Royal Astronomical Society, Vol. 370:2047-2063.

[29] Wright E. L. et al. (2009) Five-year Wilkinson Anisotropy Probe Observations: Source Catalog. The Astrophysical Journal Supplement Series, Vol. 180:283-295.

[30] Cayón L. et al. (2000) Isotropic Wavelets. A Powerful Tool to Extract Point Sources from Cosmic Microwave Background Maps. Monthly Notices of The Royal Astronomical Society, Vol. 315:757-761.

[31] González-Nuevo et al. (2006) The Mexican Hat Wavelet Family: Application to Point-Source Detection in Cosmic Microwave Background Maps. Monthly Notices of The Royal Astronomical Society 369:1603-1610.

[32] López-Caniego et al. (2007) Non-blind Catalogue of Extragalactic Sources from the Wilkinson Microwave Anisotropy Probe (WMAP) first 3-year survey data. The Astrophysical Journal Supplement Series 170:108-125

[33] López-Caniego M., Vielva P. (2012) Biparametric Adaptive Filter: Detection of Compact Sources in Complex Microwave Backgrounds. Monthly Notices of The Royal Astronomical Society, Vol. 421:2139-2154.

[34] Carvalho P., Rocha G., Hobson M. (2009) A Fast Bayesian Approach to Discrete Object Detection in Astronomical Data Sets. Monthly Notices of The Royal Astronomical Society 393:681-702.

[35] Argüeso F. et al. (2011) A Bayesian Technique for the Detection of Point Sources in CMB Maps. Monthly Notices of The Royal Astronomical Society 414:410:417.

[36] Herranz D., Sanz J.L. (2008) Matrix Filters for the Detection of Extragalactic Point Sources in Cosmic Microwave Background Images. IEEE Journal on Selected Topics in Signal Processing 2:727-734.

[37] Herranz D., et al. (2002) Filtering Techniques for the Detection of Sunyaev Zeldovich Clusters in Multifrequency Maps. Monthly Notices of The Royal Astronomical Society 336:1057-1068.

[38] Herranz D., Sanz J. L., Barreiro R., López-Caniego M. (2005) The Estimation of the Sunyaev-Zeldovich Effects with Unbiased Multifilters. Monthly Notices of The Royal Astronomical Society 356:944-954.

[39] Lanz L.F., et al. (2012) Extragalactic Point Source Detection in WMAP 7-year Data at 61 and 94 GHz. Monthly Notices of The Royal Astronomical Society, submitted. ArXiv:astro/ph:1110.6877.

[40] Carvalho P., Rocha G., Hobson M., Lasenby A. (2012) PowellSnakes II: A Fast Bayesian Approach to Discrete Object Detection in Multi-frequency Astronomical Data Sets. Monthly Notices of The Royal Astronomical Society, submitted. ArXiv:astro/ph:1112.4933.

[41] Herranz D., Argüeso F., Carvalho P. (2012) Compact Sources Detection in Multichannel Microwave Surveys: From SZ Clusters to Polarized Sources. Advances in Astronomy, 2012:1-14. doi:10.1155/2012/410965.

[42] Hall N.R., Keisler R., Knox L., et al. (2010) Angular Power Spectra of the Millimeter-wavelength Background Light from Dusty Star-forming Galaxies with the South Pole Telescope. The Astrophysical Journal, Vol. 718:632-646.

[43] Fowler J.W., Acquaviva V., Ade P.A.R., et al. (2010) The Atacama Cosmology Telescope: A Measurement of the 600 < ell < 8000 Cosmic Microwave Background Power Spectrum at 148 GHz. The Astrophysical Journal, Vol. 722:1148-1161.

[44] Planck Collaboration VIII. (2011) Planck Early results. VIII. The all-sky early Sunyaev-Zeldovich cluster sample. Astronomy and Astrophysics, Vol 536, A8:1-28.

[45] Planck Collaboration IX. (2011) Planck Early results. VIII. XMM-Newton follow-up for validation of Planck cluster candidates. Astronomy and Astrophysics, Vol 536, A9:1-20.

[46] Planck Collaboration X. (2011) Planck Early results. X. Statistical analysis of Sunyaev-Zeldovich scaling relations for X–ray galaxy clusters. Astronomy and Astrophysics, Vol 536, A10:1-14.

[47] Planck Collaboration XI. (2011) Planck Early results. XI. Calibration of the local galaxy clusters Sunyaev-Zeldovich scaling relations. Astronomy and Astrophysics, Vol 536, A11:1-14.

[48] Planck Collaboration XII. (2011) Planck Early results. XII. Cluster Sunyaev-Zeldovich optical scaling relations. Astronomy and Astrophysics, Vol 536, A12:1-10.

[49] Planck Collaboration XXVI. (2011) Planck Early results. XXVI. Detection with Planck and confirmation by XMM–Newton of PLCK G266.6-27.3, an exceptionally X–ray luminous and and massive galaxy cluster at $z \sim 1$. Astronomy and Astrophysics, Vol 536, A26:1-7.

[50] Planck Collaboration III. (2012) Planck Intermediate results. III. The relation between galaxy cluster mass and Sunyaev-Zeldovich signal. Astronomy and Astrophysics, submitted. arXiv.org//1204.2743v1.

[51] Arnaud M., Pratt G.W., Piffaretti R., Böhringer H., Croston J.H., Pointecouteau E. (2010) The universal galaxy cluster pressure profile from a representative sample of nearby systems (REXCESS) and the Y_{SZ} - M_{500} relation. Astronomy and Astrophysics, Vol 517, A92:1-20.

[52] Massardi M., Ekers R.D., Murphy T., et al. (2008) The Australia Telescope 20-GHz (AT20G) Survey: the Bright Source Sample. Monthly Notices of the Royal Astronomical Society, 384:775-802.

[53] Urry C.M., Padovani P. (1995) Unified Schemes for Radio-Loud Active Galactic Nuclei. Publication of the Astronomical Society of the Pacific, 107:803-860.

[54] De Zotti G., Ricci R., Mesa D., et al. (2005) Predictions for high-frequency radio surveys of extragalactic sources. Astronomy and Astrophysics, Vol. 431:893-903.

[55] Vieira J., Crawford T.M., Switzer E.R., et al. (2010) Extragalactic Millimeter-wave Sources in South Pole Telescope Survey Data: Source Counts, Catalog, and Statistics for an 87 Square-degree Field. The Astrophysical Journal, Vol. 719:763-783.

[56] Marriage T.A., Juin J.B., Lin Y., et al. (2011) Atacama Cosmology Telescope: Extragalactic Sources at 148 GHz in the 2008 Survey. The Astrophysical Journal, Vol. 731, A100:1-15.

[57] Kellermann K.I., and Pauliny–Toth I.I.K. (1969) The Spectra of Opaque Radio Sources. The Astrophysical Journal Letters, 155:L71-L75.

[58] Marscher A. (1980) Relativistic jets and the continuum emission in QSOs. The Astrophysical Journal, 235:386-391.

[59] Planck Collaboration XIII. (2011) Planck early results. XIII. Statistical properties of extragalactic radio sources in the Planck Early Release Compact Source Catalogue. Astronomy and Astrophysics, Vol 536, A13:1-10.

[60] Sadler E., Ricci R., Ekers R., et al. (2008) The extragalactic radio-source population at 95GHz. Monthly Notices of the Royal Astronomical Society, 385:1656-1672.

[61] Planck Collaboration XV. (2011) Planck early results. XV. Spectral energy distributions and radio continuum spectra of northern extragalactic radio sources. Astronomy and Astrophysics, Vol. 536, A15:1-56.

[62] Planck Collaboration XIV. (2011) Planck early results. XIV. ERCSC validation and extreme radio sources. Astronomy and Astrophysics, Vol. 536, A14:1-18.

[63] Tucci M., Toffolatti L., De Zotti G., and Martínez-González E. (2011). High-frequency predictions for number counts and spectral properties of extragalactic radio sources. New evidence of a break at mm wavelengths in spectra of bright blazar sources. Astronomy and Astrophysics, Vol. 533, A57:1-21.

[64] Marscher A.P. (1996) The Inner Jets of Blazars. Published in "Energy transport in radio galaxies and quasars", A.S.P. Conf. Series, eds. P.E. Hardee, A.H. Bridle and J.A. Zensus (San Francisco: Astronomical Society of the Pacific), Vol. 100:45-54.

[65] Blandford R., and Königl A. (1979) Relativistic jets as compact radio sources. The Astrophysical Journal, Vol. 232:34-48.

[66] Königl A. (1981) Relativistic jets as X-ray and gamma-ray sources. The Astrophysical Journal, Vol. 243:700-709.

[67] Marscher A.P., Gear W.K. (1985) Models for high-frequency radio outbursts in extragalactic sources, with application to the early 1983 millimeter-to-infrared flare of 3C 273. The Astrophysical Journal, Vol. 298:114-127.

[68] Planck Collaboration XVI. (2011) Planck early results. XVI. The Planck view of nearby galaxies. Astronomy and Astrophysics, Vol. 536, A16:1-14.

[69] Wang L., Rowan–Robinson M. (2009) The Imperial IRAS-FSC Redshift Catalogue. Monthly Notices of the Royal Astronomical Society, Vol. 398:109-118.

[70] Planck Collaboration VII. (2012) Planck Intermediate results. VII. Statistical properties of infrared and radio extragalactic sources from the *Planck* Early Release Compact Source Catalogue at frequencies between 100 and 857 GHz. Astronomy and Astrophysics, submitted; arXiv.org//1207.4706v1.

[71] Bethermin M., Dole H., Lagache G., Le Borgne D., and Penin A. (2011) Modeling the evolution of infrared galaxies: a parametric backward evolution model. Astronomy and Astrophysics, Vol. 529, A4:1-18.

[72] Serjeant S., and Harrison D. (2005) The local submillimetre luminosity functions and predictions from Spitzer to Herschel. Monthly Notices of the Royal Astronomical Society, Vol. 356:192–204.

[73] Oliver S.J., Wang L., Smith A.J., Altieri B., Amblard A., et al. (2010) HerMES: SPIRE galaxy number counts at 250, 350, and 500 μm. Astronomy and Astrophysics, Vol. 518, L21-L25.

[74] Clements D.L., Rigby E., Maddox S., Dunne L., Mortier A., et al. (2010) Herschel-ATLAS: Extragalactic number counts from 250 to 500 microns. Astronomy and Astrophysics, Vol. 518, L8-L12.

[75] Bethermin M., Dole H., Cousin M., and Bavouzet N. (2010) Submillimeter number counts at 250 μm, 350 μm and 500 μm in BLAST data. Astronomy and Astrophysics, Vol. 516, A43:1-15.

[76] Puget J.L., Abergel A., Bernard J.P., et al. (1996) Tentative detection of a cosmic far-infrared background with COBE. Astronomy and Astrophysics, Vol. 308, L5-L9.

[77] Hauser M.G., and Dwek E. (2001) The Cosmic Infrared Background: Measurements and Implications. Annual Review of Astronomy and Astrophysics, Vol. 39:249-307.

[78] Dole H., Lagache G., Puget J.L., et al. (2006) The cosmic infrared background resolved by Spitzer. Contributions of mid-infrared galaxies to the far-IR background. Astronomy and Astrophysics, Vol. 451:417-429.

[79] Soifer B.T., and Neugebauer G. (1991) The properties of infrared galaxies in the local universe. The Astronomical Journal, Vol. 101:354-361.

[80] Planck Collaboration XVIII. (2011) Planck early results. XVIII. The power spectrum of cosmic infrared background anisotropies. Astronomy and Astrophysics, Vol. 536, A18:1-30.

Chapter 3

Exploring the Components of the Universe Through Higher-Order Weak Lensing Statistics

Adrienne Leonard, Jean-Luc Starck,
Sandrine Pires and François-Xavier Dupé

Additional information is available at the end of the chapter

1. Introduction

Our current cosmological model, backed by a large body of evidence from a variety of different cosmological probes (for example, see [1, 2]), describes a Universe comprised of around 5% normal baryonic matter, 22% cold dark matter and 73% dark energy. While many cosmologists accept this so-called concordance cosmology – the ΛCDM cosmological model – as accurate, very little is known about the nature and properties of these dark components of the Universe.

Studies of the cosmic microwave background (CMB), combined with other observational evidence of big bang nucleosynthesis indicate that dark matter is non-baryonic. This supports measurements on galaxy and cluster scales, which found evidence of a large proportion of dark matter. This dark matter appears to be cold and collisionless, apparent only through its gravitational effects.

While dark matter is largely responsible for the growth of the largest structures in the Universe, dark energy – dominant at late times – appears to have a negative pressure, and to be responsible for an accelerated expansion of the Universe [3]. It is usually parameterised by its equation of state parameter $w = p/\rho$, where p is the pressure associated with the dark energy and ρ is its energy density. An equation of state parameter of $w = -1$ would indicate a cosmological constant, consistent with general relativistic theory, but deviations from this value would suggest a rather more exotic dark energy, and might perhaps imply a modification to our current theory of gravity.

Gravitational lensing – the deflection of light rays by massive objects due to gravitational effects, which gives rise to distortions in images of galaxies – is an ideal probe of the dark universe, as it can probe the evolution of the dark matter power spectrum in an unbiased way, and offers complementary constraints to those obtained from the CMB and other probes of large-scale structure.

Moreover, gravitational lensing probes cosmological perturbations on smaller angular scales than CMB studies, and is thus sensitive to the non-Gaussianity induced by the late-time non-linear evolution of structures such as clusters of galaxies, as well as any primordial non-Gaussianity arising, for example, due to inflation very early in the evolution of the Universe.

Traditionally, constraints from gravitational lensing are obtained by considering two-point statistics of the lensing shear field, which encodes the small elliptical distortions applied to the images of galaxies as a result of the gravitational potential field of structures along the light's path. Such two-point statistics are only weakly sensitive to the dark energy density parameter Ω_Λ[1], and depend on a degenerate combination of the amplitude of the matter power spectrum σ_8 and the matter density parameter Ω_M. In addition, two-point statistics probe only the Gaussian part of the shear field, therefore in considering such statistics alone, information about nonlinear structure evolution and primordial non-Gaussianity is lost.

Similarly, attempts to reconstruct a map of the two-dimensional projected surface mass density (the convergence κ) and three dimensional density field have often involved the use of Gaussian priors to constrain the reconstruction (for example, see [4]), thus again having limited application in studies of non-Gaussianity.

In this chapter, we present a review of recently developed gravitational lensing techniques that go beyond the standard two-point statistics, both in the arena of map-making in two and three dimensions and that of higher-order statistics of the shear field or, equivalently, the convergence field κ.

The methods we present are all based on the concept of sparse recovery, which has been found to be a very powerful tool in signal processing [5, 6]. Such methods are based on the assumption that a given image or observation can be represented sparsely in an appropriate basis (such as a Fourier or wavelet basis). Sparse priors using a wavelet basis have been used in many areas of signal processing in astronomy; of particular interest in this chapter will be the areas of denoising and in the reconstruction of the 3D density field from lensing measurements, but other applications include deconvolution and inpainting.

There is much information to be gained by considering non-Gaussian statistics and nonlinear signal processing methods, and this is an exciting and active area of research. With new surveys such as the Euclid satellite [7] coming online within the next decade, a wealth of high-quality data will soon be available. These new techniques will therefore prove invaluable in constraining the cosmological model, and allow us to better understand the nature of the primary constituents of the Universe.

2. Weak gravitational lensing

We begin with a brief overview of weak lensing theory. For a more complete description and discussion of the subject, see [8, 9]

2.1. Weak lensing theory

The basic idea underlying the theory of gravitational lensing is that massive objects distort the spacetime around them, and thus bend the path of light in their vicinity. A light ray

[1] The energy density parameters are defined by $\Omega_i = \rho_i/\rho_c$, where ρ_c is the critical density of the Universe

originating at angle β is deflected such that it appears to the observer to originate at an angular position $\boldsymbol{\theta}$, where

$$\boldsymbol{\beta} = \boldsymbol{\theta} - \boldsymbol{\alpha}(\boldsymbol{\theta}) = \boldsymbol{\theta} - \nabla\psi(\boldsymbol{\theta}), \tag{1}$$

where ψ is the two-dimensional deflection potential associated with the lens and α is the deflection angle.

For extended sources, photons from different angular positions in the source plane are deflected differently, giving rise to a distortion in the observed galaxy image, which is described – to first order – by the Jacobian of the lens equation (1):

$$\mathcal{A} = \frac{\partial\boldsymbol{\beta}}{\partial\boldsymbol{\theta}} = \delta_{i,j} - \frac{\partial^2\psi}{\partial\theta_i\partial\theta_j} = \begin{pmatrix} 1-\kappa-\gamma_1 & -\gamma_2 \\ -\gamma_2 & 1-\kappa+\gamma_1 \end{pmatrix}, \tag{2}$$

where $\kappa = \frac{1}{2}\nabla^2\psi$ is the convergence, or dimensionless surface density, and $\gamma = \gamma_1 + i\gamma_2 = |\gamma|e^{2i\phi}$ is the complex shear, which gives rise to an anisotropic elliptical distortion in the lensed image. The components of the shear are related to the potential via:

$$\gamma_1 = \frac{1}{2}\left(\partial_1^2\psi - \partial_2^2\psi\right), \quad \gamma_2 = \partial_1\partial_2\psi, \tag{3}$$

where $\partial_i = \partial/\partial\theta_i$, and the shear is related to the convergence through the relation:

$$\gamma(\boldsymbol{\theta}) = \frac{1}{\pi}\int d^2\boldsymbol{\theta}'\mathcal{D}(\boldsymbol{\theta}-\boldsymbol{\theta}')\kappa(\boldsymbol{\theta}'), \tag{4}$$

where the convolution kernel $\mathcal{D}$ is given by

$$\mathcal{D}(\boldsymbol{\theta}) \equiv \frac{\theta_2^2 - \theta_1^2 - 2i\theta_1\theta_2}{|\boldsymbol{\theta}|^4} = -\frac{1}{(\theta_1 - i\theta_2)^2}. \tag{5}$$

The convergence, in turn, can be related to the 3D density contrast $\delta(\boldsymbol{r}) \equiv \rho(\boldsymbol{r})/\bar{\rho} - 1$ by

$$\kappa(\boldsymbol{\theta}, w) = \frac{3H_0^2\Omega_M}{2c^2}\int_0^w dw' \frac{f_K(w')f_K(w-w')}{f_K(w)}\frac{\delta[f_K(w')\boldsymbol{\theta}, w']}{a(w')}, \tag{6}$$

where $\bar{\rho}$ is the mean density of the Universe, H_0 is the Hubble parameter, Ω_M is the matter density parameter, c is the speed of light, $a(w)$ is the scale parameter evaluated at comoving distance w, and

$$f_K(w) = \begin{cases} K^{-1/2}\sin(K^{1/2}w), & K > 0 \\ w, & K = 0 \\ (-K)^{-1/2}\sinh([-K]^{1/2}w) & K < 0 \end{cases}, \tag{7}$$

gives the comoving angular diameter distance as a function of the comoving distance and the curvature, K, of the Universe.

2.2. Two-point statistics

The typical gravitational shear applied to a galaxy as a result of lensing by large-scale structure in the Universe – so-called cosmic shear – is of order $|\gamma| \sim 0.01$. However, galaxies are intrinsically elliptical in shape, with a typical ellipticity of order $|\varepsilon| \sim 0.2 - 0.3$, therefore the gravitational lensing effect can only be measured statistically; under the assumption that galaxy shapes are intrinsically uncorrelated, the mean ellipticity computed over a large number of sources will yield the gravitational shear: $\langle \varepsilon \rangle \simeq \gamma$.

The most common method for constraining cosmological parameters in weak lensing studies is to use two-point statistics of the shear field. The power spectrum of the shear or convergence, $P_\kappa(\ell) = P_\gamma(\ell)$, can be related directly to the 3D matter power spectrum of density fluctuations δ by:

$$P_\kappa(\ell) = \frac{9H_0^4\Omega_M^2}{4c^4} \int dw \, \frac{W^2(w)}{a^2(w)} P_\delta \left(\frac{\ell}{f_K(w)}, w \right) , \tag{8}$$

where

$$W(w) = \int_w^\infty dw' \frac{f_K(w' - w)}{f_k(w')} p(z) \left[\frac{dz}{dw'} \right] , \tag{9}$$

is a weighting function integrated over the probability distribution of sources $p(z)$ in the sample as a function of redshift z.

When working with real data, it is often more convenient to consider statistics computed in real space, namely the shear correlation functions, which are defined in terms of the convergence power spectrum as [10, 11]:

$$\xi_\pm(\theta) = \langle \gamma_t \gamma_t \rangle \pm \langle \gamma_\times \gamma_\times \rangle = \int_0^\infty \frac{d\ell \, \ell}{2\pi} J_{0,4}(\ell\theta) P_\kappa(\ell) , \tag{10}$$

where J_n is an n-th order Bessel function of the first kind, and γ_t and $\gamma_\times$ are the tangential and cross shear components, respectively, which are defined relative to the vector connecting the two galaxies separated by angular distance θ.

2.3. Cosmological constraints

The measured two-point statistics described above can, by virtue of their relationship to the power spectrum of the underlying matter density fluctuations, be used to place constraints on cosmological parameters. These constraints may be improved if information about the distances to source galaxies is used to bin the galaxies into redshift slices, and compute the correlation functions tomographically.

Figure 1 shows cosmological parameter constraints obtained using two state-of-the-art telescopes: the Canada-France-Hawaii Telescope (CFHT), and the Hubble Space Telescope (specifically, the Cosmos survey). While the CFHT results were obtained using a 2D analysis [12], the Cosmos results show constraints obtained by both a 2D and a 3D analysis of the shear data [13]. In both cases, a large degeneracy is seen between the matter density parameter Ω_M and σ_8, the amplitude of the matter power spectrum. The CFHT results also show constraints on the dark energy equation of state parameter w. Here, again, a degeneracy between parameters is seen.

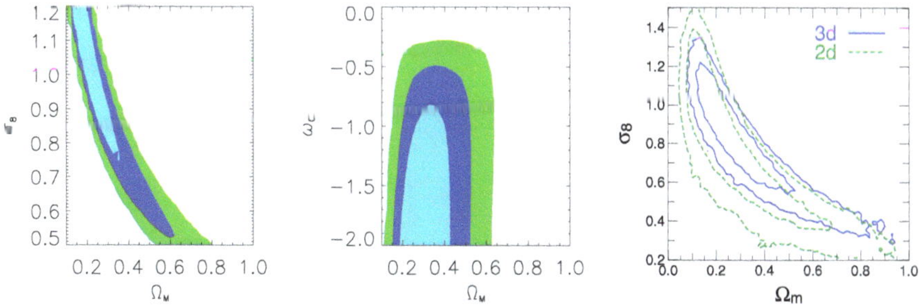

(a) Joint constraints on Ω_M and σ_8 from a 2D analysis of the combined CFHT wide and deep surveys [12].

(b) Joint constraints on Ω_M and w from a 2D analysis of the CFHT deep survey [12].

(c) Joint constraints on Ω_M and σ_8 from a 2D and 3D analysis of the Cosmos survey data [13]

Figure 1. Cosmological parameter constraints from recent weak lensing surveys. Contours are plotted at the 68% (cyan), 95% (blue) and 99.9% (green) confidence levels for the CFHT data, and at the 68% (inner) and 95% (outer) confidence levels for the Cosmos data.

While it is clear from the figures that two-point shear statistics hold a wealth of information, it is also evident that these statistics depend on degenerate combinations of the cosmological model parameters. In order to break these degeneracies, and thus more tightly constrain our cosmological model, we must therefore consider higher-order statistics of the shear or convergence. Such higher-order statistics will enable us to capture information about the non-Gaussian part of the lensing spectrum due to a combination of primordial non-Gaussianity and late-time nonlinear evolution of structures, which two-point statistics are unable to probe.

3. Scalar fields associated with the shear

The shear γ is a spin-2 field, therefore while the two-point correlations of the shear field can be reduced to a scalar quantity for parity reasons, this is not the case for higher-order moments of the shear field [14–16]. For this reason, it is useful to compute a scalar quantity from the shear field before computing higher-order statistics.

3.1. The aperture mass statistic

For this purpose, the aperture mass statistic M_{ap} [17, 18] is widely-used. The M_{ap} statistic is defined as the convolution of the convergence κ with a radially symmetrical filter function $U(|\boldsymbol{\vartheta}|)$ of width θ:

$$M_{ap}(\boldsymbol{\theta}) = \int d^2\boldsymbol{\vartheta}\ \kappa(\boldsymbol{\vartheta})U(|\boldsymbol{\vartheta}|)\ . \tag{11}$$

By considering the relationship between the shear, γ, and the convergence, one can reformulate equation (11) in terms of the measured shear as

$$M_{ap}(\boldsymbol{\theta}) = \int d^2\boldsymbol{\vartheta}\ \gamma_t(\boldsymbol{\vartheta}) Q(|\boldsymbol{\vartheta}|)\ , \tag{12}$$

where $\gamma_t(\boldsymbol{\vartheta})$ is the tangential component of the shear at position $\boldsymbol{\vartheta}$ relative to the centre of the aperture, $Q(|\boldsymbol{\vartheta}|)$ is a second radially-symmetric function, related to $U(|\boldsymbol{\vartheta}|)$ by:

$$Q(\vartheta) \equiv \frac{2}{\theta^2} \int_0^{\vartheta} \vartheta' U(\vartheta') d\vartheta' - U(\vartheta)\ , \tag{13}$$

and $U(\vartheta)$ is required to be compensated, i.e.

$$\int_0^{\vartheta_{cut}} \vartheta\ U(\vartheta)\ d\vartheta = 0\ , \tag{14}$$

with ϑ_{cut} often taken to be the radius of the aperture, θ. Furthermore, $Q(\vartheta)$ and $U(\vartheta)$ are required to go to zero smoothly at ϑ_{cut}. It is also preferable that the power spectrum of $U(\vartheta)$ is local in the frequency domain, and shows no oscillatory behaviour. This ensures that the filter function acts as a band-pass filter, allowing detection of structures at the scale of interest only.

Several authors [19, 20] have advocated a filter function of the form:

$$\begin{aligned} U(\vartheta) &= \frac{A}{\theta^2}\left(1 - \frac{b\vartheta^2}{\theta^2}\right) \exp\left(-\frac{b\vartheta^2}{\theta^2}\right)\ , \\ Q(\vartheta) &= \frac{A}{\theta^2}\frac{b\vartheta^2}{\theta^2} \exp\left(-\frac{b\vartheta^2}{\theta^2}\right)\ , \end{aligned} \tag{15}$$

where various choices for the constant b and the overall normalisation A have been used in the literature [20–22]. Of specific interest is the form used in [20] and [22], where $b = 1/2$ is chosen.[2] This form is considered to be optimal for higher-order weak lensing statistics, such as the skewness of the aperture mass statistic [19, 22].

Note that the $Q(\vartheta)$ filter function described in equation (15) shows a peak at $\vartheta = \sqrt{2}\theta$, and tends to zero as $\vartheta \to \infty$. In practice, any algorithm used to generate an aperture mass map will need to truncate this filter function at some finite radius. This will involve a trade-off between accuracy and algorithm speed, and truncation may affect the effective width of the filter.

One advantage of this method is that the aperture mass statistic can be computed directly from the shear catalogue. Moreover, as the filter acts as a band-pass filter, it is possible to boost the signal relative to the noise by considering a filter with a scale substantially larger

[2] This form is analogous to the Mexican Hat wavelet function

than the typical scale of the noise (which is typically dominant on pixel scales, depending on the binning chosen for the shear data). Indeed, optimal signal-to-noise is obtained when the filter is chosen such that its angular scale matches that of the structures we aim to detect, and its shape matches closely the expected profile of these structures.

3.2. The wavelet transform

The wavelet transform is a multiscale transform, where the wavelet coefficients of an image are computed at each position in the image at various different scales simultaneously. In one dimension, the wavelet coefficient of a function $f(x)$, evaluated at position b and scale a is defined as [23, 24]:

$$W(a,b) = \frac{1}{\sqrt{a}} \int f(x)\psi^*\left(\frac{x-b}{a}\right) dx \,, \tag{16}$$

where $\psi(x)$ is the analysing wavelet. The analysis is analogous in 2 dimensions, with $\psi(x,y) = \psi(x)\psi(y)$.

By definition, wavelets are compensated functions; i.e. the wavelet function $\psi(x)$ is constrained such that $\int_{\mathbb{R}^1} \psi(x)dx = 0$ and hence, by extension

$$\iint_{\mathbb{R}^2} \psi(x,y)dx\, dy = 0. \tag{17}$$

According to the definition in equation (16), the continuous wavelet transform of an image is therefore nothing more than the convolution of that image with compensated filter functions of various characteristic scales. If the image $f(x,y)$ is taken to be the convergence $\kappa(x,y)$, then for an appropriate choice of (radially-symmetric, local) wavelet, the wavelet transform is formally identical to the aperture mass statistic at the corresponding scales, the only difference being the choice of filter functions.

In practice in application, we use the starlet transform algorithm [6, 23–25], which simultaneously computes the wavelet transform on dyadic scales corresponding to 2^j pixels. This algorithm decomposes the convergence map of size $N \times N$ into $J = j_{\max} + 1$ sub-arrays of size $N \times N$ as follows:

$$\kappa(x,y) = c_J(x,y) + \sum_{j=1}^{j_{\max}} w_j(x,y) \,, \tag{18}$$

where $j_{\max}$ represents the number of wavelet bands (or, equivalently, aperture mass maps) considered, c_J represents a smooth (or continuum) version of the original image κ, and w_j represents the input map filtered at scale 2^j (i.e. the aperture mass map at $\theta = 2^j$ pixels).

Using the wavelet formalism to derive the aperture mass statistic presents a number of advantages. Many families of wavelet functions have been studied in the statistical literature, and all these wavelet functions could be applied to the aperture mass statistic. This allows us to tune our filter function to optimise the signal-to-noise in the resulting maps. In addition, for some specific wavelet functions, discrete and very fast algorithms exist, allowing us to compute a set of wavelet scales through the use of a filter bank with a very limited number of operations. See [6] for a full review of the different wavelet transform algorithms.

In the starlet transform algorithm, the wavelet $\psi(x,y)$ is separable and can be defined by:

$$\psi\left(\frac{x}{2},\frac{y}{2}\right) = 4\left[\varphi(x,y) - \frac{1}{4}\varphi\left(\frac{x}{2},\frac{y}{2}\right)\right], \tag{19}$$

where $\varphi(x,y) = \varphi(x)\varphi(y)$ and $\varphi(x)$ is a scaling function from which the wavelet is generated. In the case of the starlet wavelet, $\varphi(x)$ is a B3-spline:

$$\varphi(x) = \frac{1}{12}(|x-2|^3 - 4|x-1|^3 + 6|x|^3 - 4|x+1|^3 + |x+2|^3), \tag{20}$$

which is a compact function that is identically zero for $|x| > 2$.

This wavelet function has a compact support in real space, is well localized in Fourier domain, and the wavelet decomposition of an image can be obtained with a very fast algorithm (see [24] for a full description).

Figure 2 shows the starlet wavelet and aperture mass filters defined above in both real and Fourier space. Notice that the two filters presented have very similar shapes in real space, but different widths. While the starlet filter function goes to zero identically at $\vartheta = 2\theta$, and remains zero beyond this value, the M_{ap} filter function goes to zero as $\vartheta \to \infty$, and must therefore be truncated when applied in practice. Clearly at $\vartheta = 5\theta$, the M_{ap} filters are sufficiently close to zero, so that this is an appropriate truncation radius; however this will impact the computation time of the M_{ap} statistic.

In Fourier space, we show the response of the M_{ap} filter for truncation radii of $\vartheta_{cut} = \theta$ and 5θ. Notice that the shape of the response curves in Fourier space for both the M_{ap} and wavelet filter functions are similar, but that the peak of the response curve for the M_{ap} filter function shifts when ϑ_{cut} is varied. This is because the effective scale of the filter function is being changed. Also notice that in the case of $\vartheta_{cut} = \theta$, high-frequency oscillations are present in the response curve. This is a direct result of the truncation of the filter function, and will occur whenever such truncation is applied. Indeed, imperceptibly small oscillations are still present in the response curve for the filter truncated at $\vartheta_{cut} = 5\theta$.

Such oscillations are not present at all with the starlet filter function, as no truncation is applied to this function whatsoever. This gives the starlet transform the distinct advantage of being directly applicable, without any consideration needed regarding truncation radii and the associated impact on the Fourier-space response of the filter functions.

3.3. Advantages of the wavelet formalism

The aperture mass formalism tends to be the preferred method for weak lensing studies for a number of reasons. Firstly, the filter functions can be simply expressed analytically, and any associated statistics of the aperture mass can therefore be straightforwardly computed from

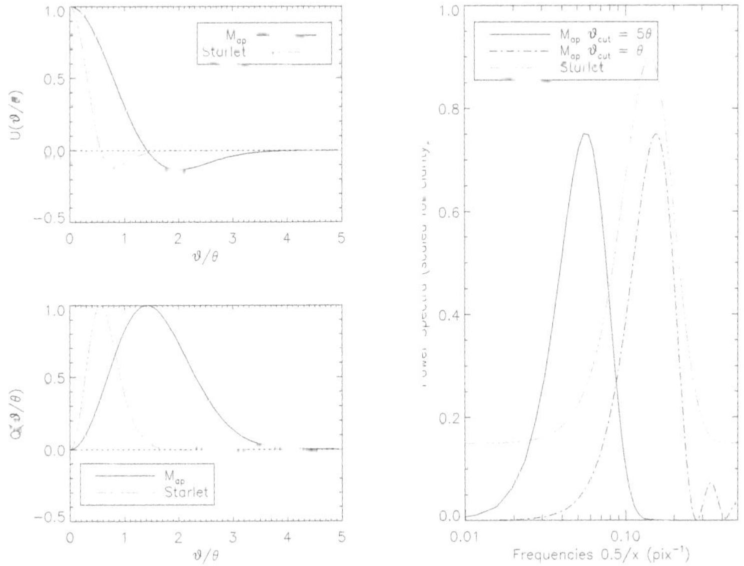

(a) M_{ap} and wavelet filter functions in real space

(b) Filter response curves in Fourier space

Figure 2. A comparison of the M_{ap} and starlet wavelet filters in real and Fourier space.

the shear catalogue directly. This avoids the need to generate an aperture mass map, which can be computationally intensive, and furthermore does not require the computation of the convergence, κ.

Computing the convergence from the shear measurements can be tricky for a number of reasons. Equation (4) describes a convolution over all 2D space. Given that we aim to invert this equation for an image of a finite size, a direct inversion in Fourier space will give rise to significant edge effects, and a leakage of power into so-called B-modes, which effectively imply a spurious (non-lensing) cross-component to the shear field γ, and usually only arise due to systematic effects in the lensing measurements.

Several methods have been devised to invert equation (4) whilst minimising these undesirable effects [26–29]. Most recently, the authors in [30] have presented a method based on a wavelet-Helmholtz decomposition, with which they demonstrate a reconstruction error at the few percent level, as compared to an error of $\sim$ 30% seen with Fourier-based methods. This implies that there is little advantage to the shear catalogue as opposed to the convergence map. Indeed, working on the convergence map and under the wavelet framework may offer some distinct advantages.

The first advantage comes directly from the nature of the wavelet transform. The aperture mass filter is a bandpass filter. Because the M_{ap} reconstruction is only computed for a discrete set of aperture scales, information on intermediate scales may be lost, particularly at both the small and large frequency extremes. In contrast, the wavelet transform retains information on all scales. The first wavelet scale is effectively a high-pass filter, retaining all the high-frequency information in the image, while the remaining wavelet scales are bandpass filtered as with the M_{ap}-filtered images. Finally, the wavelet transform retains c_J, which encodes the large-scale information in the image, and therefore consists of a smoothed version of the image (or an image with a low-pass filter applied). Figure 3(a) demonstrates this point for a wavelet transform with $J = 5$, showing the Fourier-space response of the wavelet filter as a function of scale j.

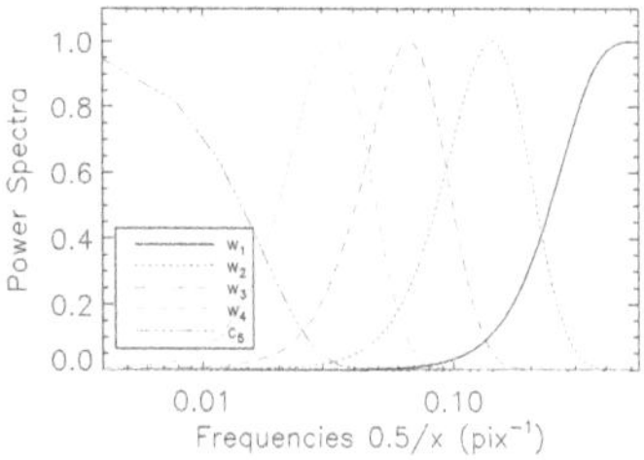

(a) The filter response of each of 5 scales of the wavelet transform

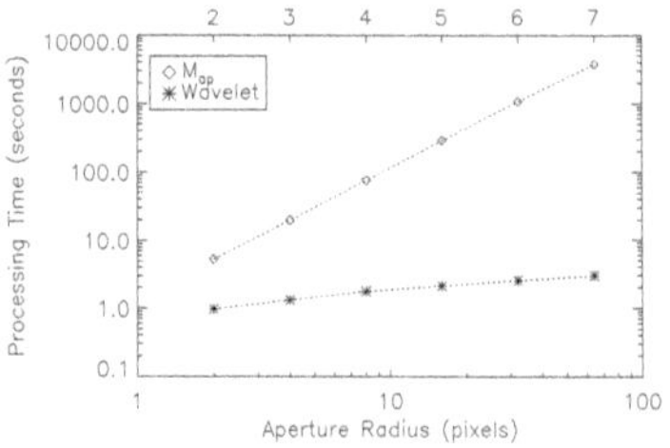

(b) Comparison of a the processing time of a brute-force aperture mass algorithm and the starlet transform algorithm

Figure 3. Illustration of the various advantages offered by the wavelet transform in terms of (a) retention of information on all scales and (b) algorithm processing time.

In addition, from a mapping perspective, the starlet wavelet transform algorithm is substantially faster than a brute-force computation of the aperture mass statistic in real space. Such a naive implementation has complexity $\propto \mathcal{O}(N^2\vartheta_{cut}^2)$, where $N \times N$ is the dimension of the image, and ϑ_{cut} is the chosen truncation radius. This scaling means that for large apertures or, equivalently, for high-resolution images, the algorithm may prove to be very time-consuming. The starlet wavelet transform algorithm is of complexity $\propto \mathcal{O}(N^2 J)$, where J is the number of scales considered, and is limited by $N \geq 2^J$. This means that the processing time for the wavelet transform algorithm is sensitive only to the number of scales considered, rather than the size of the filter functions involved, and depends linearly on this number.

In Figure 3(b), we compare the processing time for the aperture mass algorithm and the starlet transform algorithm, both programmed in C++, to analyse an image of 1024×1024 pixels on a 2×2.66GHz Intel Xeon Dual-Core processor. We consider aperture scales $\theta = [2, 4, 8, 16, 32, 64]$, which correspond to $J = j_{\max} + 1 = [2, 3, 4, 5, 6, 7]$ wavelet scales in the wavelet transform. In the aperture mass algorithm, the filters are truncated at a radius of $\vartheta_{cut} = \theta$. For the application of filters which necessitate truncation at a much larger radius, we expect the computation time to be roughly an order of magnitude longer. Even at the smallest aperture radius, the wavelet transform is $\sim 5\times$ faster than the aperture mass algorithm. At $\theta = 64$ pixels, the wavelet transform is $\sim 1200\times$ faster than the aperture mass algorithm. Note that the wavelet transform for a given value of J simultaneously computes the wavelet transform at all scales 2^j, $0 < j \leq J - 1$, in addition to the smoothed continuum map c_J, whilst the aperture mass algorithm computes the transform at a single scale θ. We note further that the computational time for the wavelet transform for $J = 7$ wavelet scales is still a factor of $\sim 2\times$ less than the computational time for the aperture mass algorithm at $\theta = 2$ pixels.

As discussed in [31], this time advantage further extends to higher-order statistics of the M_{ap}, which are typically related directly to n-point correlation functions of the shear. In recent years, tree codes have been employed to speed up computation of n-point correlation functions. Typical tree codes to compute n-point correlation functions are $\mathcal{O}(N_{gal}\log(N_{gal}))$ [32] on a shear catalogue. For a single Euclid exposure of 0.5 deg^2, we can expect $N_{gal} \sim$ 54,000 (30 galaxies/arcmin^2) - 180,000 (100 galaxies/arcmin^2). Tree codes exist that act on pixelated data [33] which run at $\mathcal{O}(N_{pix}^2 n_{bin})$ where N_{pix}^2 is the total number of pixels and n_{bin} is the number of bins in the correlation function. For a Euclid exposure, assuming pixels of 1 arcminute, we have $N_{pix}^2 = 1800$, and n_{bin} will be dependent on the required resolution of the correlation function.

The wavelet method acts on pixellated data, and is $\mathcal{O}(N^2 J)$ in computation time, so our algorithm will be comparable for computation of 2-point statistics, if the 2-point correlation function is computed on pixellated data, but if a shear catalogue is used, we will have a faster algorithm by at least an order of magnitude. For higher-order statistics, this advantage is even more pronounced. Furthermore, while optimised software is freely and publicly available to compute the wavelet transform, such optimised software is not available for n-point correlation functions.

Another advantage of the wavelet formalism is that it is possible to carry out an explicit denoising of the convergence field using thresholding based on a False Discovery Rate method. For details on this method, see [34], where the *MRLens* software package encoding this method is presented[3]. This allows one to derive robust detection levels in wavelet space, and to produce high-fidelity denoised mass maps.

In addition, wavelet-based methods offer more flexibility than aperture mass filters. Whilst we have thus far discussed only the starlet wavelet function, many other wavelet dictionaries may be used. The starlet filter seems ideal for lensing studies, due to its similarity to the aperture mass filter presented here, which was deemed to be optimal in [22]. However, different dictionaries may be optimal in different applications; for example, if one were attempting to study filamentary structure, ridgelets or curvelets might be a more appropriate basis. The vastness of the wavelet libraries and the public availability of fast algorithms to compute these transforms are major strengths of wavelet-based approaches.

4. Weak lensing beyond two-point shear statistics

As noted previously, the optimality of second-order statistics to constrain cosmological parameters depends heavily on the assumption of Gaussianity of the field. However the weak lensing field is composed, at small scales, of non-Gaussian features such as clusters of galaxies. These non-Gaussian signatures carry additional information that cannot be extracted with second-order statistics. Many studies [35–39] have shown that combining second-order statistics with higher-order statistics tightens the constraints on cosmological parameters. We now consider the higher-order statistics most commonly used in weak lensing studies aimed at detecting and constraining non-Gaussianity in the lensing field.

[3] The MRLens software, along with many other software packages for astronomical applications, is freely available here. http://cosmostat.org/software.html

4.1. Higher-order lensing statistics

As we have already noted, it is more convenient to consider higher-order statistics of the M_{ap} or wavelet transform of the convergence field, as opposed to statistics of the shear field directly. The obvious first extension to two-point statistics is to consider the three-point correlation function or, equivalently, the bispectrum, usually considered as a function of aperture or wavelet scale.

Interesting analytical results relative to the shear three-point correlation function or the convergence bispectrum have been reported (for example, see [40–43]). However, it has been shown that in using only the equilateral configuration of the bispectrum, the ability to discriminate between cosmological models is relatively poor [38]. An analytical comparison has been performed in [39] between the full bispectrum and an optimal match-filter peak count for a Euclid-like survey, and both approaches were found to provide similar results. However, as the full bispectrum calculation has a much higher complexity than other statistics, and no public software exists to compute it, we will not consider the full bispectrum here. Rather, we will restrict ourselves to statistics that are more straightforward to compute from the M_{ap} or wavelet transform maps.

4.1.1. The Skewness

The skewness of the aperture mass map, $\langle M_{ap}^3 \rangle$, is the third-order moment of the aperture mass $M_{ap}(\theta)$ and can be computed directly from shear maps filtered with different aperture mass. The skewness is a measure of the asymmetry of the probability distribution function. The probability distribution function will be more or less skewed positively depending on the abundance of dark matter haloes at the θ scale. The formalism exists to predict the skewness of the aperture mass map for a given cosmological model, which is related to the three-point correlation function or the bispectrum of density fluctuations δ [20, 37].

In [37], it is argued that the skewness as a function of scale is a preferable statistic to three-point correlation functions of the shear, as the integral relations between $\langle M_{ap}^3 \rangle$ and the bispectrum are much easier and faster to compute than the three-point correlation function. This is because the skewness is a local measure of the bispectrum, whereas the integral kernel for the three-point correlation function is a highly oscillating function with infinite support.

This statistic can equivalently be computed from the wavelet transform of the convergence map as the skewness $\langle w_j^3 \rangle$ of the wavelet band j corresponding to the aperture scale θ. This can be computed either on the noisy convergence map estimated, for example, by the Fourier space relationship between the shear and convergence – in which case the results should be comparable to the M_{ap} results – or on a denoised convergence map using the method of [34], which should provide improved results.

4.1.2. The Kurtosis

The kurtosis of the aperture mass map, $\langle M_{ap}^4 \rangle$, is the fourth-order moment of the aperture mass $M_{ap}(\theta)$ and can be computed directly from the different aperture mass maps. The kurtosis is a measure of the peakedness of the probability distribution function. The presence of dark matter haloes at a given θ scale will flatten the probability distribution function and widen its shoulders leading to a larger kurtosis. The formalism exists to predict the kurtosis of the aperture mass map for a given cosmological model, which is related to the four-point correlation function or trispectrum of the 3D density contrast [20].

Again, this statistic can be computed from the wavelet transform of the convergence map as the kurtosis $\langle w_j^4 \rangle$ of the wavelet band j corresponding to the aperture scale θ. This can be computed either on a noisy or a denoised convergence map, the latter expected to yield improved constraints.

4.1.3. Peak Counts

We define a peak as set of connected pixels above a detection threshold $\mathcal{T}$, and denote the peak count in the M_{ap} and wavelet maps by $P^{\mathcal{T}}_{M_{ap}}$ and $P^{j}_{w_j}$, respectively. If peak counting is carried out on a denoised map, the detection threshold $\mathcal{T}$ is automatically set by the denoising algorithm, and a small threshold ϵ is used when identifying peaks in the denoised maps, in order to reject spurious detections in these denoised maps.

We consider all pixels that are connected via the sides or the corners of a pixel as one structure. In a two-dimensional projected map of the convergence, we are therefore unable to discriminate between peaks due to massive halos and peaks due to projections of large-scale structures such as filaments.

While theory exists to predict cluster counts from the halo model and a cosmological model encoding the growth and evolution of structures in the Universe, there is no analytic formalism to predict the fraction of spurious detections in lensing maps arising from projections of large-scale structure. In [44], an attempt is made to derive an analytic formalism for predicting peak counts in projected maps. However, this method is based on an assumption of Gaussianity in the lensing field and, predictably, underestimates counts at the high end of the mass function, where clusters of galaxies are dominant. This means that predicting peak counts for a given cosmological model relies on considering N-body simulations generated under a range of cosmological parameters (for example, see [45]).

4.2. Optimal capture of non-Gaussianity

The question now arises: which of these statistics provides the most information about the underlying cosmology and – specifically – non-Gaussianity in the density field? In order to asses the performance of these statistics, we consider the ability of each statistic to discriminate between cosmological models using N-body simulations under a range of different cosmologies [38, 46].

To this end, we consider N-body simulations carried out for 5 different cosmological models along the $\Omega_M - \sigma_8$ degeneracy, the parameters of which are summarised in Table 1 below [46]. These simulations were carried out using the RAMSES N-body code [47], and full details of the models considered are given in [38]. For each model, 100 realizations of weak lensing maps were generated for a field of $3.95^\circ \times 3.95^\circ$, downsampled to 1024×1024 pixels ($0.23'$ per pixel). Noise was added to the simulations at a level consistent with predictions for deep space-based observations, with a number density of galaxies of $n_g = 100\,\text{gal}/\text{arcmin}^2$. This is somewhat optimistic; however, as seen in [46], increasing the noise level in the simulations does not change the conclusions of the comparison test between the higher-order statistics.

Model	Box Length (h^{-1}Mpc)	Ω_M	Ω_Λ	h	σ_8
m_1	165.8	0.23	0.77	0.594	1
m_2	159.5	0.3	0.7	0.7	0.9
m_3	152.8	0.36	0.64	0.798	0.8
m_4	145.7	0.47	0.53	0.894	0.7
m_5	137.5	0.64	0.36	0.982	0.6

Table 1. Parameters of the five cosmological models that have been chosen along the $\Omega_M - \sigma_8$ degeneracy. The simulations have 256^3 particles, h is equal to $H_0/100\,\mathrm{km\,s^{-1}\,Mpc^{-1}}$, where H_0 is the Hubble constant.

To find the best statistic, we need to characterise quantitatively for each statistic the discrimination between two different models m_1 and m_2. To do this, we consider the distribution functions for the different statistics estimated on the 100 realisations of each model. These distributions are expected to overlap, so False Discovery Rate (FDR) method is used to determine the threshold τ_1 in the distribution function of model m_1, such that the distribution function of model m_2 accounts for fewer than a fraction $\alpha = 0.05$ of the total counts. A similar threshold τ_2 is defined for the distribution function of model m_2. This is illustrated in Figure 4. The discrimination efficiency is then defined as the percentage of counts remaining under the hatched area for each model.

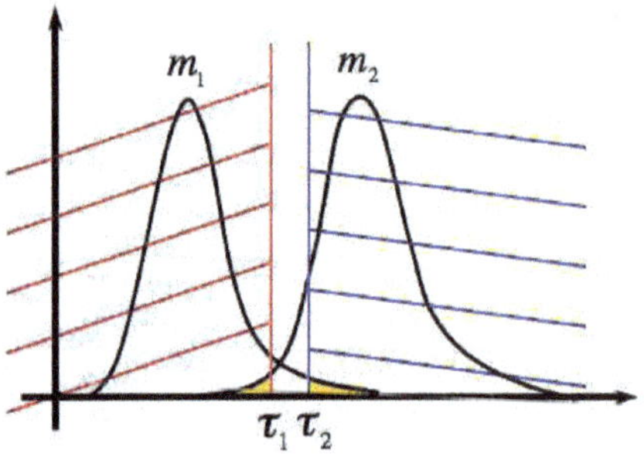

Figure 4. Illustration of the discrimination efficiency criterion. τ_1 and τ_2 are defined such that the yellow area under the curve represents 0.05 of the total area under the curves for models m_2 and m_1, respectively. The discrimination efficiency is defined as the area under the m_i curve delimited by the threshold τ_i, as a percentage of the total area under the model m_i curve.

In Table 2 below, we show the mean discrimination efficiency obtained considering all pairs of models for the higher-order statistics of interest. These are presented as a function of M_{ap} or wavelet scale, and shown for statistics computed on the aperture mass map and the wavelet transform of the noisy convergence map. The results overall are comparable, with the wavelet transform statistics appearing to offer a slight improvement over the M_{ap} statistics in all cases. This is perhaps related to the fact that, when computing the M_{ap} statistics, we truncate the filter at $\vartheta_{cut} = \theta$, which may give rise to some small systematics in the resulting maps. Furthermore, it is clear that, in all cases, the peak statistics are much more efficient at discriminating between cosmological models than either the skewness or the kurtosis.

θ_i	$\langle M_{ap}^3 \rangle$	$\langle M_{ap}^4 \rangle$	$P_{M_{ap}}^{2\sigma}$	$P_{M_{ap}}^{3\sigma}$
0.46′	04.60 %	02.30 %	39.70 %	54.95 %
0.92′	**33.40 %**	03.70 %	79.30 %	76.40 %
1.85′	03.45 %	01.45 %	**91.25 %**	**89.20 %**
3.70′	15.15 %	23.00 %	69.40 %	86.70 %
7.40′	26.95 %	**24.30 %**	4.90 %	60.50 %

(a) Mean discrimination efficiency of M_{ap} statistics

Scale	$\langle w_j^3 \rangle$	$\langle w_j^4 \rangle$	$P_{w_j}^{2\sigma}$	$P_{w_j}^{3\sigma}$
0.46′	02.00 %	01.15 %	12.05%	00.70 %
0.92′	**37.95 %**	04.75 %	86.30 %	73.05 %
1.85′	03.55 %	02.10 %	**94.40 %**	**93.85 %**
3.70′	18.25 %	25.65 %	84.05 %	87.05 %
7.40′	36.40 %	**30.90 %**	24.60 %	66.35 %

(b) Mean discrimination efficiency of wavelet statistics

Table 2. Mean discrimination efficiency (in per cent) from noisy M_{ap} reconstructions of the shear field, compared with that obtained on the wavelet transform of the noisy convergence maps.

In Table 3 below, we present the mean discrimination efficiency of statistics computed on the denoised convergence maps, as well as the peak discrimination efficiency between each pair of models in our sample. There is a clear improvement on the discrimination efficiency of all the higher-order statistics when denoising is applied to the convergence maps. This is unsurprising, as the presence of Gaussian (or near-Gaussian) noise within the shear and convergence maps will make the whole field appear more Gaussian, masking the non-Gaussian features in the map and pushing the skewness and kurtosis values closer to zero. The more noise that is present in the maps, the stronger this effect will be. It is for this reason that such a dramatic improvement is seen in these statistics when denoising is applied to the convergence maps before the statistics are computed.

The improvement seen in the discrimination efficiency of the peak statistics is also significant, and this arises due to the fact that while the peak counts in the noisy maps involve application of a simple $n\sigma$ detection threshold to the maps, the denoising algorithm involves a more sophisticated discriminant between signal and noise, again making use of the FDR method. This technique is more effective and distinguishing between signal and noise, and therefore more information about the true peaks in the map is retained using this denoising method as compared to the $n\sigma$ thresholding applied previously.

It is clear that peak statistics are much more efficient at capturing non-Gaussianity in the shear field than other higher-order statistics, and that it is advantageous to measure this statistic on denoised convergence maps. However, in order to combine constraints from peak counting with other probes, such as two-point statistics, in order to better constrain the cosmological model, it is important to be able to predict peak statistics as a function of cosmology in order to be able to compute the likelihood function for all models within the parameter space. As we have discussed, this is not possible to do analytically, due to contamination by projections of large scale structures, and it is too computationally-intensive

Scale	$\langle \tilde{w}_j^3 \rangle$	$\langle \tilde{w}_j^4 \rangle$	$P_{\tilde{w}_j}$
0.46′	53.40 %	43.20 %	68.35 %
0.92′	47.90 %	41.15 %	92.45 %
1.85′	58.80 %	44.70 %	**96.75 %**
3.70′	**63.30 %**	**48.05 %**	90.40 %
7.40′	54.90 %	40.45 %	63.45 %

(a) Mean discrimination efficiency

	m_1	m_2	m_3	m_4	m_5
m_1	x	85 %	100 %	100 %	100 %
m_2	89 %	x	92 %	100 %	100 %
m_3	100 %	92 %	x	89 %	100 %
m_4	100 %	100 %	92 %	x	98 %
m_5	100 %	100 %	100 %	98 %	x

(b) Discrimination efficiency for all models obtained using peak statistics

Table 3. Discrimination efficiency for statistics computed on denoised convergence maps.

to consider obtaining the predictions, and associated covariances, from N-body simulations if we wish to sample the parameter space completely and at high resolution.

As the quality of data available to astronomers continues to improve, it has recently become possible to consider the shear field in three dimensions, using the colours of galaxies to determine their redshifts and, therefore, distances from us. If this information may be used to deproject the lensing signal, and thus recover information about the full three-dimensional density field, it should then be possible offer predictions for peak statistics as a function of cosmology, uncontaminated by projection effects which will, in turn, allow us to improve our constraints on our cosmological model.

5. Reconstructing the density contrast in 3D

The measured shear can be related to the 2D convergence via a simple linear relationship (equation (4)), and the convergence is related to the density contrast by another linear mapping (equation (6)). We can express these relationships conveniently in matrix notation as

$$\kappa(\boldsymbol{\theta}, z) = \mathbf{Q}\delta(\boldsymbol{\theta}, z) \,, \tag{21}$$

$$\gamma(\boldsymbol{\theta}, z) = \mathbf{P}_{\gamma\kappa}\mathbf{Q}\delta(\boldsymbol{\theta}, z) \,. \tag{22}$$

The 3D lensing problem is therefore one of finding an estimator to invert equation (21) or (22) in the presence of noise. For simplicity, in what follows we assume the shear measurement noise is Gaussian and uncorrelated between redshift bins. In practice, errors in the photometric estimates of the redshifts of galaxies will introduce correlations between redshift bins. Methods have been developed to account for such errors, however, and the problem is therefore readily tractable [48].

Therefore the 3D lensing problem is effectively one of observing the density contrast convolved with the linear operator **R**, and contaminated by noise, which is assumed to be Gaussian. Formally, we can write

$$\boldsymbol{d} = \mathbf{R}\boldsymbol{s} + \varepsilon, \quad \varepsilon \sim \mathcal{N}(0, \sigma^2)\,, \tag{23}$$

where $\boldsymbol{d}$ is the observation, $\boldsymbol{s}$ the real density and ε the Gaussian noise.

5.1. Linear approaches to 3D map-making

The general idea behind linear inversion methods is to find a linear operator **H** that acts on the data vector to yield a solution that minimises some functional, such as the variance of the residual between the estimated signal and the true signal, subject to some regularisation or prior-based constraints. Two different linear approaches have been described in recent literature [4, 49].

The most competitive method is that proposed in [4], in which the authors propose a Saskatoon filter [50, 51], which combines a Wiener filter and an inverse variance filter, with a tuning parameter α introduced that allows switching between the two. This gives rise to a minimum variance filter, expressed as

$$\hat{\boldsymbol{s}}_{MV} = [\alpha\mathbf{1} + \mathbf{S}\mathbf{R}^{\dagger}\mathbf{\Sigma}^{-1}\mathbf{R}]^{-1}\mathbf{S}\mathbf{R}^{\dagger}\mathbf{\Sigma}^{-1}\boldsymbol{d}\,, \tag{24}$$

where $\mathbf{S} \equiv \langle \boldsymbol{s}\boldsymbol{s}^{\dagger} \rangle$ encodes prior information about the signal covariance, $\Sigma \equiv \langle \boldsymbol{n}\boldsymbol{n}^{\dagger} \rangle$ gives the covariance matrix of the noise, and $\mathbf{1}$ is the identity matrix.

This switching is designed to allow a balance between the increased constraining power offered by the Wiener filter over the inverse variance filter – which yields an improved signal-to-noise in the reconstruction – and the biasing that the Wiener filter imposes on the solution.

As discussed extensively in [4] and [49], linear methods give rise to a significant bias in the location of detected peaks, damping of the peak signal, and a substantial smearing of the density along the line of sight. Furthermore, the resolution attainable in the reconstructions obtained using linear methods is limited to the resolution of the data. In other words, we cannot reconstruct the density contrast field at higher resolution than the resolution of our input data which, in turn, is limited by the noise properties of the data.

Figure 5 shows the 3D reconstruction obtained in this way for a simulated cluster of galaxies at a redshift of $z_{\rm cl} = 0.25$. The tuning parameter used for this reconstruction was $\alpha = 0.05$. The cluster was simulated according to an NFW halo with $M = 10^{15}h^{-1}M_{\odot}$ and $c = 3$, and the shear data were assumed to come from a galaxy distribution given by $p(z) \propto z^2 e^{-(1.4z)^{1.5}}$ [52, 53], with a maximum redshift of $z_{max} = 2.0$ and a galaxy density of $n_g = 100\,\text{galaxies/arcmin}^2$. The simulation covers a $1^\circ \times 1^\circ$ field binned into 60×60 angular pixels, and 20 redshift bins. Shown in the figure are both the 3D rendering of the reconstruction, and a 1D plot showing the four central lines of sight through the cluster as a function of redshift. The smearing, damping and redshift bias effects are all clearly visible in the 1D plot, where the amplitude of the cluster density contrast should be $\delta \sim 36$.

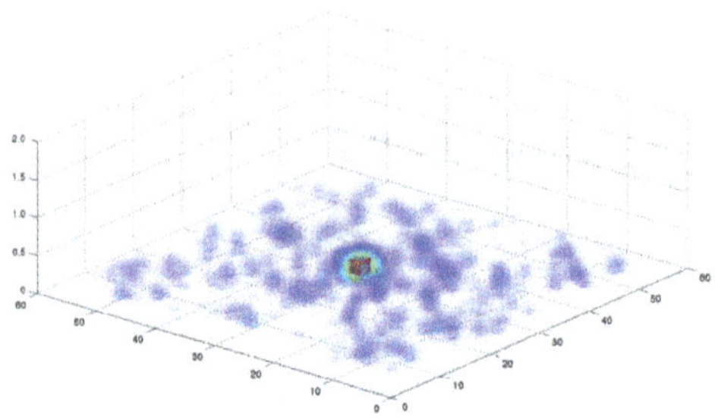

(a) 3D plot of the cluster reconstruction.

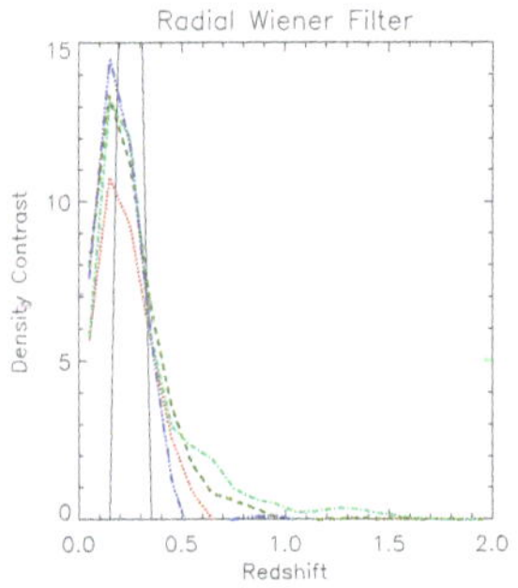

(b) The four central lines of sight through the cluster.

Figure 5. The reconstruction of a simulated cluster of galaxies using the method of [4]. Coloured broken lines in the 1D plot show the reconstruction of the four central lines of sight, while the solid line shows the true cluster density contrast.

5.2. Compressive sensing

We consider some data Y_i ($i \in [1,..,m]$) acquired through the linear system

$$Y = \mathbf{\Theta} X \,, \tag{25}$$

where $\mathbf{\Theta}$ is an $m \times n$ matrix. Compressed sensing [54, 55] is a sampling/compression theory based on the sparsity of the observed signal, which shows that, under certain conditions, one can exactly recover a k-sparse signal (a signal for which only k pixels have values different from zero, out of n total pixels, where $k < n$) from $m < n$ measurements.

This recovery is possible from undersampled data only if the sensing matrix $\mathbf{\Theta}$ verifies the *restricted isometry property* (RIP) [see 54, for more details]. This property has the effect that each measurement Y_i contains some information about the entire pixels of X; in other words, the sensing operator $\mathbf{\Theta}$ acts to spread the information contained in X across many measurements Y_i.

Under these two constraints – sparsity and a transformation meeting the RIP criterion – a signal can be recovered exactly even if the number of measurements m is much smaller than the number of unknown n. This means that, using CS methods, we will be able to outperform the well-known Shannon sampling criterion by far.

The solution X of (25) is obtained by minimizing

$$\min_X \|X\|_1 \quad s.t. \;\; Y = \mathbf{\Theta} X \,, \tag{26}$$

where the ℓ_1 norm is defined by $\|X\|_1 = \sum_i \mid X_i \mid$. The ℓ_1 norm is well-known to be a sparsity-promoting function; i.e. minimisation of the ℓ_1 norm yields the most sparse solution to the inverse problem. Many optimisation methods have been proposed in recent years to minimise this equation. More details about CS and ℓ_1 minimisation algorithms can be found in [6].

In real life, signals are generally not "strictly" sparse, but are *compressible*; i.e. we can represent the signal in a basis or frame (Fourier, Wavelets, Curvelets, etc.) in which the curve obtained by plotting the obtained coefficients, sorted by their decreasing absolute values, exhibits a polynomial decay. Note that most natural signals and images are compressible in an appropriate basis.

We can therefore reformulate the CS equation above (Equation (26)) to include the data transformation matrix $\mathbf{\Phi}$:

$$\min_{\alpha} ||\alpha||_1 \quad s.t. \ Y = \mathbf{\Theta}\mathbf{\Phi}\alpha \ , \tag{27}$$

where $X = \mathbf{\Phi}^*\alpha$, and α are the coefficients of the transformed solution X in $\mathbf{\Phi}$, which is generally referred to as the *dictionary*. Each column represents a vector (also called an *atom*), which ideally should be chosen to match the features contained in X. If $\mathbf{\Phi}$ admits a fast implicit transform (e.g. Fourier transform, Wavelet transform), fast algorithms exist to minimise Equation (27).

One problem we face when considering CS in a given application is that very few matrices meet the RIP criterion. However, it has been shown that accurate recovery can be obtained as long the mutual coherence between $\mathbf{\Theta}$ and $\mathbf{\Phi}$, $\mu_{\mathbf{\Theta},\mathbf{\Phi}} = \max_{i,k} \left|\left\langle \mathbf{\Theta}_i, \mathbf{\Phi}_k, \right\rangle\right|$, is low [56]. The mutual coherence measures the degree of similarity between the sparsifying basis and the sensing operator. Hence, in its relaxed definition, we consider a linear inverse problem $Y = \mathbf{\Theta}\mathbf{\Phi}X$ as being an instance of CS when

1. the problem is underdetermined,
2. the signal is compressible in a given dictionary $\mathbf{\Phi}$,
3. the mutual coherence $\mu_{\mathbf{\Theta},\mathbf{\Phi}}$ is low. This will happen every time the matrix $\mathbf{A} = \mathbf{\Theta}\mathbf{\Phi}$ has the effect of spreading out the coefficients α_j of the sparse signal on all measurements Y_i.

Most CS applications described in the literature are based on such a soft CS definition. Compressed sensing was introduced for the first time in astronomy for data compression [57, 58], and a direct link between CS and radio-interferometric image reconstruction was recently established in [59], leading to dramatic improvement thanks to the sparse ℓ_1 recovery [60].

The 3D weak lensing reconstruction problem can be seen to completely meet the soft-CS criteria above. The problem is underdetermined, as we seek a higher resolution than can be attained in the noise-limited observations, the matter density in the Universe is sparsely distributed, and the lensing operator spreads out the underlying density in a compressed sensing way.

In particular, for the reconstruction of clusters of galaxies, we are in a perfect situation for sparse recovery because clusters are localised in the angular domain, and are not resolved along the line of sight owing to the bin size. They can therefore be modelled as Dirac $\delta-$functions along the line of sight, while an isotropic wavelet basis can be used in the angular domain.

5.3. Results and future prospects

In [61], we present an algorithm to solve the 3D lensing problem. In this method, the 3D lensing problem is reduced to a one-dimensional problem, by taking as the data vector the (noisy) lensing convergence along each line of sight, which is related to the density contrast through Equation (21). Each line of sight can therefore be considered independently.

Clearly, a one-dimensional implementation throws away information, because we do not account at all for the correlation between neighbouring lines of sight that will arise in the presence of a large structure in the image; however, reducing the problem to a single dimension is fast and easy to implement, and allows us to test the efficacy of the algorithm using a particularly simple basis function through which we impose sparsity.

In Figure 6, we show the reconstruction obtained in this way for the simulated cluster described in section 5.1. The line of sight plot shows a clear improvement in all the target areas, reducing the bias, smearing and damping effects seen using linear methods. Small-scale noise is present, particularly at high redshifts, but this is likely due to overfitting in the algorithm used, and may be reduced by considering the whole 3D field, rather than each line of sight independently (for a full discussion of this issue, see [61]).

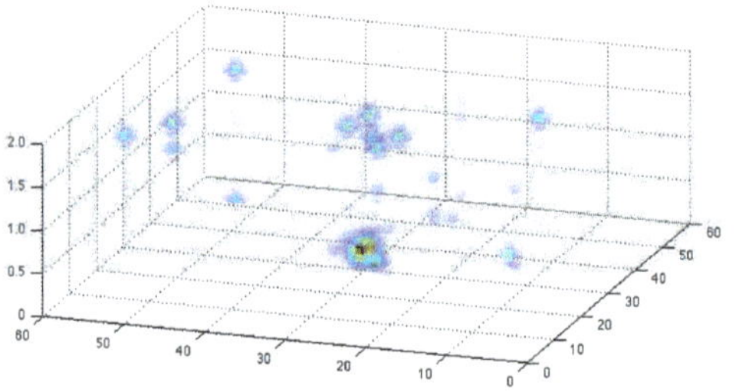

(a) 3D plot of the cluster reconstruction.

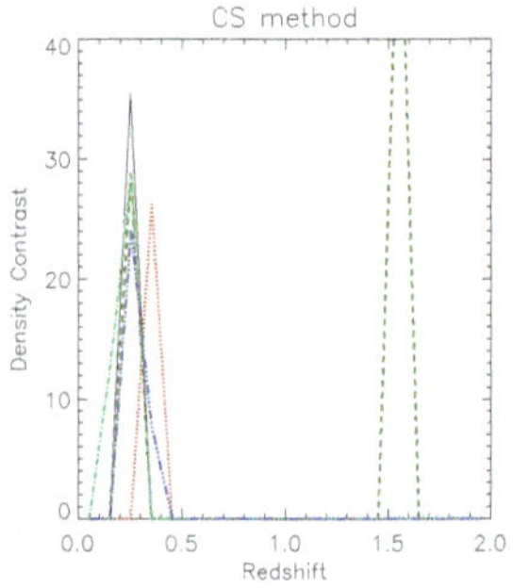

(b) The four central lines of sight through the cluster.

Figure 6. The reconstruction of a simulated cluster of galaxies using the method of [61]. Coloured broken lines in the 1D plot show the reconstruction of the four central lines of sight, while the solid line shows the true cluster density contrast.

This reconstruction was undertaken using the same resolution on the reconstruction as on the input data. However, we know that the CS approach is particularly well-adapted to dealing with ill-posed inversion problems. In order to test this, we consider a cluster at a redshifts of $z_{\rm cl} = 0.2$, simulated as before. We use $N_{\rm sp} = 20$ redshift bins in our data, but now aim to reconstruct our density contrast with a redshift resolution of $N_{\rm lp} = 25$. The results are shown in Figure 7. Here, again, we see small scale noise at high-redshift, but the overall smearing and redshift bias issues seen in the linear reconstructions is absent.

As noted previously, the one-dimensional solver employed here throws away a wealth of information about the angular correlation of the lensing signal, and is thus not optimal. Indeed, a simple algorithm based on this CS approach, but implemented as a full three-dimensional treatment, offers marked improvements in the quality of the reconstructions [62].

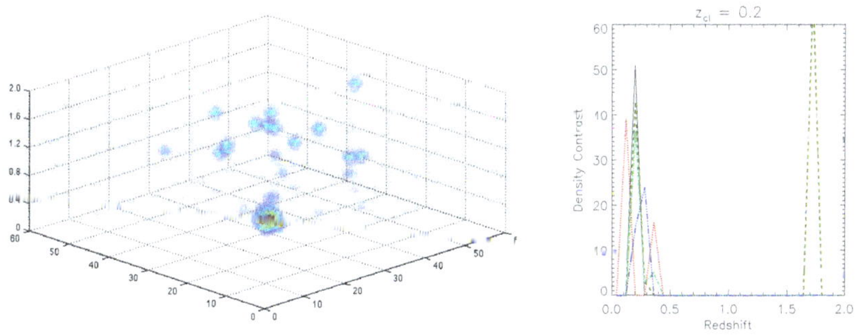

(a) 3D plot of the cluster reconstruction.

(b) The four central lines of sight through the cluster.

Figure 7. The reconstruction of a simulated cluster of galaxies at a redshift of $z_{cl} = 0.2$, and with improved resolution on the reconstruction as compared to the input data.

This is demonstrated in Figure 8, which shows reconstructions of the cluster at redshift $z_{cl} = 0.2$, with the same improved resolution in the reconstruction as before, but this time using the three-dimensional CS approach. Dramatic improvement is seen on all fronts, with the reconstructions showing no bias or redshift smearing, and very little amplitude damping, and with none of the small-scale false detections seen in the 1D CS approach.

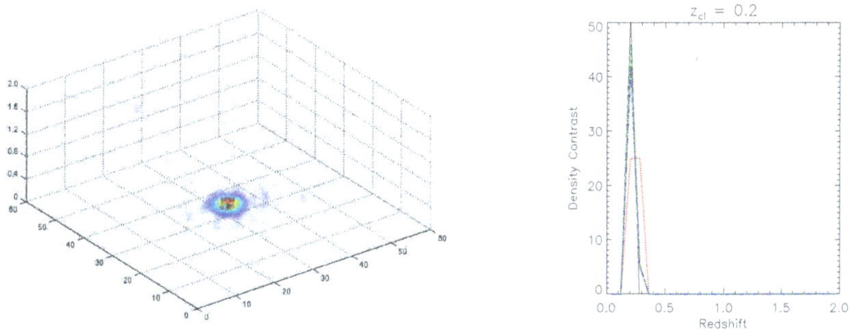

(a) 3D plot of the cluster reconstruction.

(b) The four central lines of sight through the cluster.

Figure 8. The reconstruction of a simulated cluster of galaxies at a redshift of $z_{cl} = 0.2$ using a 3D CS approach, and with improved resolution on the reconstruction as compared to the input data.

This marked improvement in the cluster reconstruction seen in the 3D CS approach represents a definite step in the right direction for weak lensing studies. There is much work to be done, however. The application of this CS approach to the 3D lensing problem is a very recent development, and many questions remain: the choice of algorithm, for example; how best to control the noise in the reconstruction; how to deal with photometric redshift errors; by what factor the reconstruction resolution might be improved as compared to the data.

Yet it is clear that this approach opens up the possibility of being able to generate accurate reconstructions of the density contrast using weak lensing measurements, and perhaps using information such as the 3D peak count – in combination with constraints from other probes – to place ever-tighter constraints on our cosmological model. This, in turn, will offer a unique insight into the nature and properties of the dark components of the Universe.

Acknowledgements

This work is supported by the European Research Council grant SparseAstro (ERC-228261). The authors would like to thank Jalal Fadili for useful discussions, and Jake VanderPlas for helpful comments and use of his code to generate the linear reconstructions.

Author details

Adrienne Leonard[1,*], Jean-Luc Starck[1],
Sandrine Pires[1] and François-Xavier Dupé[2]

* Address all correspondence to: adrienne.leonard@cea.fr

1 Laboratoire AIM (UMR 7158), CEA/DSM-CNRS-Université Paris Diderot, IRFU, SEDI-SAP, Service d'Astrophysique, Centre de Saclay, France
2 Laboratoire d'Informatique Fondamentale, Centre de Mathématique et d'Informatique, UMR CNRS, Aix-Marseille Université, Marseille, France

References

[1] D. Larson, J. Dunkley, G. Hinshaw, E. Komatsu, M. R. Nolta, C. L. Bennett, B. Gold, M. Halpern, R. S. Hill, N. Jarosik, A. Kogut, M. Limon, S. S. Meyer, N. Odegard, L. Page, K. M. Smith, D. N. Spergel, G. S. Tucker, J. L. Weiland, E. Wollack, and E. L. Wright. Seven-year Wilkinson Microwave Anisotropy Probe (WMAP) Observations: Power Spectra and WMAP-derived Parameters. *The Astrophysical Journal Supplement*, 192:16–+, 2011.

[2] G. Hinshaw, J. L. Weiland, R. S. Hill, N. Odegard, D. Larson, C. L. Bennett, J. Dunkley, B. Gold, M. R. Greason, N. Jarosik, E. Komatsu, M. R. Nolta, L. Page, D. N. Spergel, E. Wollack, M. Halpern, A. Kogut, M. Limon, S. S. Meyer, G. S. Tucker, and E. L. Wright. Five-Year Wilkinson Microwave Anisotropy Probe Observations: Data Processing, Sky Maps, and Basic Results. *The Astrophysical Journal Supplement*, 180:225–245, 2009.

[3] J. A. Frieman, M. S. Turner, and D. Huterer. Dark Energy and the Accelerating Universe. *Annual Reviews in Astronomy and Astrophysics*, 46:385–432, 2008.

[4] P. Simon, A. N. Taylor, and J. Hartlap. Unfolding the matter distribution using three-dimensional weak gravitational lensing. *Monthly Notices of the Royal Astronomical Society*, 399:48–68, 2009.

[5] M.J. Fadili and J.-L. Starck. Monotone operator splitting for optimization problems in sparse recovery. In *Proceedings of the International Conference on Image Processing, ICIP 2009, 7-10 November 2009, Cairo, Egypt*, pages 1461–1464. IEEE, 2009.

[6] J.-L. Starck, F. Murtagh, and M.J. Fadili. *Sparse Image and Signal Processing*. Cambridge University Press, 2010.

[7] A. Refregier, A. Amara, T. D. Kitching, A. Rassat, R. Scaramella, J. Weller, and f. t. Euclid Imaging Consortium. Euclid Imaging Consortium Science Book. *ArXiv:1001.0061*, 2010.

[8] P. Schneider. *Weak Gravitational Lensing*, pages 269–+. Springer, 2006.

[9] M. Bartelmann and P. Schneider. Weak gravitational lensing. *Physics Reports*, 340:291–472, 2001.

[10] P. Schneider, L. van Waerbeke, M. Kilbinger, and Y. Mellier. Analysis of two-point statistics of cosmic shear. I. Estimators and covariances. *Astronomy and Astrophysics*, 396:1–19, 2002.

[11] N. Kaiser. Weak Lensing and Cosmology. *The Astrophysical Journal*, 498:26, 1998.

[12] E. Semboloni, Y. Mellier, L. van Waerbeke, H. Hoekstra, I. Tereno, K. Benabed, S. D. J. Gwyn, L. Fu, M. J. Hudson, R. Maoli, and L. C. Parker. Cosmic shear analysis with CFHTLS deep data. *Astronomy and Astrophysics*, 452:51–61, 2006.

[13] T. Schrabback, J. Hartlap, B. Joachimi, M. Kilbinger, P. Simon, K. Benabed, M. Bradač, T. Eifler, T. Erben, C. D. Fassnacht, F. W. High, S. Hilbert, H. Hildebrandt, H. Hoekstra, K. Kuijken, P. J. Marshall, Y. Mellier, E. Morganson, P. Schneider, E. Semboloni, L. van Waerbeke, and M. Velander. Evidence of the accelerated expansion of the Universe from weak lensing tomography with COSMOS. *Astronomy and Astrophysics*, 516:A63, 2010.

[14] P. Schneider and M. Lombardi. The three-point correlation function of cosmic shear. I. The natural components. *Astronomy and Astrophysics*, 397:809–818, 2003.

[15] M. Takada and B. Jain. The Three-Point Correlation Function for Spin-2 Fields. *The Astrophysical Journal Letters*, 583:L49–L52, 2003.

[16] M. Zaldarriaga and R. Scoccimarro. Higher Order Moments of the Cosmic Shear and Other Spin-2 Fields. *The Astrophysical Journal*, 584:559–565, 2003.

[17] P. Schneider. Detection of (dark) matter concentrations via weak gravitational lensing. *Monthly Notices of the Royal Astronomical Society*, 283:837–853, 1996.

[18] P. Schneider, L. van Waerbeke, B. Jain, and G. Kruse. A new measure for cosmic shear. *Monthly Notices of the Royal Astronomical Society*, 296:873–892, 1998.

[19] L. van Waerbeke. Scale dependence of the bias investigated by weak lensing. *Astronomy and Astrophysics*, 334:1–10, 1998.

[20] M. Jarvis, G. Bernstein, and B. Jain. The skewness of the aperture mass statistic. *Monthly Notices of the Royal Astronomical Society*, 352:338–352, 2004.

[21] R. G. Crittenden, P. Natarajan, U.-L. Pen, and T. Theuns. Discriminating Weak Lensing from Intrinsic Spin Correlations Using the Curl-Gradient Decomposition. *The Astrohpysical Journal*, 568:20–27, 2002.

[22] T.-J. Zhang, U.-L. Pen, P. Zhang, and J. Dubinski. Optimal Weak-Lensing Skewness Measurements. *The Astrophysical Journal*, 598:818–826, 2003.

[23] J.-L. Starck, F. D. Murtagh, and A. Bijaoui. *Image Processing and Data Analysis*. Cambridge University Press, 1998.

[24] J.-L. Starck and F. Murtagh. *Astronomical Image and Data Analysis*. Springer, 2006. 2nd edn.

[25] J.-L. Starck, F. Murtagh, and M. Bertero. *in Handbook of Mathematical Methods in Imaging*. Springer, 2011. in press.

[26] S. Seitz and P. Schneider. A new finite-field mass reconstruction algorithm. *Astronomy and Astrophysics*, 374:740–745, 2001.

[27] C. Seitz and P. Schneider. Steps towards nonlinear cluster inversion through gravitational distortions. III. Including a redshift distribution of the sources. *Astronomy and Astrophysics*, 318:687–699, 1997.

[28] S. Seitz and P. Schneider. Cluster lens reconstruction using only observed local data: an improved finite-field inversion technique. *Astronomy and Astrophysics*, 305:383–+, 1996.

[29] C. Seitz and P. Schneider. Steps towards nonlinear cluster inversion through gravitational distortions II. Generalization of the Kaiser and Squires method. *Astronomy and Astrophysics*, 297:287–+, 1995.

[30] E. Deriaz, J.-L. Starck, and S. Pires. Wavelet Helmholtz decomposition for weak lensing mass map reconstruction. *Astronomy and Astrophysics*, 540:A34, 2012.

[31] A. Leonard, S. Pires, and J.-L. Starck. Fast calculation of the weak lensing aperture mass statistic. *Monthly Notices of the Royal Astronomical Society*, 423:3405–3412, 2012.

[32] Lucy Liuxuan Zhang and Ue-Li Pen. Fast n-point correlation functions and three-point lensing application. *New Astronomy*, 10(7):569 – 590, 2005.

[33] H. K. Eriksen, P. B. Lilje, A. J. Banday, and K. M. Gç¿Łski. Estimating n-point correlation functions from pixelized sky maps. *The Astrophysical Journal Supplement Series*, 151(1):1, 2004.

[34] J.-L. Starck, S. Pires, and A. Réfrégier. Weak lensing mass reconstruction using wavelets. *Astronomy and Astrophysics*, 451:1139–1150, 2006.

[35] F. Bernardeau, L. van Waerbeke, and Y. Mellier. Weak lensing statistics as a probe of {OMEGA} and power spectrum. *Astronomy and Astrophysics*, 322:1–18, 1997.

[36] M. Takada and B. Jain. Cosmological parameters from lensing power spectrum and bispectrum tomography. *Monthly Notices of the Royal Astronomical Society*, 348:897–915, 2004.

[37] M. Kilbinger and P. Schneider. Cosmological parameters from combined second- and third-order aperture mass statistics of cosmic shear. *Astronomy and Astrophysics*, 442:69–83, 2005.

[38] S. Pires, J.-L. Starck, A. Amara, A. Réfrégier, and R. Teyssier. Cosmological model discrimination with weak lensing. *Astronomy and Astrophysics*, 505:969–979, 2009.

[39] J. Bergé, A. Amara, and A. Réfrégier. Optimal Capture of Non-Gaussianity in Weak-Lensing Surveys: Power Spectrum, Bispectrum, and Halo Counts. *The Astrophysical Journal*, 712:992–1002, 2010.

[40] C.-P. Ma and J. N. Fry. Deriving the Nonlinear Cosmological Power Spectrum and Bispectrum from Analytic Dark Matter Halo Profiles and Mass Functions. *The Astrophysical Journal*, 543:503–513, 2000.

[41] C.-P. Ma and J. N. Fry. Halo Profiles and the Nonlinear Two- and Three-Point Correlation Functions of Cosmological Mass Density. *The Astrophysical Journal Letters*, 531:L87–L90, 2000.

[42] R. Scoccimarro and H. M. P. Couchman. A fitting formula for the non-linear evolution of the bispectrum. *Monthly Notices of the Royal Astronomical Society*, 325:1312–1316, 2001.

[43] A. Cooray and W. Hu. Weak Gravitational Lensing Bispectrum. *The Astrophysical Journal*, 548:7–18, 2001.

[44] M. Maturi, C. Angrick, F. Pace, and M. Bartelmann. An analytic approach to number counts of weak-lensing peak detections. *Astronomy and Astrophysics*, 519:A23, 2010.

[45] J. P. Dietrich and J. Hartlap. Cosmology with the shear-peak statistics. *Monthly Notices of the Royal Astronomical Society*, 402:1049–1058, 2010.

[46] S. Pires, A. Leonard, and J.-L. Starck. Cosmological constraints from the capture of non-Gaussianity in weak lensing data. *Monthly Notices of the Royal Astronomical Society*, 423:983–992, 2012.

[47] R. Teyssier. Cosmological hydrodynamics with adaptive mesh refinement. A new high resolution code called RAMSES. *Astronomy and Astrophysics*, 385:337–364, 2002.

[48] Z. Ma, W. Hu, and D. Huterer. Effects of Photometric Redshift Uncertainties on Weak-Lensing Tomography. *The Astrophysical Journal*, 636:21–29, 2006.

[49] J. T. VanderPlas, A. J. Connolly, B. Jain, and M. Jarvis. Three-dimensional Reconstruction of the Density Field: An SVD Approach to Weak-lensing Tomography. *The Astrophysical Journal*, 727:118–+, 2011.

[50] M. Tegmark. How to Make Maps from Cosmic Microwave Background Data without Losing Information. *The Astrophysical Journal Letters*, 480:L87+, 1997.

[51] M. Tegmark, A. de Oliveira-Costa, M. J. Devlin, C. B. Netterfields, L. Page, and E. J. Wollack. A High-Resolution Map of the Cosmic Microwave Background around the North Celestial Pole. *The Astrophysical Journal Letters*, 474:L77+, 1997.

[52] A. N. Taylor, T. D. Kitching, D. J. Bacon, and A. F. Heavens. Probing dark energy with the shear-ratio geometric test. *Monthly Notices of the Royal Astronomical Society*, 374:1377–1403, 2007.

[53] T. D. Kitching, A. F. Heavens, and L. Miller. 3D photometric cosmic shear. *Monthly Notices of the Royal Astronomical Society*, pages 426–+, 2011.

[54] E.J. Candès and T. Tao. Near optimal signal recovery from random projections: Universal encoding strategies? *IEEE Transactions on Information Theory*, 52(12), 2006. 5406–5425.

[55] D.L. Donoho. Compressed sensing. *IEEE Transactions on Information Theory*, 52(4), 2006. 1289–1306.

[56] E. J. Candes and Y. Plan. A probabilistic and ripless theory of compressed sensing. submitted, 2010.

[57] J. Bobin, J.-L. Starck, and R. Ottensamer. Compressed Sensing in Astronomy. *IEEE Journal of Selected Topics in Signal Processing*, 2:718–726, 2008.

[58] N. Barbey, M. Sauvage, J.-L. Starck, R. Ottensamer, and P. Chanial. Feasibility and performances of compressed sensing and sparse map-making with Herschel/PACS data. *Astronomy and Astrophysics*, 527:A102+, 2011.

[59] Y. Wiaux, L. Jacques, G. Puy, A. M. M. Scaife, and P. Vandergheynst. Compressed sensing imaging techniques for radio interferometry. *Monthly Notices of the Royal Astronomical Society*, 395:1733–1742, 2009.

[60] F. Li, T. J. Cornwell, and F. de Hoog. The application of compressive sampling to radio astronomy. I. Deconvolution. *Astronomy and Astrophysics*, 528:A31+, 2011.

[61] A. Leonard, F.-X. Dupé, and J.-L. Starck. A compressed sensing approach to 3D weak lensing. *Astronomy and Astrophysics*, 539:A85, 2012.

[62] A. Leonard, F. Lanusse, F.-X. Dupé, and J.-L. Starck. *in prep.*, 2012.

Chapter 4

Extended Theories of Gravitation and the Curvature of the Universe – Do We Really Need Dark Matter?

L. Fatibene and M. Francaviglia

Additional information is available at the end of the chapter

1. Introduction

Cosmology is a young science. Less than a century ago cosmology stopped to be a branch of philosophy and it crossbred with General Relativity to become a science. Until very recently cosmological observations were quite rough and qualitative. Until the 80s one was quite satisfied with data with error bars of few percent.

Since then a number of extremely precise surveys have been carried over producing a massive amount of very precise data. The current picture that emerged from those data is quite awkward. In order to fit observations and maintain standard GR as general framework for gravity one is forced to introduce dark sources, at least in a large amount different from the matter that can be seen in the universe and which has somehow odd behavior; see [1], [2], [3], [4].

Actually, following this direction one is led to assume that about 70% of gravitational sources in the universe is constituted by some strange kind of dark energy, closely resembling a (small and positive) cosmological constant, about 25% of gravitational sources is constituted by some kind of dark matter (for which different models have been proposed and discussed), while visible matter amounts to few percents (about 4-5% depending on the model) of the total amount of matter. It is important to notice that we do not have any direct evidence or data about dark energy and dark matter other than their supposed gravitational effects on visible matter. Moreover, the best models for dark energy and dark matter are often definitely unsatisfactory from a fundamental viewpoint; see [5].

On the other hand, it has been suggested that the description of the gravitational field given by standard GR may fail at cosmological scale and we missed something, so that a good agreement with data can be obtained by modifying the description of gravity more than adding exotic sources. In any event it is now clear that something has to be changed in our standard framework in order to understand the universe out there.

Besides these obvious considerations let us add quite a trivial remark. Our understanding of the meaning of observations is generally weak and often depending on the model. Standard GR has a good set of protocols which allow one to make predictions and tests. The theory is extremely well tested at Solar system scales, while it is known to require corrections (by adding dark sources or by modifying dynamics) at galactic, astrophysical and cosmological scale (oddly enough whenever non-vacuum solutions are considered).

However, what we observe when we measure the distance of a supernova is not clear at all. GR is a relativistic theory with a huge symmetry group, namely all spacetime diffeomorphisms. The observable quantities should be then invariant with respect to spacetime diffeomorphisms, i.e. gauge invariant. Unfortunately, due to the particular nature of diffeomorphisms and their action on the geometry of spacetime, we do not know any non-trivial quantity which is diff-invariant. Also scalars are not (unless they are constant) since the Lie derivative of a scalar with respect to a generic spacetime vector field (i.e. a generic generator of spacetime diffeomorphisms) is

$$\pounds_{\xi} f = \xi^{\mu} \partial_{\mu} f \tag{1}$$

which is in general not zero, showing that in fact the quantity is not gauge invariant. This is known since the very beginning and it is the starting point of the celebrated hole argument; see [6]. Since we do measure quantities that are not gauge invariant, the only possible explanation is that we set observational protocols which as a matter of fact break gauge invariance on a conventional basis (possibly using matter references, as suggested in [6]).

That would not be too bad, if we clearly understood the details of such conventions and gauge fixing, that we do not. Instead, standard GR mixes from its very beginning physical quantities (i.e. the gravitational field) and the observational protocols (e.g. for measuring distances and times) in the same object (namely, the metric tensor). Originally, Einstein had not many options, since at that time the only way to describe curvature was through a metric structure and general (linear) connections were still to be fully described. As a consequence it becomes very difficult to keep the two things separated as they should.

In the 70s Ehlers, Pirani and Schild (EPS) gave a fundamental contribution to the understanding of the foundations of any reasonable theory of spacetime and gravity. They proposed an axiomatic approach to gravitational theories which, instead of assuming a metric or a connection on spacetime, assumed as fundamental potentially observable quantities (namely the worldlines of particles and light rays) and derived from them the geometry of spacetime; see [7], [8]. The original project was to obtain standard GR. However, the proposal finally turned out to give us a fundamental insight about what is to be considered observable and which geometrical structures are really essential for gravity.

In particular EPS framework allows a more general geometric structure on spacetime in which standard GR comes out to be just one of many possible theories of gravitation. Moreover, a more general framework potentially allows to test which geometric structure on spacetime is actually physically realized. As a side effect, EPS has an impact on observational protocols (not all standard protocols can be trivially extended to a general extended theory).

We shall hereafter review the EPS framework, define extended theories of gravitation and attempt a rough classification of possible extended theories. Finally we shall discuss some simple application to cosmology and observational protocols.

2. EPS structures on spacetime

As already mentioned, in the early 70s Ehlers, Pirani and Schild (EPS) proposed an axiomatic framework for relativistic theories in which they showed how one can derive the geometric structure of spacetime from potentially observable quantities, i.e. worldlines of particles and light rays; see [7]. Accordingly, in the EPS framework the geometry of spacetime is not assumed but derived by more fundamental objects. We shall first briefly review EPS formalism; in the next Sections we shall discuss its consequences in gravitational theories and cosmology.

Let M be an (orientable, time orientable, connected, paracompact, smooth) m-dimensional manifold. Points in M are called *events* and M is called accordingly a *spacetime*. Let us stress that although M is chosen so that it allows global Lorentzian metrics, we do not fix any metric structure on M.

On M we consider two congruences of trajectories. Let $\mathcal{P}$ be the congruence of all possible motions of massive particles and $\mathcal{L}$ be the congruence of all possible light rays. Of course there are reasonable physical requirement to be asked about $\mathcal{P}$ and $\mathcal{L}$ since we expect they cannot be chosen to be completely generic or unrelated since we expect photons to feel the gravitational field as well as we expect matter to interact with the electromagnetic field.

If we restrict ourselves to particles and light rays passing through an event $x \in M$, we *know* that the directions of light rays form a *cone* (the *light cone*). We can express this experimental fact by asking that the directions of light rays divide spacetime directions (i.e. the projective space of T_xM) into two connected components (i.e. the directions *inside* and *outside* the light cone).

We also know that the set of vectors *inside* the light cone is topologically different from the set of vectors *outside* the light cone. If one removes the zero vector then the set of vectors *inside* the light cone disconnects into two connected components (namely, *future* and *past* directed timelike vectors), while the set of vectors *outside* the light cone keeps being connected (there is nothing like future directed spacelike vectors!).

Moreover, we know that one has two kinds of vectors tangent to light rays: the ones pointing to the future and the ones pointing to the past. Thus we assume that (once the zero vector is removed) the set of vectors tangent to light rays also splits into two connected components (namely, *future* and *past* directed vectors). Let us stress that past and future are defined *at a point* x and it does not really matter which one of them is called future or past. These three requirements are physically well founded and in the end they constrain the light cones to be *cones* without resorting to a metric structure we did not define yet.

Then we have a number of regularity conditions. We need axioms to certify that one has enough light rays to account for physical standard messaging. Let us thus assume that for any particle $P \in \mathcal{P}$ and for any event $p \in P \subset M$ there exists a neighbourhood V_p and a neighbourhood $U_p \subset V_p$ such that for any event $x \in U_p$ there are two light rays through x hitting P within V_p.

Let us remark that in Minkowski spacetime one can set $U_p = V_p = M$ and there are always two such light rays (as one can check by direct calculation remembering that particles and light rays are given as straight lines in Minkowski spacetime).

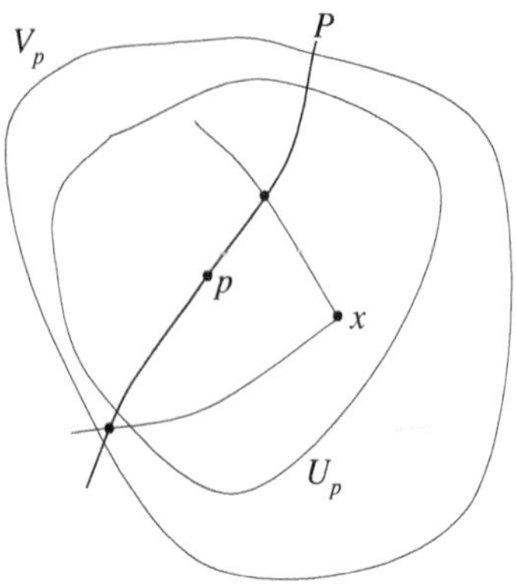

Given two particles $P, Q \in \mathcal{P}$ we can consider the family of light rays $\lambda \in \mathcal{L}$ intersecting P and Q. By the above assumption, when P and Q are close enough such family is not empty. This family of light rays does define a local one-to-one map between P and Q which is called a *message* which is denoted by $\mu : P \to Q$. If one takes the composition of a message from P to Q and a message from Q to P the resulting map $\epsilon : P \to P$ is called and *echo* of P on Q. Both messages and echoes are assumed to be smooth maps.

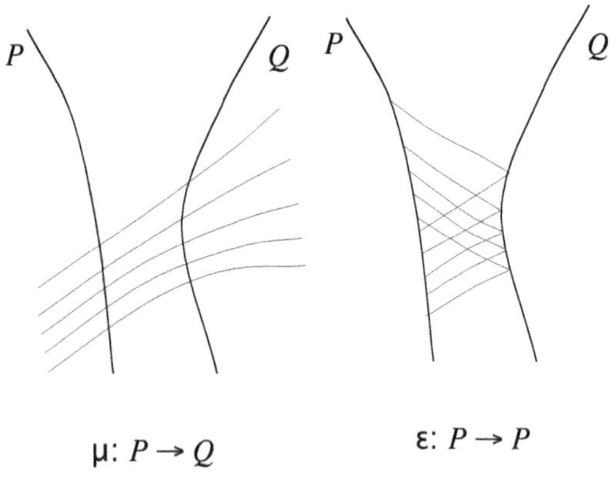

Figure 1. Messages and Echoes

Finally, we have to guarantee that there exist enough particles in $\mathcal{P}$ and light rays in $\mathcal{L}$ (which until now could be empty, as far as we know). We assume that there is a particle for each vector inside the light cone and a light ray for each vector on the light cone.

Let us now define a *clock* to be a parametrized particle, the parametrization accounting for the time maintained by the clock; [9]. For any clock $P \in \mathcal{P}$, for any event $p \in P$ one can set the parameter to be $s = 0$ at p. Using echoes one can use a number of clocks to define a special class of local coordinates, *called radar coordinates* or *parallax coodinates*. If $\dim(M) = m$ one can always choose m clocks P_i near an event $p \in M$ so that there exists a neibourhood

U_p such that for any $x \in U_p$ there is a (future directed) light ray through x then hitting the clock P_i at its parameter value s_i. The values of the parameters s_i do form a good coordinate system in U_p. We assume that the spacetime differential structure on M is the one compatible with these charts. Let us remark that parallax coordinates mimick how astronomers define positions of objects.

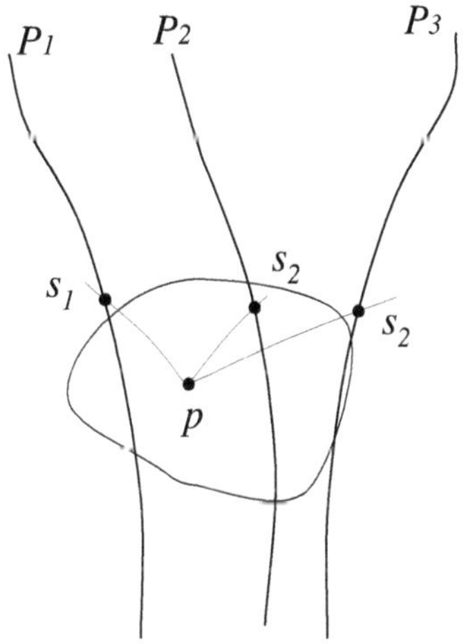

Figure 2. Parallax coordinates in dimension $m = 3$

One can show that as a consequence of these assumptions a class of Lorentzian metrics g is defined on M. Let us then fix a clock P though an event p. For any event $x \in U_p$ one has two light rays through x intersecting P, say at events $p_\pm$ which correspond to the parameter values $s_\pm$. Then we can define a local function $\Phi : U_p \to \mathbb{R} : x \mapsto -s_+ \cdot s_-$. As one can easily show, if there exists a light ray through x and p then $\Phi(x) = 0$. According to the topological assumptions made on the light cones then one can show that $\Phi(p) = 0$ and $d\Phi(p) = 0$. Then one can consider the Hessian $\partial_{\mu\nu}\Phi(p)$ as the first non-zero term in the Taylor expansion of Φ around the event p. In this case it defines a tensor field (a bilinear form)

$$g_p = g_{\mu\nu}(p)\, dx^\mu \otimes dx^\nu = \partial_{\mu\nu}\Phi(p)\, dx^\mu \otimes dx^\nu \tag{2}$$

For any light ray direction v at p one has $g(v, v) = 0$. One can also easily show that for u tangent to the clock P one has $g(u, u) < 0$.

Accordingly, g cannot be definite positive. In order not to contradict again assumptions about light cones, one can show that g is necessarily non-degenerate and Lorentzian (see [10]). Of course the the tensor g depends on the conventional choice of the clock. If one changes clock one defines a different tensor $\tilde{g}$ which is related to the previous one by a conformal transformation, namely $\tilde{g} = \varphi(x) \cdot g$ for some positive scalar field φ.

Let us now consider the set $\mathrm{Lor}(M)$ of all (global) Lorentzian metrics on M. Let us say that two metrics $g_{(1)}, g_{(2)} \in \mathrm{Lor}(M)$ are *conformally equivalent* iff there exists a positive scalar field

φ such that $\tilde{g}_x = \varphi(x) \cdot g_x$. The construction above shows that one can define out of light rays (i.e. out of the electromagnetic field) a *conformal class* of metrics $\mathfrak{C} = [g]$. Let us remark that the choice of a representative $\tilde{g} \in \mathfrak{C}$ is conventional and in fact part of the specification of the observer; conformal transformations are gauge transformations.

Notice that light cones are invariant with respect to conformal transformations; given a conformal structure $\mathfrak{C}$ on M one can define *lightlike* (*timelike* and *spacelike*, respectively) vectors being $g(v,v) = 0$ ($g(u,u) < 0$ and $g(w,w) > 0$, respectively) recovering standard notations used in GR.

Finally, we have to focus on particles. Let us first assume that we have one particle through $p \in M$ for any timelike direction and a light ray for any lightlike direction. Some constraint on particles must be set. Originally, EPS resorted to the equivalence principle and special relativity (SR) assuming that particle worldlines are geodesics of some connection $\tilde{\Gamma}$.

We cannot, for various reasons, be totally satisfied with this assumption, even if we accept of course the result. First of all relativistic theories are more fundamental than SR, which should hence be obtained in some limit from GR rather than being used to define it. Then the equivalence principle is an experimental fact and we would like to keep the possibility to test it rather than assuming it as a must. Luckily enough, one can obtain geodesic equations (together with a better insight on the nature of gravitational field) also without resorting to SR and equivalence principle. In fact, if one assumes that free fall must be described by differential equations of the second order, deterministic, covariant with respect to spacetime diffeomorphisms and with respect to arbitrary reparametrizations of worldlines those candidate equations are strongly constrained; see [11], [12]. If one then defines gravitational interaction to be the one which cannot be cancelled in a way independent of the coordinates and parametrizations then the equation uniquely determined are geodesic equations

$$\ddot{q}^\lambda + \tilde{\Gamma}^\lambda_{\alpha\beta}\dot{q}^\alpha\dot{q}^\beta = \lambda\dot{q}^\lambda \tag{3}$$

for some (global torsionless) connection $\tilde{\Gamma}(x)$ and some function $\lambda(s)$. In this way free fall is naturally associated to a connection $\tilde{\Gamma}$ and one is considering the Einstein's lift experiment as showing that *there are* observers who see a gravitational field rather than a *gedanken* experiment showing that there is a class of observers who do not (approximately) observe it.

As is well known, different connections can define the same autoparallel trajectories. In fact the connection

$$\tilde{\Gamma}'^{\alpha}_{\beta\mu} = \tilde{\Gamma}^{\alpha}_{\beta\mu} + \delta^{\alpha}_{(\beta}V_{\mu)} \tag{4}$$

defines the same geodesic trajectories as $\tilde{\Gamma}^{\alpha}_{\beta\mu}$ for any covector V_μ. In this case we say that $\tilde{\Gamma}$ and $\tilde{\Gamma}'$ are *projectively equivalent*. Accordingly, free fall corresponds to a projective class $\mathfrak{P} = [\tilde{\Gamma}]$; see [13].

Finally, we need (as we said above) a compatibility condition between the conformal class $\mathfrak{C}$ associated to light cones and the projective class $\mathfrak{P}$ associated to free fall. This is due by the simple fact that we know that light rays (and hence light cones) feel the gravitational fields as mass particles. Noticing that g-lightlike g-geodesics are conformally invariant (unlike general g-geodesics), we have then to assume that g-lightlike g-geodesics are a subset of

$\tilde{\Gamma}$-autoparallel trajectories. According to EPS-compatibility condition one can show that a representative $\tilde{\Gamma} \in \mathfrak{P}$ of the projective structure can be always (and uniquely) chosen so that there exists a covector $A = A_\mu \, dx^\mu$ such that

$$\tilde{\nabla} g = 2A \otimes g \tag{5}$$

where g is a representative of the conformal structure $g \in \mathfrak{C}$ and the covariant derivative $\tilde{\nabla}$ is the one associated to $\tilde{\Gamma}$; see [14]. Equivalently one has

$$\tilde{\Gamma}^\alpha_{\beta\mu} = \{g\}^\alpha_{\beta\mu} + (g^{\alpha\epsilon} g_{\beta\mu} - 2\delta^\alpha_{(\beta}\delta^\epsilon_{\mu)}) A_\epsilon \tag{6}$$

To summarize, by assuming particles and light rays one can define on spacetime a *EPS structure*, i.e. a triple $(M, \mathfrak{C}, \mathfrak{P})$. The conformal structure $\mathfrak{P}$ describes light cones and it is associated to light rays. Notice that having just a conformal structure one cannot yet define distances (that are not conformally invariant) and this not being a gauge covariant must resort to a *convention* which corresponds to the choice of a representative $g \in \mathfrak{C}$. On the other hand, the projective structure $\mathfrak{P}$ is associated to free fall so that one can make a canonical gauge fixing by choosing the only representative in the form (6) or, equivalently, the 1-form A.

The triple $(M, \mathfrak{C}, \tilde{\Gamma})$ (or, equivalently, the triple $(M, \mathfrak{C}, A)$) is called a *Weyl geometry* on spacetime. This setting is more general than the setting for standard GR where one has just a Lorentzian metric g determining both the conformal structure $g \in \mathfrak{C}$ and the free fall $\tilde{\Gamma} = \{g\}$ (i.e. the Levi-Civita connection uniquely associated to g). Hence standard GR is a very peculiar case of EPS framework, where there is a gauge fixing of the conformal gauge. Such a fixing is possible iff the covector $A = A_\mu dx^\mu$ is exact, i.e. $A = d\varphi$. In this case, there exists a Lorentzian metric $\tilde{g} \in \mathfrak{C}$ also determining free fall by $\tilde{\Gamma} = \{\tilde{g}\}$. When this happens the Weyl geometry $(M, [\tilde{g}], \{\tilde{g}\})$ is called a *metric Weyl geometry*. Notice that this is still more general than standard GR in the sense that the metric determining free fall and light cones is not the original g chosen to describe dynamics, but a conformal one $\tilde{g} \in [g]$. Reverting to standard GR in a sense amounts to choose φ to be a constant (so that A vanishes identically).

At this point the reader could argue that in a metric Weyl geometry one could fix the conformal metric $\tilde{g}$ at the beginning and use it to describe dynamics, thus obtaining a framework which *exactly* reproduces standard GR. We shall discuss this issue below in greater details. Now we simply notice that this is not the case. The choice of a representative of the conformal structure is, in fact, what allows us to define distances.

In fact, since astronomers do measure distances, we do have a protocol (or better a number of protocols) to measure distances. As a matter of fact, such a protocol selects (in a rather obscure way) a precise representative g' of the conformal structure which is the one that corresponds to the distances that we measure. If one metric g geometrically accounts for a given physical distance measured between two events then obviously no other conformal metric $\tilde{g}$ (i.e. no other representative of the same conformal factor) can geometrically account for the same measure (modulo constant conformal factors which can be treated as a definition of units).

In standard GR one assumes that such a representative g' *also* determines light cones and free fall. In metric Weyl geometries there is nothing ensuring that the canonical representative $\tilde{g}$ also gives us the measured distances, that as far as we can see could as well be related to any other conformally equivalent metric g. Fixing the metric that we use to calculate distances is, in the end, a choice that we can do only *a posteriori*, on the basis of observations.

At a fundamental level one can either decide to be strict on the interpretation of conformal gauge symmetry (and accordingly quantities that are not gauge invariant, such as distances, cannot be really observable) or one accepts conventional gauge fixing to define such quantities as observable, thus restricting symmetries of the system to the conformal transformations which preserves these gauge fixing. In the first case standard GR is equivalent to metric Weyl geometry (though we cannot measure distances) or, in the second case, we define distances but standard GR is not necessarily equivalent to metric Weyl geometries. Again, deciding which is the metric that really enters observational protocols is something that should not be imposed *a priori* but rather something to be tested locally.

3. Extended theories of gravitation

EPS analysis sets a number of constraints to any theoretical framework that can be called a *reasonable theory of gravitation*. Such constraints are much weaker than the strong metricity assumptions done in standard GR.

Before explicitly analyzing these constraints, let us first discuss about the interpretation of a relativistic theory. One usually chooses fundamental fields in kinematics and then considers dynamics. In gauge theories these two levels are usually quite disconnected since one is free to change fundamental field variables; this induces a change of dynamics (which is in fact assumed to be gauge covariant) and it does not affect observable quantities (which are also assumed to be gauge invariant).

However, the situation in gravitational theories is quite different. In relativistic theories, as we already discussed in the introduction, there are no known non-trivial gauge-invariant quantities. If we want to retain some connection with astrophysics and cosmology we are forced to assume as a fact that matter allows some conventional (partial) gauge fixing. Strictly speaking observables are not gauge invariant and accordingly they are not preserved under changes of fundamental field coordinates. When discussing the equivalence between different formalisms one must additionally declare how observational and measuring protocols are modified by the transformations allowed and/or chosen. For example, let us consider a metric Weyl theory in which the dynamics is described in terms of a metric g and a connection $\tilde{\Gamma}$. When a solution of field equations is found then one can determine both light cones and free fall by a single conformal representative $\tilde{g}$. Of course one can rewrite the dynamics in terms of $\tilde{g}$ only. Is this metric theory fully equivalent to GR, especially in presence of matter?

To better understand this apparently trivial question, let us recall that changing a metric g to a conformally equivalent one $\tilde{g} = \varphi \cdot g$ will not change electromagnetism but will certainly change the coupling with non-electromegnetic matter (e.g., a cosmological fluid).

We cannot answer this simple question, before considering which metric is used to define distances. In standard GR, one makes the *a priori* (unjustified) ansatz that distances would be defined by the same metric which defines free fall, i.e. in this case $\tilde{g}$. If in the original model distances were defined using g (which by the way is the only way to select a conformal gauge to write a non conformally invariant dynamics) the new model is only *similar* to standard GR but inequivalent as far as distances are concerned. If the original theory is recognized to be inequivalent to standard GR based on the Hilbert-Einstein-Palatini Lagrangian $L = \sqrt{g} g^{\mu\nu} R_{\mu\nu}(\Gamma)\, ds$, let us remark that as a matter of fact dark sources are precisely related to mismatches in observed distances...!

Now let us suppose for the sake of argument that standard GR is still a perfect theory to describe the actual universe (something that we know to be strongly questioned by actual observations). Still we believe that analyzing it within a wider framework as the one of EPS structures and Weyl geometries is in any case useful if not even necessary. If we can understand observations in this wider framework, in fact, we can better test gravity and maybe eventually show that standard GR is compatible with observations. If we assume standard GR setting and we build observational protocols for it then it may become difficult to understand which data come from assumptions and which data come instead from real physical facts, especially in a theory of gravitation in which we clearly made exceptions about gauge invariant observables.

Having said that, we see now that EPS formalism points to a Weyl geometry on spacetime in which one has a conformal structure $\mathfrak{C}$ defining light cones and a compatible (torsionless) connection $\tilde{\Gamma}$ defining free fall. Of course, whenever interaction with matter of half-integer spin is considered nothing prevents from relaxing the symmetry requirements on connections. However, until only test particles are considered matter is unaffected by torsion and one can drop it from the beginning. Then our protocols for measuring distances select a representative $g \in \mathfrak{C}$ for the conformal structure. In particular there is no reason why one should assume *a priori* that the connection $\tilde{\Gamma}$ is metric or, if such, that it is metric for the same metric one happens to have selected for distances.

Accordingly, one can use the kinematic and interpretation suggested by EPS to constrain dynamics. In a Palatini or *metric-affine* formalism the metric and connection are completely unrelated *a priori*, so that only dynamics may give their reciprocal relations. Then field equations may force a relation between the metric and the connection. That is exactly what happens in vacuum standard Palatini GR: field equations force the connection to be the Levi-Civita connection of the given metric. The same happens with some specific kind of matter, but for general matter such a feature is generally lost, and in general the connection cannot be the one associated to the original metric.

However, EPS analysis shows that the connection and the metric cannot be completely arbitrary if one wants a theory that fits fundamental principles; in fact there must be a covector A_μ for which (6) holds true. If the compatibility condition is not already imposed at the kinematical level —for example writing the theory for the fundamental fields $(g_{\mu\nu}, A_\mu)$ instead of $(g_{\mu\nu}, \tilde{\Gamma}^\alpha_{\beta\mu})$— then the only option is that field equations impose the compatibility conditions *a posteriori* as a consequence of field equations.

We shall thence call *extended theory of gravitation* any field theory for independent variables $(g_{\mu\nu}, \tilde{\Gamma}^\alpha_{\beta\mu})$ in which field equations imply the compatibility condition (6) as a consequence. In these models the geometry of spacetime is described by a Weyl geometry.

Let us call *extended metric theory of gravitation* any extended theory of gravitation in which field equations imply dynamically that the connection is a metric connection, so that in that case the geometry of spacetime is described by a *metric* Weyl geometry.

We know a class of dynamics which are in fact extended metric theories of gravitation. As is well known, any Palatini $f(\mathcal{R})$-theory is in fact an extended metric theory of gravitation; see [8], [15], [19], [20], [21]. Standard GR is a specific extended metric theory of gravitation in which field equations imply that $A_\mu = 0$ (and then $\Gamma = \{g\}$).

Of course it is well known that general Weyl geometries may have unpleasant holonomy problems in the definition of length (namely, the length of a ruler depends on the path). However, metric Weyl geometries are not affected by these problems and they are still more general than standard GR as we shall discuss hereafter for $f(\mathcal{R})$-models.

With such theories rulers cannot change length when parallel transported, although one has to be careful to notice that the metric scales can change point by point because of conformal rescaling.

3.1. Palatini $f(\mathcal{R})$-theories

In order to fix notation let us briefly review a generic $f(\mathcal{R})$-theory with matter.

Let us consider a Lorentzian metric $g_{\mu\nu}$ and a torsionless connection $\tilde{\Gamma}^{\alpha}_{\beta\mu}$ on spacetime M of dimension $m > 2$. The conformal class $[g] = \mathfrak{C}$ of metrics defines light cones. The representative $g \in \mathfrak{C}$ is chosen to define distances. The connection $\tilde{\Gamma}$ is associated to free fall and it is chosen to be torsionless since geodesic equations is insensitive to torsion.

Let us remark that in this context conformal transformations are defined to be $\tilde{g}(x) = \varphi(x) \cdot g(x)$ and they leave the connection unchanged. One is *forced* to leave the connection unchanged (as it is possible in Palatini formalism) since our connection $\tilde{\Gamma}$ is uniquely selected to describe free fall (and by the projective gauge fixing $\nabla g = 2A \otimes g$). One could say that this definition of conformal transformations preserves the interpretation of fields.

Let us restrict our analysis to dynamics induced by a Lagrangian in the form

$$L = \sqrt{g} f(\mathcal{R}) + L_m(\phi, g) \tag{7}$$

where f is a generic (analytic or *sufficiently regular*) function, ϕ is a collection of matter fields and we set $\mathcal{R} := g^{\mu\nu}\tilde{R}_{\mu\nu}$.

With this choice we are implicitly assuming that matter fields ψ minimally couple with the metric g which in turn encodes electromagnetic properties (photons and light cones). Since gravity, according to EPS formalism, is mostly inherent with the equivalence principle and free fall, that is encoded in the projective structure $\mathfrak{P}$, one should better assume that matter couples *also* with $\tilde{\Gamma}$ and investigate the more general case in which the matter Lagrangian has the form $L_m(\psi, g, \tilde{\Gamma})$. However, this case in much harder to be investigated since it entails that a second stress tensor is generated by the variational derivative

$$\sqrt{g} T^{\mu\nu}_{\alpha} = \frac{\delta L_m}{\delta \tilde{\Gamma}^{\alpha}_{\mu\nu}} \tag{8}$$

No relevant progress in this direction is still at hands, although it corresponds to an even more physically reasonable situation; only few concrete examples have been worked out insofar; see [15], [22], [23].

Let us remark that *a priori* the Ricci tensor $\tilde{R}_{\mu\nu}$ of the connection $\tilde{\Gamma}$ is not necessarily symmetric since the connection is not necessarily metric. As we said the matter Lagrangian L_m is here assumed to depend only on matter and metric (together with their derivatives up to order 1). Thus if one needs covariant derivatives of matter fields they are explicitly defined with respect to the metric field. Requiring that the matter Lagrangian does not depend on the connection $\tilde{\Gamma}$ is a standard requirement to simplify the analysis of field equation below although (as we said above) it would correspond to more reasonable physical situations. Let us here notice that what follows can be in fact extended to a more general framework; there are in fact matter Lagrangians depending on the connection $\tilde{\Gamma}$ in which field equations still imply the EPS-compatibility condition (6); see [15], [16], [17].

Field equations of (7) are

$$\begin{cases} f'(\mathcal{R})\tilde{R}_{(\mu\nu)} - \frac{1}{2}f(\mathcal{R})g_{\mu\nu} = \kappa T_{\mu\nu} \\ \tilde{\nabla}_\alpha\left(\sqrt{g}f'(\mathcal{R})g^{\beta\mu}\right) = T_\alpha^{\beta\mu} = 0 \end{cases} \tag{9}$$

We do not write the matter field equations (which will be considered as matter equations of state). The constant $\kappa = 8\pi G/c^4$ is the coupling constant between matter and gravity. The second stress tensor $T_\alpha^{\beta\mu}$ vanishes since the matter Lagrangian is assumed to be independent of the connection $\tilde{\Gamma}$. The first stress tensor $T_{\mu\nu}$ arises since the matter Lagrangian is a function of the metric

$$\sqrt{g}T_{\mu\nu} = \frac{\delta L_m}{\delta g^{\mu\nu}} \tag{10}$$

Notice that $T_{\mu\nu}$ depends both on the matter fields and on the metric g.

Under these simplifying assumptions the second field equations can be solved explicitly. Let us consider in fact a conformal transformation $\tilde{g}_{\mu\nu} = (f'(\mathcal{R}))^{\frac{2}{m-2}} \cdot g_{\mu\nu}$ (with $m > 2$). One has

$$\tilde{g}^{\mu\nu} = (f'(\mathcal{R}))^{\frac{2}{2-m}} \cdot g^{\mu\nu} \qquad \sqrt{\tilde{g}} = (f'(\mathcal{R}))^{\frac{m}{m-2}}\sqrt{g} \tag{11}$$

and then

$$\sqrt{\tilde{g}}\tilde{g}^{\beta\mu} = \sqrt{g}f'(\mathcal{R}) \cdot g^{\beta\mu} \tag{12}$$

Thus the second field equation in (9) can be recast as

$$\tilde{\nabla}_\alpha\left(\sqrt{g}f'(\mathcal{R})g^{\beta\mu}\right) = \tilde{\nabla}_\alpha\left(\sqrt{\tilde{g}}\tilde{g}^{\beta\mu}\right) = 0 \tag{13}$$

which by the Levi-Civita theorem implies

$$\tilde{\Gamma}^\alpha_{\beta\mu} = \{\tilde{g}\}^\alpha_{\beta\mu} \tag{14}$$

i.e. the connection $\tilde{\Gamma}$ is the Levi-Civita connection of the conformal metric $\tilde{g}$. Thus in these theories the connection is *a posteriori* metric and the geometry of spacetime is described by a metric Weyl geometry. As a consequence the Ricci tensor $\tilde{R}_{\mu\nu}$ is symmetric being the Ricci tensor of the metric $\tilde{g}$.

The first field equation now reads as

$$f'(\mathcal{R})\tilde{R}_{\mu\nu} - \frac{1}{2}f(\mathcal{R})g_{\mu\nu} = \kappa T_{\mu\nu} \tag{15}$$

The trace of this equation (with respect to $g^{\mu\nu}$) is so important in the analysis of these models that it is called the *master equation*. It reads

$$f'(\mathcal{R})\mathcal{R} - \frac{m}{2}f(\mathcal{R}) = \kappa T := \kappa g^{\mu\nu}T_{\mu\nu} \tag{16}$$

For any given (analytic) function f, the master equation is an *algebraic* (i.e. not differential) equation between $\mathcal{R}$ and T. Assuming that $m \neq 2$ and excluding the degenerate case in which the following holds

$$f''(\mathcal{R})\mathcal{R} + \frac{2-m}{2}f'(\mathcal{R}) = 0 \quad \Rightarrow f'(\mathcal{R}) = \frac{m}{2}C_1\mathcal{R}^{\frac{m-2}{2}} \quad \Rightarrow f(\mathcal{R}) = C_1\mathcal{R}^{\frac{m}{2}} + C_2 \tag{17}$$

we see that the function $F(\mathcal{R}, T) := f'(\mathcal{R})\mathcal{R} - \frac{m}{2}f(\mathcal{R}) - \kappa T$ is also analytic and can be generically (i.e. except a discrete set of values for $\mathcal{R}$) solved for $\mathcal{R} = r(T) = \kappa\hat{r}(T)$.

In vacuum or for purely electromagnetic matter obeying Maxwell equations, one has $T = 0$, i.e. the trace T of $T_{\mu\nu}$ is zero and $\mathcal{R}$ takes a constant value from a discrete set $\mathcal{R} \equiv \rho \in \{\rho_0, \rho_1, \dots\}$ that of course depends on f. In this vacuum (as well as in purely electromagnetic) case the field equations simplify to

$$\tilde{G}_{\mu\nu} = \tilde{R}_{\mu\nu} - \frac{1}{2}\tilde{R}\tilde{g}_{\mu\nu} = \left(\frac{2-m}{2m}(f'(\rho))^{\frac{2}{2-m}}\rho\right)\tilde{g}_{\mu\nu} = \Lambda(\rho)\tilde{g}_{\mu\nu} \tag{18}$$

Accordingly, vacuum (or purely electromagnetic) Palatini $f(\mathcal{R})$-theories are generically equivalent to Einstein models with cosmological constant and the possible value of the cosmological constant is chosen in a discrete set which depends on the function f. This is known as *universality theorem* for Einstein equations (see [18]). The meaning of this result is not to be overestimated; the equivalence is important but one has a huge freedom in choosing the function f (which depends on countable infinite parameters) so that any value for the cosmological constant can be in principle attained. Let us stress once more that this includes all cases in which matter is present but the trace $T = 0$ as it happens for the electromagnetic field.

Accordingly, the physics described by Palatini $f(\mathcal{R})$-theories in vacuum is not richer than standard GR physics with cosmological constant. Still one should notice that in these vacuum $f(\mathcal{R})$-theories free fall is given by $\tilde{g}$ while in standard GR it is given by g (while distances are defined by g in both cases); however, the conformal factor $\varphi = (f'(\rho))^{\frac{2}{2-m}}$ is constant and it does not affect geodesics and it can be compensated by a change of units.

However, when *real matter* is present the situation is completely different. In this more general case, we have that $\mathcal{R} = r(T)$ depends on $x \in M$. The first field equation becomes then

$$\tilde{G}_{\mu\nu} = \tilde{R}_{\mu\nu} - \frac{1}{2}\tilde{R}\tilde{g}_{\mu\nu} = \kappa\left(\frac{1}{f'(r(T))}\left(T_{\mu\nu} - \frac{1}{m}Tg_{\mu\nu}\right) + \frac{2-m}{2m}\hat{r}(T)g_{\mu\nu}\right) = \kappa\tilde{T}_{\mu\nu} \tag{19}$$

so that a Palatini $f(\mathcal{R})$-theory with *real* matter behaves like standard GR with a strongly modified source stress tensor. Naively speaking, one can reasonably hope that the modifications dictated by the choice of the function f can be chosen to fit observational data.

In a sense, whenever $T \neq 0$ the presence of standard visible matter ψ (assumed to generate, through the matter Lagrangian $L_m(g,\psi)$, an energy momentum stress tensor $T_{\mu\nu}$) would produce by gravitational interaction with $\tilde{\Gamma}$ (i.e. with the Levi-Civita connection of the conformal metric $\tilde{g} = f'(T) \cdot g$) a kind *effective* energy momentum stress tensor $\tilde{T}_{\mu\nu}$ in which standard matter ψ is seen to exist together with *dark (virtual)* matter generated by the gauging of the rulers imposed by the T-dependent conformal transformations on g. In a sense, the *dark side* of Einstein equations can be mimicked by suitably choosing f and L_m, as a curvature effect induced by $T = g^{\mu\nu}T_{\mu\nu} \neq 0$.

3.2. Equivalence with Brans-Dicke theories

Let us hereafter briefly review the mathematical equivalence between Palatini $f(\mathcal{R})$-theories and Brans-Dicke theories and discuss about how physical is such an equivalence. Let us hereafter restrict to the case in dimension $m = 4$.

A *Brans-Dicke* theory is a theory for a metric $g_{\mu\nu}$ and a scalar field φ. The dynamics is described by a Lagrangian in the following form

$$L_{BD} = \sqrt{g}\left[\varphi R - \frac{\omega}{\varphi}\nabla_\mu\varphi\nabla^\mu\varphi + U(\varphi)\right] + L_m(g,\psi) \tag{20}$$

where ω is a real parameter and $U(\varphi)$ is a potential function.

Field equations for such a theory are

$$\begin{cases} \varphi(R_{\mu\nu} - \frac{1}{2}Rg_{\mu\nu}) = \kappa T_{\mu\nu} + \nabla_{\mu\nu}\varphi + \Box\varphi g_{\mu\nu} + \frac{\omega}{\varphi}\left(\nabla_\mu\varphi\nabla_\nu\varphi - \frac{1}{2}\nabla_\alpha\varphi\nabla^\alpha\varphi g_{\mu\nu}\right) + \frac{1}{2}Ug_{\mu\nu} \\ R = \frac{\omega}{\varphi^2}\nabla_\alpha\varphi\nabla^\alpha\varphi - 2\frac{\omega}{\varphi}\Box\varphi - U'(\varphi) \end{cases} \tag{21}$$

If one considers now the field equation (19) for a Palatini $f(\mathcal{R})$-theory and writes them for the original metric $g_{\mu\nu} = \varphi^{-1} \cdot \tilde{g}_{\mu\nu}$ and the conformal factor $\varphi = f'(\mathcal{R})$ field equation reads as

$$\varphi R_{\mu\nu} = \nabla_{\mu\nu}\varphi + \frac{1}{2}\Box\varphi g_{\mu\nu} - \frac{3}{2\varphi}\nabla_\mu\varphi\nabla_\nu\varphi + \frac{1}{4}\varphi\hat{r}(T)g_{\mu\nu} + \kappa\left(T_{\mu\nu} - \frac{1}{4}Tg_{\mu\nu}\right) \tag{22}$$

while the master equation reads as

$$\varphi R = 3\Box\varphi - \frac{3}{2\varphi}\nabla_\alpha\varphi\nabla^\alpha\varphi + \kappa T + 2f \tag{23}$$

Within the framework for $f(\mathcal{R})$-theory, one can generically invert the definition of the conformal factor

$$\varphi = f'(\mathcal{R}) \quad \Rightarrow \mathcal{R} = \sigma(\varphi) \tag{24}$$

and define a potential function

$$U(\varphi) = -\varphi\sigma(\varphi) + f(\sigma(\varphi)) \qquad (U'(\varphi) = -\sigma'\varphi - \sigma + f'\sigma' = -\sigma) \tag{25}$$

Then one has a manifest correspondence between a Palatini $f(\mathcal{R})$-theory and a Brans-Dicke theory with the potential $U(\varphi) = -\varphi\sigma(\varphi) + f(\sigma(\varphi))$ and $\omega = -\frac{3}{2}$. This correspondence holds at the level of field equations (and solutions) but it can be shown at the level of action principles as well; see [20].

This equivalence is sometimes used against $f(\mathcal{R})$-theories since Brans-Dicke theories go to standard GR for $\omega \to \infty$ and the value $\omega = -\frac{3}{2}$ is ruled out by standard tests in the solar system, e.g. by precession of perihelia of Mercury. In view of the correspondence shown above the same tests would rule out $f(\mathcal{R})$-theories as well.

Letting aside the fact that tests rule out Brans-Dicke theories *without potential*, there is a further aspect that we believe is worth discussing here. In Brans-Dicke theory the gravitational interaction is mediated by a scalar field as well as the metric field. That means that g determines light cones, free fall and distances while the scalar field φ just participates to the dynamics.

In the corresponding Palatini $f(\mathcal{R})$-theory the metric g defines distances, it defines light cones (as well as $\tilde{g}$ does), but free fall is described by $\tilde{g}$ *not* by g!

The standard tests (as the precession of perihelia of Mercury) which rule out Brans-Dicke theories (see e.g. [24]) simply do not apply to the corresponding $f(\mathcal{R})$-theory since, in the two different models, Mercury moves along the geodesics of two different metrics. In Brans-Dicke theories it moves along the geodesics of a metric g which can be expanded in series of ω^{-1} around the standard Schwarzschild solution of standard GR; in the corresponding $f(\mathcal{R})$-theory it moves along geodesics of a different metric $\tilde{g}$ which, being in vacuum and in view of universality theorem, is a Schwarzschild-AdS solution; see also [25]. Since it is reasonable to assume a value for the cosmological constant which has no measurable effect at solar system scales, Mercury can be assumed move with good approximation along geodesics of the standard Schwarzschild metric and, despite the mathematical equivalence, $f(\mathcal{R})$-theories pass the tests while Brans-Dicke does not.

This is a pretty neat example in which a mathematical equivalence between two field theories is broken by the interpretation of the theories since the physical assumptions are not preserved by the transformation mapping one framework into the other; see [26], [27].

4. Extended cosmologies

Let us now apply to a cosmological situation the discussion above for a general Palatini $f(\mathcal{R})$-model.

In a cosmological setting let us assume that the matter stress tensor $T_{\mu\nu}$ is the energy momentum tensor of a (perfect) fluid

$$T_{\mu\nu} := p g_{\mu\nu} + (p+\rho) u_\mu u_\nu \tag{26}$$

where $u^\alpha u^\beta g_{\alpha\beta} = -1$ and we set ρ for the fluid density and p for its pressure. Matter field equations are assumed to provide a relation between pressure and density under the form $p = w\rho$ for some (constant) w. Then one has

$$\begin{aligned}\tilde{T}_{\mu\nu} =& \frac{1}{f'}\left(T_{\mu\nu} - \frac{1}{m} T g_{\mu\nu}\right) + \frac{2-m}{2m}\hat{r}(T) g_{\mu\nu} = \\ =& (\rho+p)(f')^{\frac{m}{2-m}} \tilde{u}_\mu \tilde{u}_\nu + \left(\frac{p+\rho}{m}(f')^{\frac{m}{2-m}} - \frac{m-2}{2m}\hat{r}(f')^{\frac{2}{2-m}}\right)\tilde{g}_{\mu\nu} = \\ =& (\tilde{\rho}+\tilde{p})\tilde{u}_\mu \tilde{u}_\nu + \tilde{p}\tilde{g}_{\mu\nu}\end{aligned} \tag{27}$$

where we set $\tilde{u}_\mu = (f')^{\frac{1}{m-2}} u_\mu$ and

$$\begin{cases} \tilde{\rho} = \frac{m-1}{m}(p+\rho)(f')^{\frac{m}{2-m}} + \frac{m-2}{2m}\hat{r}(f')^{\frac{2}{2-m}} \\ \tilde{p} = \frac{p+\rho}{m}(f')^{\frac{m}{2-m}} - \frac{m-2}{2m}\hat{r}(f')^{\frac{2}{2-m}} \end{cases} \tag{28}$$

Thus the effect of a Palatini $f(\mathcal{R})$-dynamics is to modify the fluid tensor representing sources into another stress tensor which is again in the form of a (perfect) fluid, with modified pressure and density. This can be split quite naturally (though of course non-uniquely) into three fluids with

$$\begin{cases} \tilde{\rho}_1 = \rho \\ \tilde{p}_1 = p \end{cases} \qquad \begin{cases} \tilde{\rho}_2 = \frac{m-1}{m}(p+\rho)(f')^{\frac{m}{2-m}} - \rho \\ \tilde{p}_2 = \frac{p+\rho}{m}(f')^{\frac{m}{2-m}} - p \end{cases} \qquad \begin{cases} \tilde{\rho}_3 = \frac{m-2}{2m}\hat{r}(f')^{\frac{2}{2-m}} \\ \tilde{p}_3 = -\frac{m-2}{2m}\hat{r}(f')^{\frac{2}{2-m}} = -\tilde{\rho}_3 \end{cases} \tag{29}$$

The first fluid accounts for what we see as visible matter and it has standard equation of states $p_1 = w_1\rho_1$ with $w_1 = w$, i.e. the same state equation chosen for the visible matter. The third fluid has equation of states in the form $p_3 = w_3\rho_3$ with $w_3 = -1$, i.e. it is a quintessence field.

For the second fluid, taking into account the equation of state $p = w\rho$ of visible matter, one can set

$$\begin{cases} \tilde{\rho}_2 = \frac{m-1}{m}(f')^{\frac{m}{2-m}} p + \left(\frac{m-1}{m}(f')^{\frac{m}{2-m}} - 1\right)\rho = \left(\frac{m-1}{m}(f')^{\frac{m}{2-m}}(w+1) - 1\right)\rho \\ \tilde{p}_2 = \left(\frac{1}{m}(f')^{\frac{m}{2-m}} - 1\right)p + \frac{1}{m}(f')^{\frac{m}{2-m}}\rho = \left(\frac{1}{m}(f')^{\frac{m}{2-m}}(w+1) - w\right)\rho \end{cases} \tag{30}$$

which corresponds to an equation of state of the form $p_2 = w_2\rho_2$ with

$$w_2 = \frac{\frac{1}{m}(f')^{\frac{m}{2-m}}(w+1) - w}{\frac{m-1}{m}(f')^{\frac{m}{2-m}}(w+1) - 1} \tag{31}$$

Within the standard viewpoint this kind of matter is quite puzzling. Its equation of state is changing in time (since in cosmology $f'(r((m-1)p(t) - \rho(t)))$ is a function of time).

It is reasonable to assume that at present time visible matter is dominated by dust ($w = 0$) and $m = 4$, in which case we have

$$w_2^{\text{dust}} = \frac{1}{3 - 4(f')^2} \tag{32}$$

Of course, the splitting of the fluid is not canonical or unique. In particular the second fluid can be further split in different components (for example in order to isolate components which are dominant in various regimes).

This is probably the main reason to consider Palatini $f(\mathcal{R})$-theories as good as models also for cosmology: although we assumed only dust at fundamental level, from the gravitational viewpoint that behaves effectively as a more general fluid the characteristics of which depend on the extended gravitational theories chosen, i.e. on f.

Morever, let us also remark that this simple toy model can be easily tested and falsified by current data and it makes predictions about near future surveys. In the standard ΛCDM one assumes a cosmological constant Λ which is here modeled by the third fluid. Thus in order to fit data one has to fix the current value for f', which in turn fixes the current equation of state for the CDM dark matter which is also observed. Of course one can consider other reasonable models considering more realistic and finer descriptions of visible matter. Near future surveys will provide data about the evolution of the cosmological constant in time allowing in principle to observe $f'(t)$ directly.

Let us now set $m = 4$, $w = 0$ and impose the cosmological principle ansazt, i.e. homogeneity and isotropy. Again should we impose it for g or $\tilde{g}$? It is fortunate that this does not matter at all! If one does that for $\tilde{g}$ assuming the form

$$\tilde{g} = -d\tilde{t}^2 - \tilde{a}^2(\tilde{t})\left(\frac{dr^2}{1-Kr^2} + r^2 d\Omega^2\right)$$

then also g is homogeneous and isotropic, i.e. in the form

$$g = -dt^2 - a^2(t)\left(\frac{dr^2}{1-Kr^2} + r^2 d\Omega^2\right)$$

provided one rescales the cosmological time with the conformal factor (which depends only on time)

$$d\tilde{t} = \sqrt{f'}\,dt \qquad \Rightarrow \tilde{t}(t) = \int \sqrt{f'}\,dt$$

and rescales the Friedmann-Lemaître-Robertson-Walker (FLRW) scale factor accordingly

$$\tilde{a}(\tilde{t}) = \sqrt{f'}\,a(t)$$

The equations for the scale factor are the celebrated Friedmann equations

$$\begin{cases} \frac{\dot{\tilde{a}}^2+K}{\tilde{a}^2} = \frac{\kappa}{3}\tilde{\rho} = \frac{\kappa}{12f'}\left(\hat{r}(\rho) + \frac{3\rho}{f'}\right) \\ \frac{\ddot{\tilde{a}}}{\tilde{a}} = -\frac{\kappa}{6}\left(\tilde{\rho} + 3\tilde{p}\right) = \frac{\kappa}{12f'}\left(\hat{r}(\rho) - \frac{3\rho}{f'}\right) \end{cases}$$

For a given f (and the associated $\hat{r}(\rho)$) these are two equations for the two unknowns $\tilde{a}(\tilde{t})$ and $\rho(t)$ which in principle should be determined as functions of t.

There is no much one can say in general without specifying f. Nevertheless, one can still notice that the worldlines $\gamma : s \mapsto (t_0 + s, r_0, \theta_0, \phi_0)$ are always geodesics (something that depends on the cosmological principle, not on Friedmann equations). Also the curves $\bar{\gamma} : s \mapsto (t_0, r_0 s, \theta_0, \phi_0)$ are geodesic trajectories and their length is thence related to spacial distances at time t_0.

Let us thus consider a point $(t_0, r_0, \theta_0, \phi_0)$ representing for example a galaxy, and let us suppose we want to compute its distance from us. If we defined distances by $\tilde{g}$ (as one would probably do in scalar tensor theories) such a distance would be given by

$$\tilde{d} = \tilde{a}(\tilde{t}_0) r_0 \int_0^1 \frac{ds}{\sqrt{1 - Ks^2 r_0^2}}$$

However, we defined distances by using g. Accordingly, one has

$$d = a(t_0) r_0 \int_0^1 \frac{ds}{\sqrt{1 - Ks^2 r_0^2}} = \frac{1}{\sqrt{f'}}\tilde{d}$$

Then these $f(\mathcal{R})$-theories have an extra time-dependent mismatch in measuring distances. Being the conformal factor dependent on time, it would affect non-trivially the measured acceleration of faraway galaxies.

To the best of our knowledge such a possible effect has not only been totally ignored in interpreting raw data, but it has not been discussed or proved to vanish.

For example, in this context the *measured* acceleration of the universe (which is defined to be the acceleration of galaxies per unit of distance) would be

$$\frac{\ddot{a}}{a} = \Phi^2 \frac{\ddot{\tilde{a}}}{\tilde{a}} - \frac{\dot{\Phi}}{\Phi}\frac{\dot{a}}{a} - \frac{\ddot{\Phi}}{\Phi} + \left(\frac{\dot{\Phi}}{\Phi}\right)^2$$

where we set $\Phi^2(t) = f'$ for the conformal factor.

It is therefore not difficult to find whole classes of functions f for which a solution in the FLRW form is allowed $\ddot{\tilde{a}}/\tilde{a}$ is negative (corresponding to an ever slower expansion) while $\ddot{a}/a$ is positive (corresponding to an accelerating expansion). When this were the case part of the effect of dark energy would be explained as a simple aberration of distance measurement.

Whether for some f this can fit experimental data better than the acceleration $\ddot{\tilde{a}}/\tilde{a}$ is something to be discussed on the observational ground. We just remark on a fundamental ground that extrapolating *our terrestrial current* rulers to 10 billion years ago and 10 billion light years away (in a theory in which geometry is dynamical and measurement protocols depends on all sorts of physical assumptions on the behavior of electromagnetic and gravitational fields) could be slightly hasty.

Of course we are not claiming these effect to be real. However, they are plausible and hence they should be considered in data analysis (and possibly eventually shown to be null). They were not introduced by *ad hoc* argument. On the contrary they are quite natural in metric extended theories of gravitation.

5. Conclusions and perspectives

The astrophysical and cosmological observations of the last decade clearly point to a deep reconsideration of standard scenarios based on standard GR, either on the source side or on the gravitational dynamics; or both. Basically, all observations about gravity in non-vacuum situations need to be somehow corrected.

If one decides to keep stuck to standard gravitational dynamics, then observations force us to modify the matter energy momentum tensor by adding dark sources. If one decides to modify gravity dynamics, the family of different available (covariant, variational, ...) theories is huge. Moreover, one variational model for gravity usually may support (for example when the model contains more than one metric) many inequivalent definitions of observational protocols. It is quite natural that in such a huge family one can find (many) models which fit the observations.

Thus usually one has to choose which of these two ways is preferable. In any event, one should reconsider foundations of gravitational theories from a more general perspective. EPS formalism provides us with such a reconsideration. It clearly shows that on spacetime coexist a conformal structure (associated to light rays and defining light cones and causality), a (torsionless) projective structure (associated to particles and free fall), and a metric structure (associated to our definition of clocks and rulers). These three structures can be assumed to be *a priori* independent, provided that dynamics forces *a posteriori* some compatibility conditions. The metric structure should also define the conformal structure and the projective structure can be represented by an affine (torsionless) connection so that lightlike geodesics of the metric structure are *also* autoparallel curves of the connection. This framework strongly constrains possible dynamics and it leads to *extended theories of gravitation*. Standard GR is a model within this extended family of *reasonable* gravitational models.

In extended gravitational theories one can also recast the fields so that an extended gravitational theory looks like standard GR with additional effective sources. This scenario thoroughly agrees with observations as long as dark matter and dark energy will be detected only through their gravitational effects. This scenario is reasonable also in view of cosmological observations which clearly suggest that the spacetime geometrical structure at large scale might be substantially different from the simple standard GR that we observe at Solar System scale.

Even if in the end standard GR were the correct theory and dark energy and matter will be understood at a fundamental level, this wider framework would be fundamental. It provides an extended framework in which one could test *directly* the assumptions of standard GR on an experimental basis without resorting to uncertain approximations.

In this paper we reviewed EPS formalism and defined *extended theories of gravitation* and *extended metric theories of gravitation*. Then we showed that Palatini $f(\mathcal{R})$-theories provide a family of such metric extended theories of gravitation.

If we restrict and apply $f(\mathcal{R})$-theories to cosmology we showed that matter naturally induces effective sources which can naturally modelled by fluid energy momentum source tensors which at least qualitatively present the main features of dark source models used phenomenologically to fit data. A (running) cosmological constant naturally emerges as well as a fluid with a running equation of states which depends explicitly of the $f(\mathcal{R})$ dynamics chosen. We also briefly discussed how one should define distances (as well as velocities and acceleration parameteres) in this extended framework.

Acknowledgements

We acknowledge the contribution of INFN (Iniziativa Specifica NA12) the local research project *Metodi Geometrici in Fisica Matematica e Applicazioni* (2011) of Dipartimento di Matematica of University of Torino (Italy). This paper is also supported by INdAM-GNFM.

Author details

L. Fatibene[1,2,*] and M. Francaviglia[1,2]

* Address all correspondence to: lorenzo.fatibene@unito.it

1 Department of Mathematics, University of Torino, Italy
2 INFN- Sezione Torino, Iniziativa Specifica Na12, Italy

References

[1] D. J. Mortlock, R. L. Webster, Mon. Not. Roy. Astron. Soc. 319 872 (2000)

[2] A. G. Riess et al. [Supernova Search Team Collaboration], Astron. J. 116 1009 (1998);
S. Perlmutter et al. [Supernova Cosmology Project Collaboration], Astrophys. J. 517 565 (1999);
J. L. Tonry et al. [Supernova Search Team Collaboration], Astrophys. J. 594 1 (2003)

[3] D. N. Spergel et al. [WMAP Collaboration], Astrophys. J. Suppl. 148 175 (2003);
C. L. Bennett et al., Astrophys. J. Suppl. 148 1 (2003);
M. Tegmark et al. [SDSS Collaboration], Phys. Rev. D 69 103501 (2004)

[4] U. Seljak et al. [SDSS Collaboration], Phys. Rev. D 71, 103515 (2005)

[5] S. M. Carroll, W. H. Press and E. L. Turner, Ann. Rev. Astron. Astrophys. 30, 499 (1992).

[6] C. Rovelli, *What is observable in classical and quantum gravity?*, Class. Quantum Grav. 8: 297, 1991;

[7] J. Elhers, F.A. E. Pirani, A. Schild, *The Geometry of free fall and light propagation in Studies in Relativity*, Papers in honour of J. L. Synge 6384 (1972)

[8] M. Di Mauro, L. Fatibene, M.Ferraris, M.Francaviglia, *Further Extended Theories of Gravitation: Part I* , Int. J. Geom. Methods Mod. Phys. Volume: 7, Issue: 5 (2010); gr-qc/0911.2841

[9] V. Perlick, *Characterization of standard clocks by means of light rays and freely falling particles*, Gen. Rel. Grav. 19, 1059-1073 (1987)

[10] J.L. Synge, *Relativity: the special theory*, (Amsterdam, 1956)

[11] L. Fatibene, M. Francaviglia, G. Magnano, *On a Characterization of Geodesic Trajectories and Gravitational Motions*, (in press) Int. J. Geom. Meth. Mod. Phys.; arXiv:1106.2221v2 [gr-qc]

[12] N. Dadhich, *Universal Velocity and Universal Force*, Physics News, 39, 20-25 (2009); arXiv:1003.2359v1 [physics.gen- ph]

[13] J.A.Schouten, *Ricci-Calculus: An Introduction to Tensor Analysis and its Geometrical Applications*, Springer Verlag (1954)

[14] N. Dadhich, J.M. Pons, *Equivalence of the Einstein-Hilbert and the Einstein-Palatini formulations of general relativity for an arbitrary connection*, (to appear on GRG); arXiv:1010.0869v3 [gr-qc]

[15] L. Fatibene, M.Ferraris, M.Francaviglia, S.Mercadante, *Further Extended Theories of Gravitation: Part II*, Int. J. Geom. Methods Mod. Phys. 7, (5) (2010), pp. 899-906; gr-qc/0911.284

[16] T.P. Sotiriou, *$f(R)$ gravity, torsion and non-metricity*, Class. Quant. Grav. 26 (2009) 152001; gr-qc/0904.2774

[17] T.P. Sotiriou, S. Liberati, *Metric-affine f(R) theories of gravity*, Annals Phys. 322 (2007) 935-966; gr-qc/0604006

[18] A. Borowiec, M. Ferraris, M. Francaviglia, I. Volovich, *Universality of Einstein Equations for the Ricci Squared Lagrangians*, Class. Quantum Grav. 15, 43-55, 1998

[19] S. Capozziello, V.F. Cardone, A. Troisi, JCAP 08 001 (2006).

[20] A. De Felice, S. Tsujikawa, *$f(R)$ Theories*, http://www.livingreviews.org/lrr-2010-3

[21] G.J. Olmo, *Palatini Approach to Modified Gravity: f(R) Theories and Beyond*, Int. J. Mod. Phys.D 20, 413-462 (2011); arXiv:1101.3864v1 [gr-qc]

[22] G.J. Olmo, P. Singh, *Covariant Effective Action for Loop Quantum Cosmology a la Palatini*, Journal of Cosmology and Astroparticle Physics 0901:030, 2009; arXiv:0806.2783

[23] L.Fatibene, M.Francaviglia, S. Mercadante, *Matter Lagrangians Coupled with Connections*, Int. J. Geom. Methods Mod. Phys. 7(5) (2010), 1185-1189; arXiv:0911.2981

[24] S. Weinberg, *Gravitation and Cosmology: Principles and Applications of the General Theory of Relativity*, Wiley, New York (a.o.) (1972). XXVIII, 657 S. : graph. Darst.. ISBN: 0-471-92567-5.

[25] G.J.Olmo, *Nonsingular black holes in quadratic Palatini gravity*, JCAP 1110 (2011) 018; arXiv:1112.0475v2 [gr-qc]

[26] S. Capozziello, M.F. De Laurentis, M. Francaviglia, S. Mercadante, Found. of Physics 39 1161 (2009)

[27] V. Faraoni and E. Gunzig, Int. J. Theor. Phys. 38 217 (1999)

Chapter 5

Uncertainties in Dark Matter Indirect Detection

Katie Auchettl and Csaba Balázs

Additional information is available at the end of the chapter

1. Introduction

Astrophysical observations interpreted in the standard (ΛCDM) cosmological framework indicate that about eighty percent of matter in the Universe is non-luminous. This dark matter poses a major problem for particle physics: no known particle explains its inferred properties. Observations are most consistent with the assumption that dark matter is composed of weakly interacting massive particles. The discovery of these particles is vital to validate the prevailing dark matter paradigm. In this work, we examine the uncertainties affecting the astrophysical discovery of dark matter particles via secondary cosmic ray emission.

Before trying to discover dark matter particles, we should know some of their properties. These properties of dark matter are reconstructed from astrophysical observations, most of which (including the galactic rotational velocities, galactic structure formation, and weak gravitational lensing) indicate that dark matter particles have mass and are present in large numbers around us. Measurements of the cosmic microwave background radiation and the abundance of light elements further suggests that dark matter is not composed of baryonic particles (that is quarks). Since electromagnetic interactions imply light emission, we are led to conclude that dark matter particles may only interact with ordinary matter weakly, either via the standard W and Z bosons, or via an unknown force. Since they are electrically neutral, the simplest assumption is that dark matter particles are their own anti-particles. Their diffuse distribution, inferred from their gravitational effects, indicates that they probably interact weakly with each other. To be present over the observed distance scales, dark matter particles have to be stable on the timescale of the age of the Universe. Lastly, the observed large scale structure indicates that dark matter is cold - its particles are non-relativistic at its present temperature.

Based on the above properties, dark matter particles are being searched for in three major types of experiments. First, since the CERN Large Hadron Collider (LHC) in Geneva was built specifically to explore the electroweak sector of the standard particle model, it is an obvious place for trying to create dark matter particles. While we are in (almost) full control of this experiment, without knowing the exact mass and interaction strength between ordinary and dark matter we can only hope that the energy and luminosity of the LHC are high enough to produce the latter. Second, because it appears that the Solar System is immersed in a high flux stream of dark matter particles, it is natural to try to detect collisions between them and well shielded nuclei in underground laboratories. These experiments have the potential to probe interactions between dark and ordinary matter even beyond the reach of the LHC. However, these experiments are also limited by the unknown mass, interaction strength, flux and velocity distribution of the dark matter particles. Finally, perhaps the most general and unconstrained way to discover dark matter particles is to find traces of their annihilation or decay products in cosmic rays bombarding Earth.

This last type of experiment is called indirect dark matter detection and it is a sensible way to find dark matter particles if they are either their own anti-particles or the matter-antimatter asymmetry in the dark sector is not pronounced. In this case, via weak interactions, dark matter particles self annihilate into standard ones and create secondary cosmic rays. Alternatively, if dark matter decays into standard particles (with the lifetime of about 10^{26} sec or more) its decay products can contribute to the secondaries. The most promising detection modes are the photon final states or the ones that contribute to anti-matter cosmic rays, such as positrons or anti-protons. In the last few years several anomalies were found in the cosmic positron fluxes by the PAMELA and Fermi-LAT satellites, which could be the first glimpses of dark matter.

However, various factors make indirect detection of dark matter challenging and less straightforward than we would like it to be. First, the immense cosmic ray background originating from ordinary astrophysical sources makes it hard to find the signal contributed by dark matter. Next, in many cases sources of the cosmic ray background are not known or not understood well enough. Finally, an important source of uncertainty and the main subject of our study is the cosmic ray propagation through the galaxy. This propagation is described by the diffusion equation; an equation with many unknown parameters. We review the state of this field of research. We show that using state of the art numerical codes, CPU intensive statistical inference, and the latest cosmic ray observations, the most important of these propagation parameters can be determined with a certain precision. Then we show how to propagate these uncertainties into recent cosmic ray measurements of Fermi-LAT and PAMELA. In the light of these findings we quantify the statistical significance of the present hints of signals in dark matter indirect detection. Finally, we contrast the experimental standing with some of the theoretical dark matter models proposed in the recent literature to explain cosmic ray 'anomalies'.

2. Experimental status of cosmic electrons and positrons

Experiments detecting cosmic rays near Earth have been finding various unexpected deviations from theoretical predictions over the last twenty years. The local flux of high energy positrons is notoriously anomalous as reported by the

- TS [1],
- AMS [2],

- CAPRICE [3],
- MASS [4], and
- HEAT [5, 6]

collaborations. Measurements of the PAMELA satellite stirred great interest by showing an unmistakable rise of the local e^+/e^- fraction, that deviates significantly from theoretical predictions for $E_{e^+} > 10$ GeV [7]. The combined experimental and theoretical uncertainties do not seem to account for such a large excess [8–11].

The summed flux of electrons and positrons also indicates an anomalous excess of observation over theory, as measured by the

- AMS [12],
- PPB-BETS [13], and
- HESS [14, 15]

collaborations. The Fermi Large Area Telescope (LAT) satellite confirmed the excess of the $e^- + e^+$ flux for energies over 100 GeV [16, 17]. The Fermi-LAT results are consistent with those of the PAMELA collaboration, which measured the cosmic ray electron flux up to 625 GeV [18]. To date the Fermi-LAT data are the most precise indication of such an anomaly in the electron-positron spectrum. The Fermi-LAT data differ by several standard deviations from the theoretical calculations encoded in GALPROP by [19].

3. The problem of cosmic ray background calculation

Between 2008 and 2011, of the order of a thousand papers were devoted to explaining the difference between the experimental measurements of Fermi-LAT and PAMELA and theoretical calculations. Speculation ranged from the modification of the cosmic ray propagation through supernova remnants, to dark matter annihilation. A concise summary of this literature with detailed references can be found in [20] and [21].

Before drawing conclusions from the electron-positron anomaly, however, one has to carefully examine the status of the theoretical understanding of Galactic cosmic rays. Unfortunately, even the origin of the cosmic rays is uncertain. The theory describing the propagation of cosmic ray particles from their birthplace through the Milky Way is based on the diffusion-convection model. The quantitative description of propagation is facilitated by the transport equation. This is a partial differential equation for each cosmic ray species which requires fixing the distribution of initial sources and the boundary conditions. Specifying the initial source distribution is a source of significant uncertainty in these calculations. The local cosmic ray fluxes are obtained as the self-consistent solutions of the set of transport equations. Obtaining these solutions is challenging due to the large number of free parameters such as the convection velocities, spatial diffusion coefficients, and momentum loss rates.

In the rest of this chapter we show how to determine those uncertainties of the electron-positron cosmic ray flux that originate from the propagation parameters of the diffusion equation. First we find the set of propagation parameters that the electron-positron flux is most sensitive to. Then we extract the values of these propagation parameters from cosmic ray data (different from the Fermi-LAT and PAMELA measurements). Based on the values of the propagation parameters most favored by the data we calculate theoretical predictions for the electron-positron fluxes and compare these to Fermi-LAT and PAMELA. By calculating the difference between our predictions and the observed fluxes we are able to isolate the anomalous part of the cosmic e^- and e^+ fluxes.

Similar results have been published in the literature before. However, our results supersede these in two important aspects. First, we show that when analyzed in the framework of the standard propagation model there exists a statistically significant tension between the e^-, e^+ and the rest of the charged cosmic ray fluxes. Second, unlike anyone else before us, we isolate the anomalous contribution within the e^- and e^+ spectrum together with its theoretical uncertainty.

Our analysis uses more charged cosmic ray spectral data points than similar studies before us such as [22]. Unlike us [23] use gamma ray data when extracting the background, however this may bias the analysis since gamma rays originating from anomalous electrons or positrons are not part of the background. Our numerical treatment, similar to that of [24], is more complete than the one of [25–27].

Our statistical analysis can be considered as an extension of [24] since we calculate the e^-, e^+ background with a theoretical uncertainty. Ref. [24] use 76 spectral data points, while we use 219 which gives us a significant edge over their analyses. The parameters that we freely vary are somewhat different from those of [24]. Before we choose the parameter space, we analyzed the sensitivity of the electron and positron spectrum to the parameters to maximize the efficiency of our parameter extraction. Our choice and treatment of the nuisance parameters also differs from [24]. Finally, we use a different scanning technique from the one they use.

4. Cosmic ray propagation through the Galaxy

The propagation of charged particles through the Galaxy can be well described using the diffusion-convection model [28]. This model assumes that the charged particles propagate homogeneously within a defined region of diffusion (similar to the leaky box propagation model), while taking the effects of energy loss into account. The diffusive region is assumed to be a solid flat cylinder defined with a radius R and a half-height of L. Its shape is such that it encloses the Galactic plane which confines charged cosmic rays to the Galactic magnetic fields inside it, while cosmic rays outside are free to stream away. The solar system in this diffusive region is defined in cylindrical coordinates as $\vec{r}_\odot = (8.33\text{ kpc}, 0\text{ kpc}, 0\text{ kpc})$ [29]. The phase-space density $\psi_a(\vec{r}, p, t)$ of a particular cosmic ray species a at time t, Galactic position $\vec{r}$ and with momentum p can be determined by solving the cosmic ray transport equation, which has the general form [30]

$$\begin{aligned}\frac{\partial \psi_a(\vec{r}, p, t)}{\partial t} &= Q_a(\vec{r}, p, t) + \nabla \cdot (D_{xx} \nabla \psi_a - \vec{V} \psi_a) \\ &+ \frac{\partial}{\partial p} \left(p^2 D_{pp} \frac{\partial}{\partial p} \frac{1}{p^2} \psi_a \right) - \frac{\partial}{\partial p} \left(\dot{p} \psi_a - \frac{p}{3} (\nabla \cdot \vec{V}) \psi_a \right) - \frac{1}{\tau_f} \psi_a - \frac{1}{\tau_r} \psi_a. \end{aligned} \quad (1)$$

If the time-scale of cosmic ray propagation (which is of the order of 1 Myr at 100 GeV energies) is much longer than the typical time scales of the galactic collapse of dark matter and the variation in the propagation conditions, then one can assume that the steady state condition holds. In this case, the left hand side of equation 1 can be set to zero and the time dependence of all quantities can be dropped. For our analysis we focus on a simplified version of the transport equation which to a first order approximation is sufficient to describe the propagation of electrons, positrons or anti-protons through the Galaxy and their corresponding spectrum at Earth:

$$0 = Q_a(\vec{r}, E) + K(E)\,\nabla^2\psi_a + \frac{\partial}{\partial E}\Big(b(E)\,\psi_a - K_{\rm EE}(E)\,\psi_a\Big) - \frac{\partial}{\partial z}\left(\mathrm{sign}(z)\,V_C\,\psi_a\right), \qquad (2)$$

where E is the energy of the secondary particle species a. To ensure that on the outer surface of the cylinder the cosmic ray density vanishes, boundary conditions are imposed. Similarly, outside of the diffusive region, these boundary condtions allow the particles to freely propagate and escape. This ensures that the modelling is consistent with the physical picture described above. One also imposes the symmetric condition $\partial\psi_a/\partial r(r=0) = 0$ at $r = 0$. In momentum space, null boundary conditions are imposed.

The transport of cosmic ray species through turbulent magnetic fields, the energy losses experienced by these particles due to Inverse Compton scattering (ICS), synchrotron radiation, Coulomb scattering or bremsstrahlung and their re-acceleration due to their interaction with moving magnetised scattering targets in the Galaxy is defined by the spatial diffusion coefficient $K(E)$, the energy loss rate $b(E)$ and the diffusive re-acceleration coefficient $K_{\rm EE}(E)$ respectively. The effect of Galactic winds propagating vertically from stars in the Galactic disk can be incorporated by defining the convective velocity V_C. The source of the cosmic rays is defined by $Q_a(\vec{r}, E)$ in equation 2, with a standard source term resulting from the annihilation of dark matter which can be written as:

$$Q_a(\vec{r}, E) = \frac{1}{2}\frac{dN_a}{dE}\langle\sigma_a v\rangle_0 \left(\frac{\rho_g(\vec{r})}{m_\chi}\right)^2. \qquad (3)$$

Here $\langle\sigma_a v\rangle_0$ corresponds to the thermally averaged annihilation cross section of the relevant species, and $\rho_g(\vec{r})$ is the energy density of dark matter in the Galaxy. The energy distribution of the secondary particle a is defined as dN_a/dE and is normalised per annihilation. This formula applies to self-conjugated annihilating dark matter. In the case of non-self-conjugated dark matter, or of multicomponent dark matter, the quantities in Eq. (3) should be replaced as follows, where an index i denotes a charge state and/or particle species (indeed any particle property, collectively called "component") and $f_i = n_i/n$ is the number fraction of the i-th component:

$$m_\chi \to \sum_i f_i m_i \qquad \text{(mean mass)}, \qquad (4)$$

$$\langle\sigma_a v\rangle \to \sum_{ij} f_i f_j \langle\sigma_{a,ij} v_{ij}\rangle \qquad \text{(mean cross section times relative velocity)}, \qquad (5)$$

$$dN_a/dE \to \frac{\sum_{ij} f_i f_j \sigma_{a,ij} v_{ij}\,(dN_{a,ij}/dE)}{\sum_{ij} f_i f_j \sigma_{a,ij} v_{ij}}, \qquad \text{(annihilation spectrum per annihilation)}. \qquad (6)$$

The spatial diffusion coefficient $K(E)$ is assumed to have the form

$$K(E) = K_0 \, v^{\eta} \left(\frac{R}{\text{GeV}} \right)^{\delta}, \tag{7}$$

where v is the speed (in units of c) and $R = p/eZ$ is the magnetic rigidity of the cosmic ray particles. Here Z is the effective nuclear charge of the particle and e is the absolute value of its electric charge (if we considered particles other than electrons, positrons, protons or anti-protons then this quantity would be different from 1). At low energies the behaviour of the cosmic rays as they diffuse is controlled using the parameter η. Traditionally one will set $\eta = 1$ but departures from this traditional value to other values (either positive or negative) have been suggested. More detailed treatments allow one to incorporate spatial dependence into the diffusion coefficient ($K(\vec{r}, E)$) and the influence of particle motion on the diffusion of these particles (which leads to anisotropic diffusion).

Synchrotron radiation and Inverse Compton scattering are position dependent phenomena. Synchrotron radiation arises from the interaction of a charged particle with Galactic magnetic fields and thus it depends on the strength of the magnetic field which changes in the Galaxy. Similarly, inverse Compton scattering is dependent on the distribution of background light which varies in the Galaxy. If one neglects this position dependence of energy losses in the Galactic halo and assumes that all energy losses can be described using a relationship that is proportional to E^2 (which is only valid if one neglects energy losses, such as Coulomb losses and bremsstrahlung, and considers only inverse Compton scattering for electrons with relatively low energy - Thomson scattering regime), then the energy loss rate can be parametrized as

$$b(E) = b_0 \, E^2. \tag{8}$$

In more detailed treatments the spatial dependence associated with the energy loss rate $b(\vec{r}, E)$ would be considered and a more general energy dependence relationship for the energy loss rate would be obtained. Additionally Coulomb losses ($dE/dt \sim$ const) and bremsstrahlung losses ($dE/dt \sim bE$) could also be taken into account. These losses can be calculated using functions dependent on position and energy as well as gas, interstellar radiation and magnetic field distributions [31].

Finally, the diffusive re-acceleration coefficient $K_{\text{EE}}(E)$ is usually parametrized as

$$K_{\text{EE}}(E) = \frac{2}{9} v_{\text{A}}^2 \frac{v^4 \, E^2}{K(E)}, \tag{9}$$

where v_{A} is the Alfvén speed.

A propagator, or Green's function G, is used to describe the evolution of the cosmic ray that originates from a source Q at $\vec{r}_S$ with energy E_S through the diffusive halo and reaches the Earth at point $\vec{r}$ with energy E. This allows the general solution for Eq. (2) to be written as

$$\psi_a(\vec{r}, E) = \int_E^{m_X} dE_S \int d^3 r_S \, G(\vec{r}, E; \vec{r}_S, E_S) \, Q_a(\vec{r}_S, E_S). \tag{10}$$

The differential flux is related to the solution in Eq. (10) via

$$\frac{d\Phi_a}{dE} = \frac{v(E)}{4\pi} \psi_a(\vec{r}, E). \tag{11}$$

For the propagation of protons or anti-protons in the Galactic halo, additional terms in Eq. (2) should be introduced to account for spallations on the gas in the disk.

5. Statistical framework

We use standard Bayesian parameter inference to determine the statistically favored regions of the propagation parameters $P = \{p_1, ..., p_N\}$ that the electron and positron cosmic ray fluxes are the most sensitive to. For full mathematical details we refer the reader to [21]. Here we only highlight the main concepts used.

Using the experimental data $D = \{d_1, ..., d_M\}$ and their corresponding theoretical predictions $T = \{t_1(P), ..., t_M(P)\}$, as the fuction of the parameters, we construct the likelihood function

$$\mathcal{L}(D|P) = \prod_{i=1}^{M} \frac{1}{\sqrt{2\pi}\sigma_i} \exp(-\chi_i^2(D,P)/2). \tag{12}$$

Here

$$\chi_i^2(D,P) = \left(\frac{d_i - t_i(P)}{\sigma_i} \right)^2, \tag{13}$$

and σ_i are the corresponding combined theoretical and experimental uncertainties. Assuming flat priors $\mathcal{P}(P)$, we then construct the posterior probability distribution

$$\mathcal{P}(P|D) = \mathcal{L}(D|P)\mathcal{P}(P)/\mathcal{E}(D). \tag{14}$$

At this stage the value of the evidence $\mathcal{E}(D)$ is unknown. After integrating over the whole parameter space its value can be recovered as

$$\mathcal{E}(D) = \int \mathcal{L}(D|P)\mathcal{P}(P) \prod_{j=1}^{N} dp_j. \tag{15}$$

More relevant to our purpose is the adaptive scan of the likelihood function during this integration which will give us the shape of the posterior distribution over the relevant part of the parameter space (where the likelihood is the highest). Having this shape we can calculate the probability density of a certain theoretical parameter p_i acquiring a given value by marginalization

$$\mathcal{P}(p_i|D) = \int \mathcal{P}(P|D) \prod_{i \neq j=1}^{N} dp_j. \tag{16}$$

Here the integral is carried out over the full range of the parameters. We can also determine Bayesian credibility regions $\mathcal{R}_x$ for each of the parameters:

$$x = \int_{\mathcal{R}_x} \mathcal{P}(p_i|D)\, dp_i. \tag{17}$$

The above relation expresses the fact that x % of the total posterior probability lies within the region $\mathcal{R}_x$.

After examining the electron and positron fluxes for parameter sensitivity we found that the relevant parameters are:

$$P = \{\gamma^{e^-}, \gamma^{nucleus}, \delta_1, \delta_2, D_{0xx}\}. \tag{18}$$

These parameters enter into the diffusion calculation as follows; γ^{e^-} and $\gamma^{nucleus}$ are the primary electron and nucleus injection indices parameterizing an injection spectrum without a break; δ_1 and δ_2 are spatial diffusion coefficients below and above a reference rigidity ρ_0; and D_{0xx} determines the normalization of the spatial diffusion coefficient.

We treat the normalizations of all charged cosmic ray fluxes as theoretical nuisance parameters:

$$P_{nuisance} = \{\Phi^0_{e^-}, \Phi^0_{e^+}, \Phi^0_{\bar{p}/p}, \Phi^0_{B/C}, \Phi^0_{(SC+Ti+V)/Fe}, \Phi^0_{Be-10/Be-9}\}. \tag{19}$$

We discuss other statistical and numerical issues, such as the choice of priors, the systematic uncertainties, sampling and convergence, in detail in [21].

6. Experimental data used in this analysis

In our statistical analysis we use 219 of the most recent data points corresponding to five different types of cosmic ray experiment. A majority of these data points (114 in total) come from electron-positron related experiments while the other 105 are made up of Boron/Carbon, anti-proton/proton and (Sc+Ti+V)/Fe and Be-10/Be-9 cosmic ray flux measurements. If any of the energy ranges of the experiments overlap, the most recent experimental data point was chosen in that energy range.

There are three main experiments that have measured electrons and positrons over different decades of energy. These experiments include AMS by [12], Fermi-LAT by [17] and HESS by

[14, 15]. The AMS collaboration reported an excess in positrons with energies greater than 10 GeV [12], while the HESS collaboration measured a significant steepening of the electron plus photon spectrum above one TeV as measured by HESS's atmospheric Cherenkov telescope (ACT). Using the Large Area Telescope (LAT) on the Fermi Satellite, the Fermi-LAT collaboration released a high precision measurement of the $e^+ + e^-$ spectrum for energies from 7 GeV to 1 TeV [17], extending the energy range of their previously published results. We defined our $e^+ + e^-$ spectrum by using data from these three experiments. The PAMELA collaboration recently released their measurement of the e^- only spectrum [18] confirming the behaviour of the $e^+ + e^-$ spectrum as measured by Fermi-LAT. The energy ranges that these data points were selected over are listed in Table 1.

Measured flux	Experiment	Energy (GeV)	Number of data points
	AMS [12]	0.60 - 0.91	3
$e^+ + e^-$	Fermi-LAT [17]	7.05 - 886	47
	HESS [14, 15]	918 - 3480	9
$e^+/(e^+ + e^-)$	PAMELA [32]	1.65 - 82.40	16
e^-	PAMELA [18]	1.11 - 491.4	39
anti-proton/proton	PAMELA [33]	0.28 - 129	23
	IMP8 [34]	0.03 - 0.11	7
	ISEE3 [35]	0.12 - 0.18	6
Boron/Carbon	[36]	0.30 - 0.50	2
	HEAO3 [37]	0.62 - 0.99	3
	PAMELA [38]	1.24 - 72.36	8
	CREAM [39]	91 - 1433	3
(Sc+Ti+V)/Fe	ACE [40]	0.14 - 35	20
	SANRIKU [41]	46 - 460	6
	[42]	0.003 - 0.029	3
	[43]	0.034 - 0.034	1
	[42]	0.06 - 0.06	1
Be-10/Be-9	ISOMAX98 [44]	0.08 - 0.08	1
	ACE-CRIS [45]	0.11 - 0.11	1
	ACE [46]	0.13 - 0.13	1
	AMS-02 [47]	0.15 - 9.03	15

Table 1. In this table we have listed the cosmic ray experiments and the energy ranges of the corresponding data points that we selected for our analysis. There are two sets of cosmic ray data listed in this table. Electron positron flux related experiments make up the first five lines of the table, while all other experiments make up the rest of the table. On these two sets of data, we perform a Bayesian analysis to highlight the tension between the two sets of data.

Apart from measuring the electron positron sum, collaborations such as PAMELA have measured the differential positron fraction $e^+/(e^+ + e^-)$ between energies of 1.5 and 100 GeV [7]. If one assumes that all secondary positrons are produced during the propagation of cosmic rays in the Galaxy, one would expect the positron fraction to decrease, however, the observed fraction increases for energies greater than 10 GeV.

How primary and secondary cosmic rays are produced and transported throughout the Galaxy can be studied by using cosmic-ray particles such as anti-protons. One requires a large number of measurements with good statistics over a large energy range to produce detailed anti-proton spectra for study. Anti-proton spectra obtained from previous balloon borne experiments such as CAPRISE98 [3] and HEAT [48] had very low statistics, but recently the PAMELA satellite experiment [33] released a high-quality measurement of the anti-proton/proton flux ratio for an energy range of 1-100 GeV. This spectrum confirmed the behaviour of the anti-proton/proton ratio as observed by previous experiments.

Additionally one can use stable secondary to primary cosmic ray ratios such as Boron/Carbon and (Sc+Ti+V)/Fe ratio to study the variation experienced by cosmic rays as they propagate through the Galaxy. These ratios are particularly sensitive to the properties of cosmic ray propagation as the element in the numerator is produced by a different mechanism to the element that defines the denominator. Primary cosmic rays are produced by the original source of the cosmic rays such as a supernova remnant, while the secondary cosmic rays are generated by the interaction of their primaries with the interstellar medium [49]. Ratios that are defined by a numerator and denominator which are produced by the same mechanism, such as a primary/primary or a secondary/secondary cosmic ray ratio, have a low sensitivity to any variation in the propagation parameters. Analysing Galactic Boron/Carbon and (Sc+Ti+V)/Fe ratios allows one to determine the amount of interstellar material transversed by the primary cosmic ray and its energy dependence [49].

Unstable isotopes such as Beryllium-10/Beryllium-9 are also beneficial to analyse as they produce a constraint on the time it takes for a cosmic ray to propagate through the Galaxy [50]. Various experiments such as ISOMAX98 [44], ACE-CRIS [45], ACE [46] and AMS-02 [47] have all measured Be-10/Be-9 data with varying statistics.

In Table 1 we state over what energy range and from which experiment we selected the data points that define our spectrum of anti-proton/proton, Boron/Carbon, (Sc+Ti+V)/Fe and Be-10/Be-9 ratio.

For energies below $E < 10$ GeV, solar magnetic and coronal activities perturb the low energy part of the cosmic ray spectrum. This is called solar modulation and has an important role in determining the observed spectral shape(s) of cosmic rays measured at earth [51, 52]. Solar modulation is accounted for in GALPROP by using a force field approximation. It should be noted that this is an approximation and does not include important influences such as the structure of the heliospheric magnetic field. To incorporate these effects into our analysis we vary the value of the modulation potential in GALPROP. Following Gast & Schael (2009), we also assume that the positively and negatively charged cosmic rays are modulated differently by solar activities (charge-sign dependent modulation). This charge dependent modulation has a significant effect on positrons and its effect on the anti-proton/proton ratio can be comparable to the experiment's statistical uncertainties. The modulation effect on heavy nuclei such as B, C, Sc, Ti, V, Fe and Be is mild even though these nuclei have a higher positive charge compared to the proton. The reason for this is that the modulation potential is proportional to the charge-to-mass ratio and these heavy nuclei have a much lower charge-to-mass ratio than the proton therefore the effect is minute. Regardless, as we use the ratio of their fluxes, most of the effect of solar modulation on these nuclei is cancelled, thus we can safely absorb this modulation effect into the systematic uncertainties of the experiment. To be able to compare experimental data we set the modulation potential in GALPROP for positrons (electrons) to the value determined by Gast & Schael (2009), $\phi+ = 442(2)$ MeV. Previous work by Usoskin et al. (2011) showed that the time dependence of the

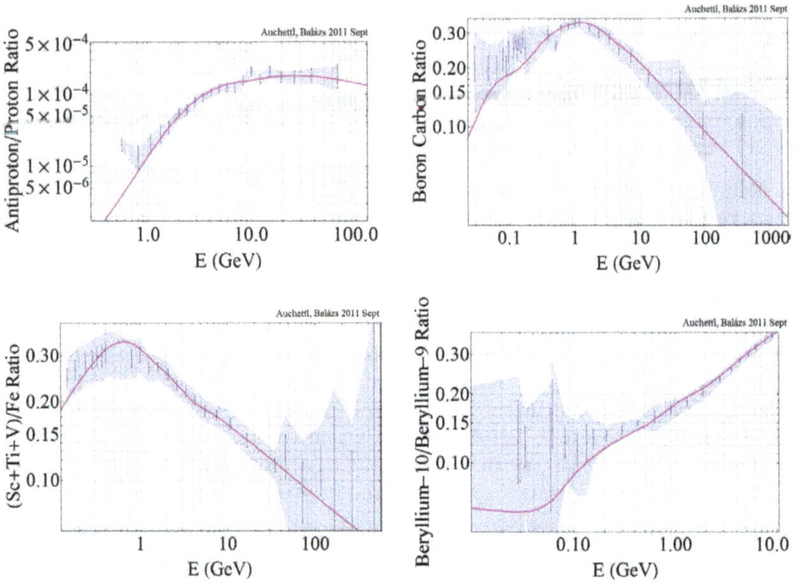

Figure 1. Best fit curves plotted against non-electron-positron related data. These curves were calculated using the most probable parameter values obtained from the peak values of the posterior probabilities (inferred from $\bar{p}/p$, B/C, (Sc+Ti+V)/Fe and Be data) shown in red in Fig. 2. The best fit curves pass through the estimated systematic error bands, shown in gray.

solar modulation potential was not substantial over the period of PAMELA's data taking, and approximately the same average value for the potential can be used for Fermi-LAT.

7. Results

7.1. The presence or absence of a cosmic ray anomaly

There has been a plethora of experiments which have hinted at the existence of an anomaly in the electron-positron spectrum. The most notable measurements are the Fermi-LAT electron positron sum and the PAMELA positron fraction. Ref. [53] and [24] have all questioned the reality of the anomaly in the PAMELA and Fermi-LAT data as well as the absence of an anomaly in the anti-proton flux. Ref. [24] suggested that one requires only to readjust the diffusion parameters that define the propagation model as encoded in GALPROP to reproduce the Fermi-LAT data. This conclusion is highlighted clearly in figure 8 of [24], where their propagation model obtained from the best fit to 76 cosmic ray spectral data points agrees well with the Fermi-LAT data. Interestingly, the corresponding positron fraction obtained from their best fit does not agree with the PAMELA data, indicating that one cannot fit the PAMELA data by simply adjusting the parameters of the propagation model. This indicates to us that the anomaly observed in cosmic electron-positron data is real and rather than adjusting the propagation parameters, one has to perform a detailed investigation of its existence and characteristics.

One of our important results is that just by adjusting the parameters in Eq.(18) and (19) it is possible to generate a theoretical prediction which is fully consistent with the Fermi-LAT $e^- + e^+$ flux measurement. Similarly, we found that both the PAMELA positron fraction and the electron flux can be reproduced by theory. This means that the theory has enough flexibility to accommodate the experimentally measured fluxes.

If one assumes that all types of cosmic ray data (electron-positron related and non-electron-positron related) can be described well using a single set of propagation parameters it quickly becomes obvious that one cannot fit the data simultaneously. To analyse this behaviour, we first divided the cosmic ray data listed in table 1 into two groups: electron and/or positron fluxes as measured by AMS, Fermi, HESS and PAMELA and non-electron and/or positron fluxes such as anti-proton/proton, Boron/Carbon, (Sc+Ti+V)/Fe, Be-10/Be-9. We then attempted to fit only the second group of cosmic ray data, that is, excluding the electron-positron related data. We obtain a χ^2 per degree of freedom of 0.34 from this fit and as a consequence the corresponding best fit curves each pass through all the estimated systematic error bands shown in grey in Fig. 1. When we apply the best-fit parameters to the electron and/or positron flux data, however, we obtain a χ^2 per degree of freedom of 24. Similarly, when we do the converse, i.e. find the best-fit propagation parameters using only the electron-positron related flux, we obtain an excellent χ^2 per degree of freedom of 1.0, but the best-fit parameters gave a larger χ^2 per degree of freedom (3.1) for the non-electron-positron data. As we have a large number (105) of data points the deviations observed between these two sets of data is significant which signals statistically significant tension between electron-positron and non-electron-positron measurements. These results support the conclusion highlighted in Fig. 7 and 8 of [24], that one requires something more than simply adjusting the propagation parameters to accommodate for the cosmic ray anomaly.

This tension between the electron-positron and non-electron-positron measurements was further investigated by performing an independent Bayesian analysis on the two groups of data. This allows us to extract the values of the propagation parameters as preferred by the different sets of data. Interestingly, one can derive information about the propagation parameters of the electron-positron related data from the non-electron-positron related data. This arises due to a number of reasons. Firstly, the value of some propagation parameters such as D_{0xx} is highly dependent on the species that one is modelling the propagation of. Secondly, to model the propagation of cosmic rays one uses the transport equation (equation 1). In this equation a large number of processes, including nuclear fragmentation and decay, are incorporated, which directly affects the predicted secondary electron-positron flux. Thirdly, as the energy density of cosmic rays is comparable to the energy density of the interstellar radiation field and the local magnetic field, different cosmic ray species will influence the dynamics of each species non-negligibly.

As a consequence, even if no electron-positron related data is used in our fit one can still constrain some of the propagation parameters of the electron-positron data. Unfortunately this method does not constrain the value of injection indices sufficiently, so in order to fix these parameters we have to include a minimal amount of information about the electron-positron related fluxes in our analysis. We selected data points from the $e^- + e^+$ spectrum for four reasons: (1) these points cover the largest energy range; (2) before setting out to find the optimal parameter value, within uncertainties the end points of the $e^- + e^+$ spectrum agree with theoretical predictions; (3) for low energies the effect of

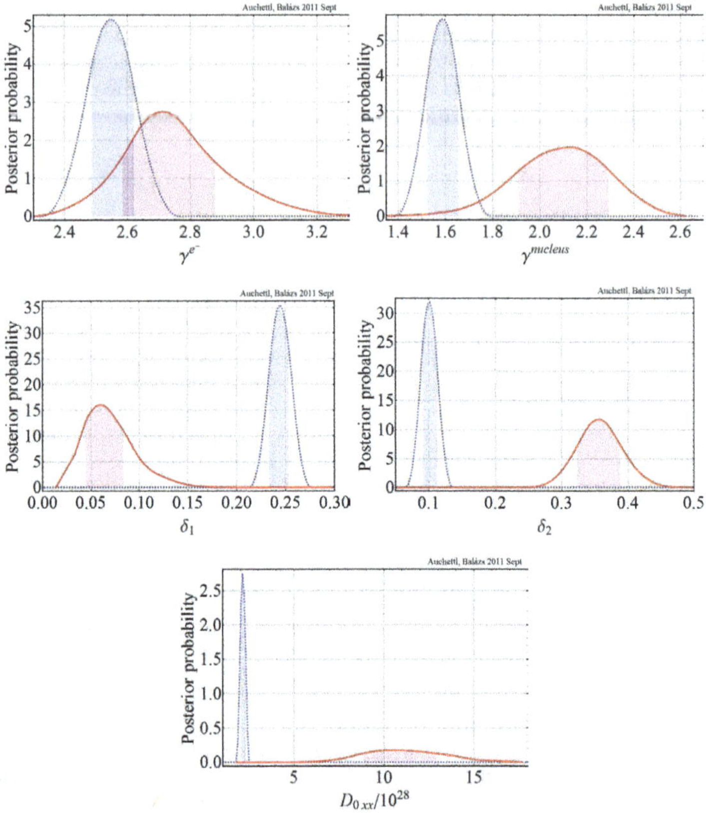

Figure 2. Marginalized posterior probability distributions of propagation parameters listed in Eq.(18). The likelihood functions containing electron and/or positron flux data are plotted as blue dashed lines while the likelihood functions for the rest of the comic ray data are plotted as solid red lines. The 68 % credibility regions are highlighted by the shaded areas of the posteriors. In the lower three frames it is evident that there exists a statistically significant tension between the electron-positron data and the rest.

solar modulation on this data is minor; and (4) the theoretical prediction for the $e^- + e^+$ is insensitive to the value of the propagation parameter for mid-range energies (this is highlighted by the distinct bow-tie shape of the theoretical uncertainty band).

In addition to using non-$e^{\pm}$ related data points (i.e. $\bar{p}/p$, B/C, (Sc+Ti+V)/Fe and Be data), we also selected four $e^{\pm}$ related data points to use in our analysis. This included the lowest energy data point from AMS, the highest energy data point from HESS and the 19.40 GeV and 29.20 GeV data points of Fermi-LAT. We checked that this selection of $e^{\pm}$ related data points does not bias the final conclusion and the results that we obtained with this selection are robust.

In figure 2 we have plotted the marginalized posterior probability densities of our selected propagation parameters as obtained by completing a Bayesian analysis on the two sets of data. The blue dashed curve represents the likelihood functions generated for the electron-positron related data (AMS, Fermi-LAT, HESS, and PAMELA), while the red solid curves represent the likelihood functions obtained for the rest of the cosmic ray data (anti-proton/proton, Boron/Carbon, (Sc+Ti+V)/Fe, Be-10/Be-9) listed in table 1. The 68% credibility regions of the likelihood functions are highlighted by the shaded areas of figure 2 and table 2 lists the numerical values of these credibility regions as well as the best-fit values of each propagation parameter. By looking at figure 2 it is obvious that the electron-positron related data and the non-electron-positron related data are inconsistent with the hypothesis that the model of cosmic ray propagation and/or the sources encoded in GALPROP provide a sufficient theoretical description.

parameter	Fit for the $e^{\pm}$ related data		Fit for the rest of the data	
	best fit value	68% Cr range	best fit value	68% Cr range
γ^{e^-}	2.55	{2.45, 2.60}	2.71	{2.54, 2.92}
$\gamma^{nucleus}$	1.60	{1.51, 1.69}	2.10	{1.88, 2.92}
δ_1	0.24	{0.23, 0.26}	0.06	{0.04, 0.08}
δ_2	0.10	{0.08, 0.12}	0.35	{0.32, 0.39}
D_{0xx} [$\times 10^{28}$]	2.17	{1.85, 2.19}	11.49	{8.86,13.48}

Table 2. Best fit values of the propagation parameters and their 68 % credibility ranges. Numerical values are shown for both fits: including the electron-positron related cosmic ray data only, and including the rest of the data.

For the posterior densities of the electron and nucleus injection indices γ^{e^-} and $\gamma^{nucleus}$ shown in the first two frames of figure 2, there is a mild but tolerable tension between the two data sets. In the final three frames of figure 2 the posterior densities for δ_1, δ_2 and D_{0xx} are shown. These frames indicate a statistically significant tension between the two sets of data as the 68 % credibility regions of each set for the two spatial diffusion coefficients δ_1 and δ_2, as well as D_{0xx} do not overlap each other. Although not shown it is easily extrapolated that not even the 99% credibility regions of these posteriors will overlap. As a consequence we can conclude that if one adjusts the values of the cosmic ray parameters, one can indeed obtain a good fit for either the electron-positron related data or to the rest of the data individually, however, you cannot obtain a good fit for both sets of data simultaneously.

This tension can be interpreted to mean that the data measured by the PAMELA and Fermi-LAT collaborations is affected by new physics that is unaccounted for by the propagation model and/or cosmic ray sources encoded in GALPROP. Based on simple theoretical arguments, the observed behaviour of the PAMELA positron fraction is unexpected. If one attempts to fit this data by simply adjusting the value of the propagation parameters this will lead to a bad fit of the non-electron-positron related data. One also expects that the anomaly in the PAMELA $e^+/(e^+ + e^-)$ would also produce an observable anomaly in other electron-positron related data such as the Fermi-LAT $e^+ + e^-$ and the PAMELA e^- spectra. This conclusion agrees with the argument of [24] that "secondary positron production in the general ISM is not capable of producing an abundance that rises with energy".

The observed tension in our data is dramatically increased when one incorporates the recently released PAMELA e^- flux [18] in our electron-positron related data. For consistency we checked the result that we would obtain if we excluded this new electron flux data into our analysis. We noticed that the tension we observe is significantly milder if it is not included. This, and the effect of using a larger amount of data compared to previous studies, suggests why the tension we observe was not detected by authors such as [24].

7.2. The size of the anomaly

As we conclude that new physics is buried within the electron-positron fluxes, we now attempt to extract from the data the size of this new physics signal. Assuming that the new physics affects only the electron-positron fluxes but its influence on the rest of the cosmic ray data is negligible, we can determine the central value and credibility regions of the cosmic ray propagation parameters from the unbiased data: anti-proton/proton, Boron/Carbon, (Sc+Ti+V)/Fe, Be-10/Be-9 to generate a background prediction for all cosmic ray data including the electron-positron fluxes. Once we calculate the theoretical background prediction we can subtract this background from the electron-positron data and determine if a statistically significant signal can be extracted.

To do this we calculate the prediction for the PAMELA and Fermi-LAT electron-positron fluxes by using the central values of the propagation parameters determined using $\bar{p}/p$, B/C, (Sc+Ti+V)/Fe, Be-10/Be-9. Then using all the scanned values of all five propagation parameters lying within the 68 % credibility region we generate a 1-σ uncertainty band for the background around this central value. In figure 3 we overlay the uncertainty background in gray over the Fermi-LAT electron+positron and the PAMELA electron and positron fraction fluxes. For the Fermi-LAT and PAMELA e^- the statistical and systematic uncertainties were combined in quadrature, while as the PAMELA $e^+/(e^+ + e^-)$ only had statistical uncertainties, we scaled these uncertainties using $\tau = 0.2$ to produce the experimental error bands. The magenta bands correspond to our background predictions, while the green dashed lines and band correspond respectively to the central value and the 1-σ uncertainty of the calculated anomaly.

In figure 3 one can see that our background prediction deviates from the data at energies below $\approx$ 10 GeV and above 100 GeV. For this analysis we focus on the deviation between the background and the data for energies greater than 100 GeV, while for the deviation observed at low energies we leave this to future research, but we note that this deviation could arise from inadequacies of the propagation model. Based on our background prediction we obtain a weak but statistically significant anomaly signal which we interpret as the presence of a new physics in the Fermi-LAT electron+positron flux. A similar conclusion can be drawn about the PAMELA positron fraction when taking the difference between the central values of the data and the background, but due to sizeable uncertainties of the PAMELA measurement we cannot claim a statistically significant deviation. [1]

To determine the size of the new physics signal in the electron positron data we subtract the central value of the corresponding background prediction from the central value of the data. The 1-σ uncertainty band of the signal is obtained by combining the experimental

[1] The Fermi-LAT collaboration recently presented a preliminary measurement of the positron fraction [54] which confirmed the results of PAMELA. At first glance this measurement appears to have smaller systematic uncertainties than that of PAMELA. If the officially published Fermi-LAT measurement has systematic errors of approximately the same size as the statistical errors of PAMELA then the data will also deviate from our background unveiling an anomalous signal in the positron fraction.

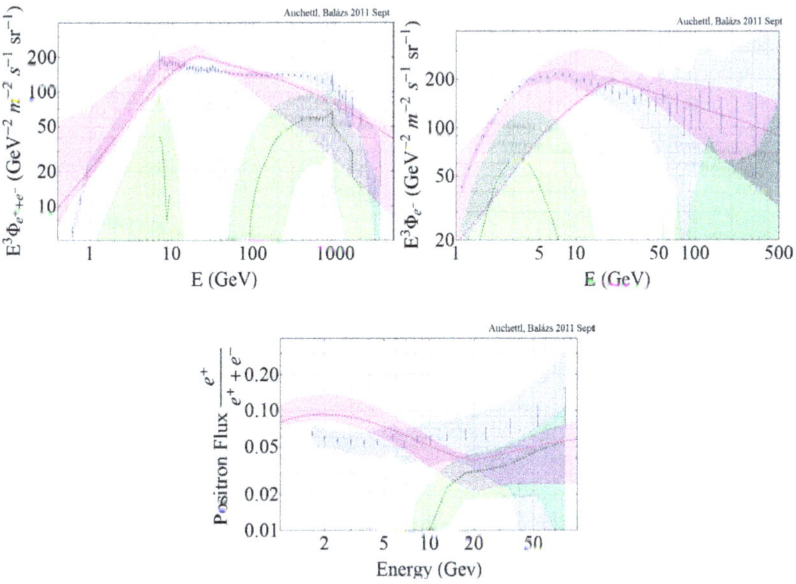

Figure 3. The anomalous signal we extracted for various electron-positron fluxes. The green dotted curves (marking the central values) and bands (showing 68 % credible intervals) correspond to the extracted size of the anomaly. The data points correspond to the spectra measured by Fermi-LAT and PAMELA. Combined statistical and systematic uncertainties are shown (by the gray bands) for Fermi-LAT and PAMELA e^-, while (τ = 0.2) scaled statistical uncertainties are shown for PAMELA $e^+/(e^+ + e^-)$. Overlaid in magenta is our background prediction (central value curve and 68 % credible intervals).

and background uncertainties quadratically. In Fig. 3 the results for the electron-positron anomaly are shown. Based on our background predictions, we obtain a non-vanishing anomalous signal for the Fermi-LAT $e^+ + e^-$ flux, while for the PAMELA data we cannot claim the presence of a statistically significant anomaly due to the large uncertainties of the data.

7.3. The source of the anomaly

Since the publication of the PAMELA positron fraction [7], there have been numerous publications that have speculated on the origin of the discrepancy between the theoretical prediction of electron-positron spectra and the experimental data. Based on the available evidence we can only postulate on the origin of this deviation. An obvious guess would be that the model used to describe the propagation of electrons and positrons in our Galaxy is insufficient in some respect, which if correct would mean that there exists no anomalous signal in the data. One such reasonable effect which is not incorporated in the two dimensional GALPROP calculation is the spectral hardening of cosmic ray spectra due to the presence of non-steady sources. To confirm these possibilities it would be an interesting exercise to repeat our analysis using different calculation tools such as DRAGON by [55], USINE by [56], PPPC4DMID by [57] or the code of [58].

If one assumes that the propagation model satisfactorily describes the propagation of cosmic rays through our Galaxy it is only natural to suspect that local effects are modifying the distribution of electrons and positrons. The lack of sources included in the GALPROP calculation seem to confirm this suspicion. There have been a plethora of papers that account for this anomaly by proposing various new sources of cosmic rays. There are two major categories of new cosmic ray sources that have been proposed. The first involves known astrophysical objects with uncertain parameters such as supernova remnants, pulsars, or various other objects in the Galactic centre, while the second involves more exotic astronomical and/or particle physics phenomena such as dark matter. Literature discussing these cases is extensively cited by [21].

For energies greater than 100 GeV, energy losses such as inverse Compton scattering of interstellar dust and cosmic microwave background light or synchrotron radiation become important. These effects result in a relatively short lifetime of the electron and positron while simultaneously this causes a decrease in the intensity of these particles as energy increases. As a result it is hypothesised that a large number of the electrons and positrons detected at Earth with an energy above 100 GeV come from individual sources within a few kilo-parsecs of Earth [51, 52]. Random fluctuations in the injection spectrum and spatial distribution of these nearby sources can produce detectable differences between the predicted background and the most energetic part of the observed electron and positron spectrum. This deviation could indicate the presence of new physics arising from either an astrophysical object(s) or dark matter.

If the size of the anomalous signal can be isolated from the experimental data then, regardless of the origin of the anomaly, the source will have to produce a signal with those characteristics. In Fig. 4 we compare our extracted signal to a few randomly selected attempts from the literature to match this anomaly. The first frame features the spectrum of electrons and positrons unaccounted for from local supernovae as calculated by [59] . The top right frame shows the contribution from additional primary cosmic ray sources such as pulsars or annihilation of particle dark matter as calculated by [52]. The bottom left frame contains the predictions of [60] for anomalous electron-positron sources from dark matter annihilations, while the last frame shows the dark matter annihilation contributions calculated by [61].

If the theoretical uncertainty of a new cosmic ray source and its contribution to cosmic ray measurements at Earth is unknown it can be difficult to draw any conclusion about its contribution to our isolated signal. In the case where the theoretical uncertainty of a new cosmic ray source is known it usually tends to be of significant size that it can prevent us from judging whether it is a valid explanation of our signal. Regardless, we can select a few scenarios that are more likely to be favoured than some others based on the present amount of information we have obtained from our analysis. With more data it will be possible to reduce the size of the uncertainty of our signal, while with more detailed calculations we can produce a more precise prediction of the cosmic ray spectrum as measured at Earth. This may enable the various suggestions of the source of the electron-positron anomaly to be confirmed or ruled out.

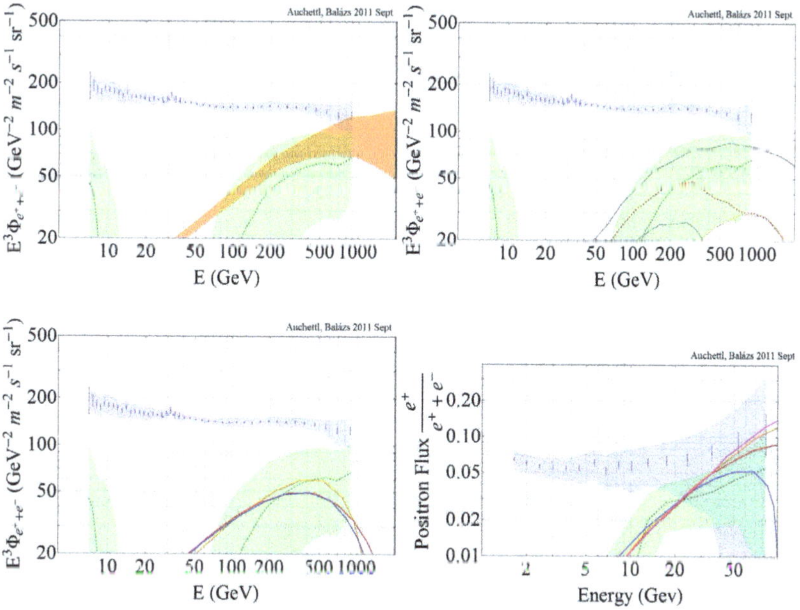

Figure 4. Comparing the detected cosmic ray flux (data points and gray band) and the signal extracted in this work (green dotted curve and band) to potential explanations of the electron-positron cosmic ray anomaly (solid curves). The various theoretical predictions come from [59], [52], [60] and [61]. Currently the comparison is fairly inconclusive but with more data it will be possible to shrink the uncertainty in the determination of the signal. Then various suggestions can be confirmed or ruled out.

8. Conclusions

Motivated by the possibility of new physics contributing to the measurements of PAMELA and Fermi-LAT, we subjected a wide range of cosmic ray observations to a Bayesian likelihood analysis. In the context of the propagation model coded in GALPROP, we found a significant tension between the $e^-\backslash e^+$ related data and the rest of the cosmic ray fluxes. This tension can be interpreted as the failure of the model to describe all the data simultaneously or as the effect of a missing source component.

Since the PAMELA and Fermi-LAT data are suspected to contain a component unaccounted for in GALPROP, we extracted the preferred values of the cosmic ray propagation parameters from the non-electron-positron related measurements. Based on these parameter values we calculated background predictions, with uncertainties, for PAMELA and Fermi-LAT. We found a deviation between the PAMELA and Fermi-LAT data and the predicted background even when uncertainties, including systematics, were taken into account. Interpreting this as an indication of new physics we subtracted the background from the data isolating the size of the anomalous component.

The signal of new physics in the electron+positron spectrum was found to be non-vanishing within the calculated uncertainties. Thus the use of 219 cosmic ray spectral data points within the Bayesian framework allowed us to confirm the existence of new physics effects in the electron+positron flux in a model independent fashion. Using the statistical techniques we were able to extract the size, shape and uncertainty of the anomalous contribution in the $e^- + e^+$ cosmic ray spectrum. We briefly compared the extracted signal to some theoretical results predicting such an anomaly.

Author details

Katie Auchettl and Csaba Balázs

Monash University, Australia

References

[1] R. L. Golden et al. Observations of cosmic ray electrons and positrons using an imaging calorimeter. *Astrophys. J.*, 436:769–775, 1994.

[2] J. Alcaraz et al. Leptons in near earth orbit. *Phys. Lett.*, B484:10–22, 2000.

[3] M. Boezio et al. The cosmic-ray anti-proton flux between 3-GeV and 49- GeV. *Astrophys. J.*, 561:787–799, 2001.

[4] C. Grimani et al. Measurements of the absolute energy spectra of cosmic-ray positrons and electrons above 7-GeV. *Astron. Astrophys.*, 392:287–294, 2002.

[5] S. W. Barwick et al. Measurements of the cosmic-ray positron fraction from 1- GeV to 50-GeV. *Astrophys. J.*, 482:L191–L194, 1997.

[6] J. J. Beatty et al. New measurement of the cosmic-ray positron fraction from 5-GeV to 15-GeV. *Phys. Rev. Lett.*, 93:241102, 2004.

[7] Oscar Adriani et al. An anomalous positron abundance in cosmic rays with energies 1.5-100 GeV. *Nature*, 458:607–609, 2009.

[8] T. Delahaye et al. Galactic secondary positron flux at the Earth. *Astron. Astrophys.*, 501:821–833, 2009.

[9] T. Delahaye, J. Lavalle, R. Lineros, F. Donato, and N. Fornengo. Galactic electrons and positrons at the Earth:new estimate of the primary and secondary fluxes. *Astron. Astrophys.*, 524:A51, 2010.

[10] Philipp Mertsch. Cosmic ray backgrounds for dark matter indirect detection. *arXiv:1012.4239*, 2010.

[11] Timur Delahaye, Fiasson Armand, Martin Pohl, and Pierre Salati. The GeV-TeV Galactic gamma-ray diffuse emission I. Uncertainties in the predictions of the hadronic component. *arXiv:1102.0744*, 2011.

[12] M. Aguilar et al. The Alpha Magnetic Spectrometer (AMS) on the International Space Station. I: Results from the test flight on the space shuttle. *Phys. Rept.*, 366:331–405, 2002.

[13] S. Torii et al. High-energy electron observations by PPB-BETS flight in Antarctica. *arXiv:0809.0760*, 2008.

[14] F. Aharonian et al. The energy spectrum of cosmic-ray electrons at TeV energies. *Phys. Rev. Lett.*, 101:261104, 2008.

[15] F. Aharonian et al. Probing the ATIC peak in the cosmic-ray electron spectrum with H.E.S.S. *Astron. Astrophys.*, 508:561, 2009.

[16] Aous A. Abdo et al. Measurement of the Cosmic Ray e+ plus e- spectrum from 20 GeV to 1 TeV with the Fermi Large Area Telescope. *Phys. Rev. Lett.*, 102:181101, 2009.

[17] M. Ackermann et al. Fermi LAT observations of cosmic-ray electrons from 7 GeV to 1 TeV. *Phys. Rev.*, D82:092004, 2010.

[18] O. Adriani et al. The cosmic-ray electron flux measured by the PAMELA experiment between 1 and 625 GeV. *Phys. Rev. Lett.*, 106(20):201101, 2011.

[19] A. W. Strong and I. V. Moskalenko. Propagation of cosmic-ray nucleons in the Galaxy. *Astrophys. J.*, 509:212–228, 1998.

[20] Pasquale D. Serpico. Astrophysical models for the origin of the positron 'excess', 2011.

[21] Katie Auchettl and Csaba Balazs. Extracting the size of the cosmic electron-positron anomaly. *Astrophys.J.*, 749:184, 2012.

[22] A. Putze, L. Derome, and D. Maurin. A Markov Chain Monte Carlo technique to sample transport and source parameters of Galactic cosmic rays. II. Results for the diffusion model combining B/C and radioactive nuclei. *Astronomy and Astrophysics*, 516:A66+, June 2010.

[23] Tongyan Lin, Douglas P. Finkbeiner, and Gregory Dobler. The Electron Injection Spectrum Determined by Anomalous Excesses in Cosmic Ray, Gamma Ray, and Microwave Signals. *Phys. Rev.*, D82:023518, 2010.

[24] R. Trotta et al. Constraints on cosmic-ray propagation models from a global Bayesian analysis. *The Astrophysical Journal*, 729(2):106, 2010.

[25] D. Maurin, F. Donato, R. Taillet, and P. Salati. Cosmic Rays below Z=30 in a diffusion model: new constraints on propagation parameters. *Astrophys. J.*, 555:585–596, 2001.

[26] David Maurin, Richard Taillet, and Fiorenza Donato. New results on source and diffusion spectral features of Galactic cosmic rays: I- B/C ratio. *Astron. Astrophys.*, 394:1039–1056, 2002.

[27] D. Maurin, A. Putze, and L. Derome. Systematic uncertainties on the cosmic-ray transport parameters. Is it possible to reconcile B/C data with $\delta = 1/3$ or $\delta = 1/2$? *Astronomy and Astrophysics*, 516:A67+, June 2010.

[28] (ed.) Ginzburg, V.L., V.A. Dogiel, V.S. Berezinsky, S.V. Bulanov, and V.S. Ptuskin. *Astrophysics of cosmic rays*. North Holland, 1990.

[29] S. Gillessen, F. Eisenhauer, S. Trippe, T. Alexander, R. Genzel, et al. Monitoring stellar orbits around the Massive Black Hole in the Galactic Center. *Astrophys.J.*, 692:1075–1109, 2009.

[30] Andrew W. Strong, Igor V. Moskalenko, and Vladimir S. Ptuskin. Cosmic-ray propagation and interactions in the Galaxy. *Ann. Rev. Nucl. Part. Sci.*, 57:285–327, 2007.

[31] A.W. Strong, I.V. Moskalenko, T.A. Porter, G. Johannesson, E. Orlando, et al. The GALPROP Cosmic-Ray Propagation Code. 2009.

[32] O. Adriani et al. A statistical procedure for the identification of positrons in the PAMELA experiment. *Astropart.Phys.*, 34:1–11, 2010.

[33] O. Adriani et al. PAMELA results on the cosmic-ray antiproton flux from 60 MeV to 180 GeV in kinetic energy. *Phys. Rev. Lett.*, 105:121101, 2010.

[34] Igor V. Moskalenko, Andrew W. Strong, Jonathan F. Ormes, and Marius S. Potgieter. Secondary anti-protons and propagation of cosmic rays in the galaxy and heliosphere. *Astrophys.J.*, 565:280–296, 2002.

[35] K. E. Krombel and M. E. Wiedenbeck. Isotopic composition of cosmic-ray boron and nitrogen. *Astrophys. J.*, 328:940–953, May 1988.

[36] J. A. Lezniak and W. R. Webber. The charge composition and energy spectra of cosmic-ray nuclei from 3000 MeV per nucleon to 50 GeV per nucleon. *Astrophys. J.*, 223:676–696, July 1978.

[37] J. J. Engelmann, P. Ferrando, A. Soutoul, P. Goret, and E. Juliusson. Charge composition and energy spectra of cosmic-ray nuclei for elements from Be to NI - Results from HEAO-3-C2. *Astronomy and Astrophysics*, 233:96–111, July 1990.

[38] E. Mocchiutti et al. The PAMELA Experiment: Preliminary Results after Two Years of Data Taking. In *21st European Cosmic Ray Symposium (ECRS 2008)*, Proceeding of 21st European Cosmic Ray Symposium, pages 396–401, 2008.

[39] H. S. Ahn et al. Measurements of cosmic-ray secondary nuclei at high energies with the first flight of the CREAM balloon-borne experiment. *Astropart. Phys.*, 30:133–141, 2008.

[40] A. J. Davis, R. A. Mewaldt, W. R. Binns, E. R. Christian, A. C. Cummings, J. S. George, P. L. Hink, R. A. Leske, T. T. von Rosenvinge, M. E. Wiedenbeck, and N. E. Yanasak. On the low energy decrease in galactic cosmic ray secondary/primary ratios. In R. A. Mewaldt, J. R. Jokipii, M. A. Lee, E. Möbius, & T. H. Zurbuchen , editor, *Acceleration and Transport of Energetic Particles Observed in the Heliosphere*, volume 528 of *American Institute of Physics Conference Series*, pages 421–424, September 2000.

[41] M. Hareyama. SUB-Fe/Fe ratio obtained by Sanriku balloon experiment. In *International Cosmic Ray Conference*, volume 3 of *International Cosmic Ray Conference*, pages 105–+, August 1999.

[42] M. E. Wiedenbeck and D. E. Greiner. A cosmic-ray age based on the abundance of Be-10. *Astrophysical Journal*, 239:L139–L142, August 1980.

[43] M. Garcia-Munoz, T. G. Guzik, S. H. Margolis, J. A. Simpson, and J. P. Wefel. The Energy Dependence of Cosmic-Ray Propagation at Low Energy. In *International Cosmic Ray Conference*, volume 9 of *International Cosmic Ray Conference*, pages 195–+, 1981.

[44] T. Hams, L. M. Barbier, M. Bremerich, E. R. Christian, G. A. de Nolfo, S. Geier, H. Goebel, S. K. Gupta, M. Hof, W. Menn, R. A. Mewaldt, J. W. Mitchell, S. M. Schindler, M. Simon, and R. E. Streitmatter. $^{10}Be/^{9}Be$ ratio up to 1.0 GeV/nucleon measured in the ISOMAX 98 balloon flight. In *International Cosmic Ray Conference*, volume 5 of *International Cosmic Ray Conference*, pages 1655–+, August 2001.

[45] A. J. Davis, R. A. Mewaldt, W. R. Binns, E. R. Christian, A. C. Cummings, J. S. George, P. L. Hink, R. A. Leske, T. T. von Rosenvinge, M. E. Wiedenbeck, and N. E. Yanasak. On the low energy decrease in galactic cosmic ray secondary/primary ratios. *AIP Conference Proceedings*, 528(1):421–424, 2000.

[46] N. E. Yanasak, M. E. Wiedenbeck, W. R. Binns, E. R. Christian, A. C. Cummings, A. J. Davis, J. S. George, P. L. Hink, M. H. Israel, R. A. Leske, M. Lijowski, R. A. Mewaldt, E. C. Stone, and T. T. von Rosenvinge. Cosmic ray time scales using radioactive clocks. *Advances in Space Research*, 27:727–736, 2001.

[47] J. Burger. Cosmic ray physics with AMS. *European Physical Journal C*, 33:941–943, 2004.

[48] A. S. Beach, J. J. Beatty, A. Bhattacharyya, C. Bower, S. Coutu, M. A. Duvernois, A. W. Labrador, S. McKee, S. A. Minnick, D. Müller, J. Musser, S. Nutter, M. Schubnell, S. Swordy, G. Tarlé, and A. Tomasch. Measurement of the Cosmic-Ray Antiproton-to-Proton Abundance Ratio between 4 and 50 GeV. *Physical Review Letters*, 87(26):A261101+, December 2001.

[49] J. T. Childers and M. A. Duvernois. Expected boron to carbon at TeV energies. In *International Cosmic Ray Conference*, volume 2 of *International Cosmic Ray Conference*, pages 183–186, 2008.

[50] A. G. Malinin. Astroparticle physics with AMS-02. *Phys. Atom. Nucl.*, 67:2044–2049, 2004.

[51] Timur Delahaye et al. Anti-Matter in Cosmic Rays : Backgrounds and Signals. *arXiv:0905.2144*, 2009.

[52] D. Grasso et al. On possible interpretations of the high energy electron- positron spectrum measured by the Fermi Large Area Telescope. *Astropart. Phys.*, 32:140–151, 2009.

[53] Boaz Katz, Kfir Blum, and Eli Waxman. What can we really learn from positron flux 'anomalies'? *Monthly Notices of the Royal Astronomical Society*, 405(3):1458–1472, 2009.

[54] Warrit Mitthumsiri. Cosmic-ray positron measurement with the fermi-lat using the earth's magnetic field. In *Fermi Symposium*, Fermi Symposium, May 2011.

[55] Luca Maccione, Carmelo Evoli, Daniele Gaggero, Giuseppe Di Bernardo, and Dario Grasso. DRAGON: A public code to compute the propagation of high-energy Cosmic Rays in the Galaxy., 2010.

[56] David Maurin, Antje Putze, Laurent Derome, Richard Taillet, Fernando Barao, Fiorenza Donato, Pierre Salati, and CÃl'line Combet. USINE - a galactic cosmic-ray propagation code., 2011.

[57] Marco Cirelli, Gennaro Corcella, Andi Hektor, Gert Hutsi, Mario Kadastik, et al. PPPC 4 DM ID: A Poor Particle Physicist Cookbook for Dark Matter Indirect Detection. *JCAP*, 1103:051, 2011.

[58] I. Buesching, A. Kopp, M. Pohl, and R. Shlickeiser. A New Propagation Code for Cosmic Ray Nucleons. In *International Cosmic Ray Conference*, volume 4 of *International Cosmic Ray Conference*, pages 1985–+, July 2003.

[59] Markus Ahlers, Philipp Mertsch, and Subir Sarkar. On cosmic ray acceleration in supernova remnants and the FERMI/PAMELA data. *Phys. Rev.*, D80:123017, 2009.

[60] Lars Bergstrom, Joakim Edsjo, and Gabrijela Zaharijas. Dark matter interpretation of recent electron and positron data. *Phys. Rev. Lett.*, 103:031103, 2009.

[61] Ilias Cholis, Douglas P. Finkbeiner, Lisa Goodenough, and Neal Weiner. The PAMELA Positron Excess from Annihilations into a Light Boson. *JCAP*, 0912:007, 2009.

Chapter 6

Extending Cosmology: The Metric Approach

Sergio Mendoza

Additional information is available at the end of the chapter

1. Introduction

In this chapter it is reviewed a possible physical scenario for which the introduction of a fundamental constant of nature with dimensions of acceleration into the theory of gravity makes it possible to extend gravity in a very consistent manner. In the non-relativistic regime a MOND-like theory with a modification in the force sector is obtained. This description turns out to be the the weak-field limit of a more general metric relativistic theory of gravity. The mass and length scales involved in the dynamics of the whole universe require small accelerations which are of the order of Milgrom's acceleration constant and so, it turns out that this relativistic theory of gravity can be used to explain the expansion of the universe. In this work it is explained how to build that relativistic theory of gravity in such a way that the overall large-scale dynamics of the universe can be treated in a pure metric approach without the need to introduce dark matter and/or dark energy components.

Cosmological and astrophysical observations are generally explained introducing two unknown mysterious dark components, namely dark matter and dark energy. These ad hoc hypothesis represent a big cosmological paradigm, since they arise due to the fact that Einstein's field equations are forced to remain unchanged under certain observed astrophysical phenomenology.

A natural alternative scenario would be to see whether viable cosmological solutions can be found if dark unknown entities are assumed non-existent. The price to pay with this assumption is that the field equations of the theory of gravity need to be extended and so, new Friedmann-like equations will arise. The most natural approach to extend gravity arises when a metric extension $f(R)$ is introduced into the theory [see e.g. 9, and references therein].

In a series of recent articles, Bernal, Capozziello, Cristofano & de Laurentis [4], Bernal, Capozziello, Hidalgo & Mendoza [5], Carranza et al. [10], Hernandez et al. [17, 18], Mendoza et al. [22, 23] have shown how relevant the introduction of a new fundamental physical constant $a_0 \approx 10^{-10} \mathrm{m/s}^2$ with dimensions of acceleration is in excellent agreement with different phenomenology at many astrophysical mass and length sizes, from solar-system to extragalactic and cosmological scales. The introduction of the so called Milgrom's acceleration constant a_0 in a description of gravity means that any gravitational field produced by a certain distribution of mass (and hence energy) needs to incorporate the acceleration a_0 together with Newton's gravitational constant G and the speed of light c in the description of gravity.

In section 2 it is shown, through a description of an extended Newtonian gravity scenario, the advantages of working with a modification of gravity dependent on the mass and lengths associated with the dimensions and masses of the sources that generate the gravitational field, and not with the dynamical acceleration they produce on test particles. Section 3 describes how it is possible to build a metric theory of gravity which generalises the extended Newtonian description mentioned in section 2 and section 4 interconnects this extended relativistic description of gravity with a metric description of gravity for which the energy-momentum tensor appears in the gravitational field's action. On section 5 we use the developed theory of gravity for cosmological applications in a dust universe and see how it is a coherent representation of gravity at cosmological scales. Finally on section 6, we discuss the consequences of the developed approach of gravity and some of the future developments of the theory.

2. Extended Newtonian gravity

Milgrom [26, 27, 29] constructed a MOdified Newtonian Dynamics (MOND) theory, based on the introduction of a fundamental constant of nature $a_0 = 1.2 \times 10^{-10} \mathrm{m\,s^{-2}}$ in such a way that the acceleration experienced by a test particle on a gravitational field produced by a point mass source M is such that:

$$a = \begin{cases} -\frac{GM}{r^2}, & \text{for} \quad a \gg a_0, \\ -\frac{\sqrt{a_0 GM}}{r}, & \text{for} \quad a \ll a_0, \end{cases} \tag{1}$$

where r is the radial distance to the central mass. In other words, for accelerations $a \gg a_0$, Newtonian gravity is recovered and new MONDian effects are expected to appear for accelerations $a \lesssim a_0$. The strong $a \ll a_0$ MONDian regime means that Kepler's third law is not valid since for a circular orbit about the central mass M, the acceleration $a = v/r$, where v is velocity of the test mass, and so $v = (a_0 GM)^{1/4} \propto M^{1/4}$, which is the Tully-Fisher relation [see e.g. 33] for the case of a spiral galaxy and is the same relation experienced by wide-open binaries [17] and by the tail of the "rotation curve" in globular clusters [15, 16].

In order to interpolate from the strong $a \gg a_0$ Newtonian regime to the weak $a \ll a_0$ one, the traditional MONDian approach is to construct a somewhat built-by-hand interpolation function $\mu(y)$ in such a way that

$$a\mu(y) = -\frac{GM}{r^2}, \tag{2}$$

where

$$\mu(y) = \begin{cases} 1, & \text{for} \quad y \gg 1, \\ y, & \text{for} \quad y \ll 1, \end{cases} \quad \text{and} \quad y := \frac{a}{\mathrm{a}_0}.$$

The usual approach to MOND as expressed by equation (2) means that Newton's 2nd law of mechanics needs to be modified [see e.g. 1]. As explained by Mendoza et al. [23], a better physical approach can be constructed if the modification is made in the force (gravitational) sector. Indeed, by the use of Buckingham's theorem of dimensional analysis [cf. 35], the gravitational acceleration experienced by a test particle is given by

$$a = \mathrm{a}_0 g(x), \tag{3}$$

where the dimensionless quantity

$$x := \frac{l_M}{r}, \tag{4}$$

and a mass-length scale

$$l_M := \left(\frac{GM}{\mathrm{a}_0} \right)^{1/2}. \tag{5}$$

The length l_M plays an important role in the description of the theory and is such that when $l_M \gg r$, the strong Newtonian regime of gravity is recovered and when $l_M \ll r$ the weak MONDian regime of gravity appears. As such, the dimensionless acceleration (or *transition function*) $g(x)$ is such that:

$$\frac{a}{\mathrm{a}_0} = g(x) := \begin{cases} x^2, & \text{when} \quad x \gg 1, \\ x, & \text{when} \quad x \ll 1. \end{cases} \tag{6}$$

In general terms, a mass distribution whose length is much greater than its associated mass-length l_M is in the MONDian regime (since $x \ll 1$) and a mass distribution whose length is much smaller than its mass-length scale is in the Newtonian regime (since $x \gg 1$). The case $x = 1$ can roughly be thought of as the point where the transition from the Newtonian to the MONDian regime occurs.

A general transition function $g(x)$ was built by [23] taking Taylor expansion series about the correct MONDian and Newtonian limits, yielding:

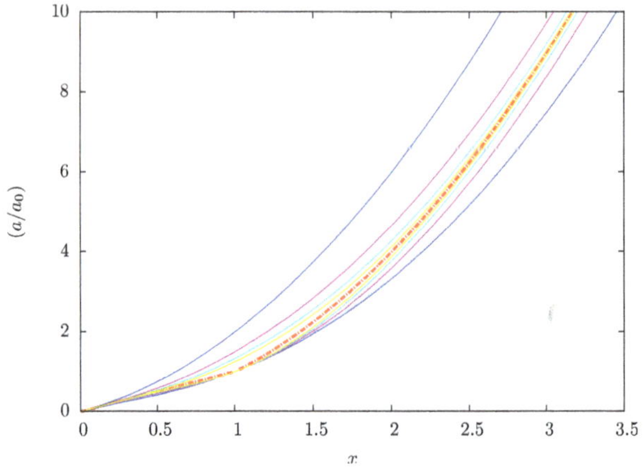

Figure 1. The figure taken from Mendoza et al. [23] shows the acceleration function a in units of Milgrom's acceleration constant a_0 as a function of the parameter x. The thick dash-dot curve is the extreme limiting value $n \to \infty$, i.e. $a/a_0 = x$ for $x \leq 1$ and $a/a_0 = x^2$ for $x \geq 1$. The curves above and below this extreme acceleration line represent values of $n = 4, 3, 2, 1$, for the minus and plus signs of equation (7) respectively. The extreme limiting curve has a kink at $x = 1$.

$$g(x) = x\,\frac{1 \pm x^{n+1}}{1 \pm x^n}. \tag{7}$$

This non-singular function converges to the correct expected limits of equation (6) for any value of the parameter $n \geq 0$. As shown in Figure 1, the transition function $g(x)$ rapidly converges to the limit "step function"

$$g(x)\bigg|_{n\to\infty} = \begin{cases} x, & \text{for } 0 \leq x \leq 1, \\ x^2, & \text{for } x \geq 1, \end{cases} \tag{8}$$

when $n \gtrsim 3$. The parameter n needs to be found empirically by astronomical observations. The value found by Mendoza et al. [23] for the rotation curve of our galaxy is $n \gtrsim 3$ and the one found by Hernandez & Jiménez [15], Hernandez et al. [16, 17] is $n \gtrsim 8$, with a minus sign selection on the numerator and denominator on the right hand side of equation (7). These authors have shown that a large value of n is coherent with solar system motion of planets, rotation curves of spiral galaxies, equilibrium relations of dwarf spheroidal galaxies and their correspondent relations in globular clusters, the Faber-Jackson relation and the fundamental plane of elliptical galaxies as well as with the orbits of wide binary stars. The $n = 3$ model in which a small, but measurable transition is obtained, has also been tested on earth and moon-like experiments by Meyer et al. [25] and Exirifard [12] respectively, showing that it is coherent with such precise measurements. In fact, these experiments also validate all $n \geq 3$ models.

Care must be taken when the introduction of a new fundamental constant of nature with dimensions of acceleration a_0 is made. In fact, the introduction of a_0 does not impose any causality arguments such as the ones given by the velocity of light c. In fact, one may think of a_0 as a fundamental constant needed to transit from one gravity regime to another. In this respect for example, instead of using a_0 as a fundamental constant, one may define

$$\Sigma_0 := a_0/G = 1.8\,\mathrm{kg\,m^{-2}}, \tag{9}$$

as the new fundamental constant of nature. The constant Σ_0, with dimensions of surface mass density, enters in the description of the gravitational theory in such a way that equations (3) and (5) are given by:

$$a = -G\Sigma_0 g\left(l_M/r\right), \qquad l_M := \left(M/\Sigma_0\right). \tag{10}$$

and the acceleration in the full MONDian regime and the corresponding Tully-Fisher relation are

$$a = -G\frac{\left(\Sigma_0 M\right)^{1/2}}{r}, \qquad v = G^{1/2}\Sigma_0^{1/4}M^{1/4}. \tag{11}$$

Also, a more manageable extended fundamental quantity, directly measurable through the Tully-Fisher relation, can be defined:

$$\epsilon_M := a_0 G = 8.004\times10^{-21}\mathrm{m^4\,s^{-4}\,kg^{-1}}, \tag{12}$$

with dimensions of velocity to the fourth over mass, for which

$$a = -\frac{\epsilon_M}{G}g\left(l_M/r\right), \qquad l_M := \left(G^2\,M/\epsilon_0\right). \tag{13}$$

With this, the acceleration of a test particle in the full MOND regime and the Tully-Fisher relation are:

$$a = -\frac{\left(\epsilon_M\,M\right)^{1/2}}{r}, \qquad v = \left(\epsilon_M\,M\right)^{1/4}. \tag{14}$$

The choice of a new fundamental constant of nature has many ways in which it can be introduced into the theory [35]. In this work, the use of a_0 is kept as it is traditionally done, but we note the fact that ϵ_0 is the best fundamental constant to use since it is directly measured through the flattened rotation curves of spiral galaxies.

The extended Newtonian model of gravity presented in this section is equivalent with MOND on spherical and cylindrical symmetry but deviates considerable from it for systems away from this symmetry [23]. As we have already shown, there are however many advantages of using this approach, the most objective meaning that the modification is made on the force sector and not a modification on the dynamics.

3. Relativistic metric extension

Finding a relativistic theory of gravity for which one of its non-relativistic limits converges to MOND yields usually strange assumptions and/or complicated ideas [see e.g. 3, 6, 30]. A good first approach was provided by a slight modification of Einstein's field equations by Sobouti [36], but the attempt is not complete.

In order to find an elegant and simple theory of gravity for which a MONDian solution is found, Bernal, Capozziello, Hidalgo & Mendoza [5] used a correct dimensional metric interpretation of Hilbert's gravitational action S_{f} for a point mass source M generating the gravitational field, in such a way that:

$$S_{\mathrm{f}} = -\frac{c^3}{16\pi G L_M^2} \int f(\chi) \sqrt{-g}\, \mathrm{d}^4 x, \tag{15}$$

which slightly differs from its traditional form (see e.g. [7, 9, 37]) since the following dimensionless quantity has been introduced:

$$\chi := L_M^2 R, \tag{16}$$

where R is Ricci's scalar and L_M defines a length fixed by the parameters of the model: The explicit form of the length L has to be obtained once a certain known limit of the theory is taken, usually a non-relativistic limit. Note that the definition of χ gives a correct dimensional character to the action (15), something that is not completely clear in all previous works dealing with a metric description of the gravitational field. For $f(\chi) = \chi$ the standard Einstein-Hilbert action of general relativity is obtained.

On the other hand, the matter action has its usual form,

$$S_{\mathrm{m}} = -\frac{1}{2c} \int \mathcal{L}_{\mathrm{m}} \sqrt{-g}\, \mathrm{d}^4 x, \tag{17}$$

with $\mathcal{L}_{\mathrm{m}}$ the matter Lagrangian density of the system. The null variations of the complete action, i.e. $\delta \left(S_{\mathrm{H}} + S_{\mathrm{m}} \right) = 0$, yield the following field equations:

$$f'(\chi)\,\chi_{\mu\nu} - \frac{1}{2} f(\chi) g_{\mu\nu} - L_M^2 \left(\nabla_\mu \nabla_\nu - g_{\mu\nu}\Delta\right) f'(\chi) = \frac{8\pi G L_M^2}{c^4} T_{\mu\nu}, \tag{18}$$

where the dimensionless Ricci tensor $\chi_{\mu\nu}$ is given by:

$$\chi_{\mu\nu} := L_M^2 R_{\mu\nu}, \tag{19}$$

and $R_{\mu\nu}$ is the standard Ricci tensor. The Laplace-Beltrami operator has been written as $\Delta := \nabla^\alpha \nabla_\alpha$ and the prime denotes derivative with respect to its argument. The energy-momentum tensor $T_{\mu\nu}$ is defined through the following standard relation: $\delta S_{\mathrm{m}} = -(1/2c)\, T_{\alpha\beta}\, \delta g^{\alpha\beta}$. In here and in what follows, we choose a $(+,-,-,-)$ signature for the metric $g_{\mu\nu}$ and use Einstein's summation convention over repeated indices.

The trace of equation (18) is:

$$f'(\chi)\,\chi - 2f(\chi) + 3L_M^2\, \Delta f'(\chi) = \frac{8\pi G L_M^2}{c^4} T, \tag{20}$$

where $T := T^\alpha_\alpha$.

In order to search for a MONDian solution, Bernal, Capozziello, Hidalgo & Mendoza [5] analysed the problem in two ways. First by performing an order of magnitude approach to the problem, and second, by doing a full perturbation analysis. Since the second technique is merely to fix constants of proportionality of the problem, their order of magnitude approach and its consequences are discussed in the remain of this section. Also, since we are interested at the moment on a point mass distribution generating a stationary spherically symmetric space-time, the trace equation (20) contains all the relevant information relating the field equations. At this point it is also useful to assume a power law form for the function

$$f(\chi) = \chi^b. \tag{21}$$

An order of magnitude approach to the problem means that $\mathrm{d}/\mathrm{d}\chi \approx 1/\chi$, $\Delta \approx -1/r^2$ and the mass density $\rho \approx M/r^3$. With this, the trace (20) takes the following form:

$$\chi^b\,(b-2) - 3bL_M^2 \frac{\chi^{(b-1)}}{r^2} \approx \frac{8\pi G M L_M^2}{c^2 r^3}. \tag{22}$$

Note that the second term on the left-hand side of equation (22) is much greater than the first term when the following condition is satisfied:

$$Rr^2 \lesssim \frac{3b}{2-b}. \tag{23}$$

At the same order of approximation, Ricci's scalar $R \approx \kappa = R_c^{-2}$, where κ is the Gaussian curvature of space and R_c its radius of curvature and so, relation (23) essentially means that

$$R_c \gg r. \tag{24}$$

In other words, the second term on the left-hand side of equation (22) dominates the first one when the local radius of curvature of space is much grater than the characteristic length r. This should occur in the weak-field regime, where MONDian effects are expected. For a metric description of gravity, this limit must correspond to the relativistic regime of MOND.

Under assumption (24), equation (22) takes the following form:

$$R^{(b-1)} \approx -\frac{8\pi GM}{3bc^2 r L_M^{2(b-1)}}. \tag{25}$$

We now recall the well known relation followed by the Ricci scalar at second order of approximation at the non-relativistic level [19]:

$$R = -\frac{2}{c^2}\nabla^2\phi = +\frac{2}{c^2}\nabla \cdot \boldsymbol{a}, \tag{26}$$

where the negative gradients of the gravitational potential ϕ provide the acceleration $\boldsymbol{a} := -\nabla\phi$ felt by a test particle on a non-relativistic gravitational field. At order of magnitude, equation (26) can be approximated as

$$R \approx -\frac{2\phi}{c^2 r^2} \approx \frac{2a}{c^2 r}. \tag{27}$$

Substitution of this last equation on relation (25) gives

$$\begin{aligned} a &\approx -\frac{c^2 r}{2L_M^2}\left(\frac{8\pi GM}{3bc^2 r}\right)^{1/(b-1)}, \\ &\approx -c^{(2b-4)/(b-1)} r^{(b-2)/(b-1)} L_M^{-2} (GM)^{1/(b-1)}. \end{aligned} \tag{28}$$

This last equation converges to a MOND-like acceleration $a \propto 1/r$ if $b - 2 = -(b-1)$, i.e. when $b = 3/2$. Also, at the lowest order of approximation, in the extreme non-relativistic limit, the velocity of light c should not appear on equation (28) and so, the only way this condition is fulfilled is that L_M depends on a power of c, i.e.

$$L_M^{-2} \propto c^{(4-2b)/(b-1)} = c^2, \quad \text{and so,} \qquad L_M \propto c^{-1}. \tag{29}$$

As discussed by Bernal, Capozziello, Hidalgo & Mendoza [5], the length L_M must be constructed by fundamental parameters describing the theory of gravity and since the only two characteristic lengths of the problem are the mass-length l_M and the gravitational radius

$$r_g = \frac{GM}{c^2}, \tag{30}$$

then the correct dimensional form of the length L_M is given by

$$L_M = \zeta\, r_g^{\alpha} l_M^{\beta}, \qquad \text{with} \qquad \alpha + \beta = 1, \tag{31}$$

where the constant of proportionality ζ is a dimensionless number that can be found by a full perturbation analysis technique and is given by [5]:

$$\zeta = \frac{2\sqrt{2}}{9}, \tag{32}$$

Substituting equation (31) and the value $b = 3/2$ into relation (29), it then follows that

$$\alpha = \beta = 1/2, \qquad \text{i.e.} \qquad L_M \approx r_g^{1/2} l_M^{1/2}. \tag{33}$$

If we now substitute this last result and the value $b = 3/2$ in equation (28) we get:

$$a \approx -\frac{(a_0 GM)^{1/2}}{r}, \tag{34}$$

which is the traditional form of MOND for a point mass source (see e.g. [2, 28, 29] and references therein). Also, the results of equation (34) in (27) mean that

$$R \approx \frac{r_g}{l_M} \frac{1}{r^2}, \tag{35}$$

and so, inequality (24) is equivalent to

$$l_M \gg r_g. \tag{36}$$

The regime imposed by equation (36) is precisely the one for which MONDian effects should appear in a relativistic theory of gravity. This is an expected generalisation of the results presented in section 2. Note that in the weak field limit regime for which $l_M \ll r$ together with equation (36) yields $r \gg l_M \gg r_g$. In this connection, we also note that Newton's theory of gravity is recovered in the limit $l_M \gg r \gg r_g$.

In exactly the same way as it was done to build the transition function for the case of extended Newtonian gravity in section 2, a general function $f(\chi)$ can be constructed:

$$f(\chi) = \chi^{3/2} \frac{1 \pm \chi^{p+1}}{1 \pm \chi^{3/2+p}} \to \begin{cases} \chi^{3/2}, & \text{for } \chi \ll 1, \\ \chi, & \text{for } \chi \gg 1. \end{cases} \tag{37}$$

In other words, general relativity is recovered when $\chi \gg 1$ in the strong field regime and the relativistic version of MOND with $\chi^{3/2}$ is recovered for the weak field regime of gravity when $\chi \ll 1$ (see Figure 2). The unknown parameter $p \geq -1$ needs to be calibrated with astronomical observations, in an analogous form as the calibration of the parameter n in equation (7) was done. This is a much harder task and a matter of future research. However, since the non-relativistic approach to gravity explained in section 2 means that the transition from the Newtonian to the MONDian regimes of gravity is very sharp, it most probably means that the function $f(\chi) = \chi$ for $\chi \geq 1$ and that $f(\chi) = \chi^{3/2}$ for $\chi \leq 1$, but this has to be tested by some astronomical observations.

The mass dependence of χ and L_M mean that Hilbert's action (15) is a function of the mass M. This is usually not assumed, since that action is thought to be purely a function of the geometry of space-time due to the presence of mass and energy sources. However, it was Sobouti [36] who first encountered this peculiarity in the Hilbert action when dealing with a metric generalisation of MOND. Following the remarks by Sobouti [36] and Mendoza & Rosas-Guevara [24] one should not be surprised if some of the commonly accepted notions, even at the fundamental level of the action, require generalisations and re-thinking. An extended metric theory of gravity goes beyond the traditional general relativity ideas and in this way, we need to change our standard view of its fundamental principles.

4. $F(R,T)$ connection

For the description of gravity shown in section 3 it follows that an adequate way of writing up the gravitational field's action is given by:

$$S_f = -\frac{c^3}{16\pi G} \int \frac{f(\chi)}{L_M^2} \sqrt{-g}\, d^4x. \tag{38}$$

The function L_M is a function of the mass of the system and in general terms it is a function of the space-time coordinates. For the particular case of a spherically symmetric space-time

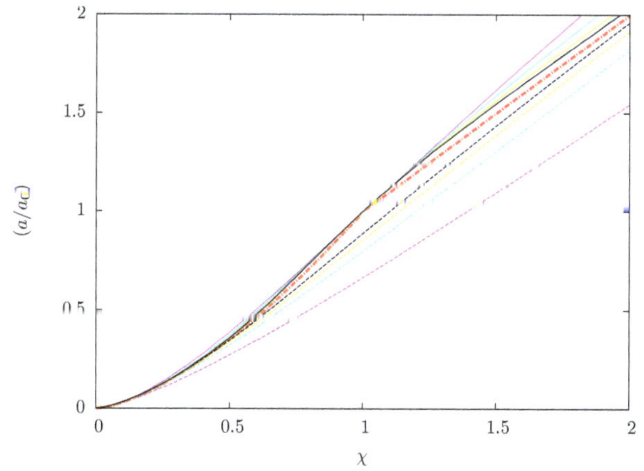

Figure 2. The figure shows the transition function $f(\chi)$, as a function of the dimensionless Ricci scalar χ, for different regimes of gravity, converging to $f(\chi) = \chi$ for $\chi \gg 1$ (general relativity) and to $f(\chi) = \chi^{3/2}$ for $\chi \ll 1$ (a relativistic regime with MOND as its weak field limit) -see equation (37). The thick dash-dot curve is the extreme limiting value $p \to \infty$, i.e. $f(\chi) = \chi^{3/2}$ for $\chi \leq 1$ and $f(\chi) = \chi$ for $\chi > 1$. The curves above and below this extreme function represent values of $p = 3,\ 2,\ 1,\ 0$ for the minus and plus signs of equation (7) respectively. The extreme limiting curve has a kink at $\chi = 1$.

it coincides with the mass of the central object generating the gravitational field as expressed in equations (31) and (33). Generally speaking what the meaning of M would be for a particular distribution of mass and energy needs further research, beyond the scope of this book chapter. Nevertheless one expects that for dust systems with spherically symmetric distributions, the function M would be given by the standard mass-energy relation [see e.g. 31]:

$$M := \frac{4\pi}{c^2} \int T\, r^2\, \mathrm{d}r, \tag{39}$$

In very general terms, the definition of M in this last equation means that M would not be invariant. However, in some particular systems with high degree of symmetry it is possible to make this quantity invariant. For example, in the case of a spherically symmetric spacetime produced by a point mass that quantity is simply the "Schwarzschild" mass of the point mass generating the gravitational field. In the cosmological case it is also possible to define it as an invariant quantity as discussed in section 5.

The field equations produced by the null variations of the addition of the field's action $S_{\mathrm{f}} + S_{\mathrm{m}}$ can be constructed in the following form. Harko et al. [14] have built an $F(R,T)$ theory of gravity, so making the natural identification:

$$F(R,T) := \frac{f(\chi)}{L_M^2}, \tag{40}$$

it is possible to use all their results for our particular case expressed in equation (40). For example, the null variations of the complete action $S_{\rm f} + S_{\rm m}$ for the particular case of equation (40) is given by Harko et al. [14]:

$$\begin{aligned}\left(\frac{f_R}{L_M^2}\right) R_{\mu\nu} - \frac{1}{2L_M^2} f\, g_{\mu\nu} + \left[g_{\mu\nu}\Delta - \nabla_\mu \nabla_\nu\right]\left(\frac{f_R}{L_M^2}\right) = \\ \frac{8\pi G}{c^4} T_{\mu\nu} - \left(\frac{f}{L_M^2}\right)_T \left[T_{\mu\nu} + \Theta_{\mu\nu}\right],\end{aligned} \tag{41}$$

and its trace is given by:

$$\frac{f_R R}{L_M^2} - \frac{2f}{L_M^2} + 3\Delta\left(\frac{f_R}{L_M^2}\right) = \frac{8\pi G}{c^4} T - \left(\frac{f}{L_M^2}\right)_T \left[T + \Theta\right], \tag{42}$$

where the subscripts R and T stand for the partial derivatives with respect to those quantities, i.e.

$$\Big(\;\Big)_R := \frac{\partial}{\partial R}, \qquad \text{and} \qquad \Big(\;\Big)_T := \frac{\partial}{\partial T}. \tag{43}$$

The tensor $\Theta_{\mu\nu}$ is such that $\Theta_{\mu\nu}\delta g^{\mu\nu} := g^{\alpha\beta}\delta T_{\alpha\beta}$ and for the case of an ideal fluid it can be written as [14]:

$$\Theta_{\mu\nu} = -2T_{\mu\nu} - p g_{\mu\nu}. \tag{44}$$

Note that equation (41) or (42) converge to the field (18) and trace (20) relations as discussed in section 3 when one considers a point mass generating the gravitational field, i.e. when $L_M = \text{const.}$ and so $\partial/\partial R = L_M^2 \partial/\partial \chi$.

In general terms, the $F(R,T)$ theory described by Harko et al. [14] produces non-geodesic motion of test particles since:

$$\begin{aligned}\nabla^\mu T_{\mu\nu} = \left(\frac{f}{L_M^2}\right)_T \left\{\frac{8\pi G}{c^4} - \left(\frac{f}{L_M^2}\right)_T\right\}^{-1} \times \\ \left[\left(T_{\mu\nu} + \Theta_{\mu\nu}\right)\nabla^\mu \ln\left(\frac{f}{L_M^2}\right)_T (R,T) + \nabla^\mu \Theta_{\mu\nu}\right] \neq 0,\end{aligned} \tag{45}$$

and as such the geodesic equation has a force term:

$$\frac{d^2x^\mu}{ds^2} + \Gamma^\mu_{\nu\lambda} u^\nu u^\lambda = \lambda^\mu, \tag{46}$$

where the four-force

$$\lambda^\mu := \frac{8\pi G}{c^4} \left(\rho c^2 + p\right)^{-1} \left[\frac{8\pi G}{c^4} + \left(\frac{f}{L_M^2}\right)_T\right]^{-1} \times \\ \left(g^{\mu\nu} - u^\mu u^\nu\right) \nabla_\nu p, \tag{47}$$

is perpendicular to the four velocity $\mathrm{d}x^\alpha/\mathrm{d}s$. As discussed by Harko et al. [14], the motion of test particles is geodesic, i.e. $\lambda^\mu = 0$ and/or $\nabla^\alpha T_{\alpha\beta} = 0$, (i) for the case of a pressureless $p = 0$ (dust) fluid and (ii) for the cases in which $F_T(R,T) = 0$.

In what follows we will see how all the previous ideas can be applied to a Friedmann-Lemaître-Robertson-Walker dust universe and so, the divergence of the energy momentum tensor in equation (45) is null. It is worth noting that this condition on the energy-momentum tensor for many applications needs to be zero, including applications to the universe at any epoch.

5. Cosmological applications

There are many good and interesting attempts to explain many cosmological observations using modified theories of gravity [see e.g. 32, and references therein], however these theories are not generally fully consistent with the gravitational anomalies shown at galactic and extragalactic scales discussed in sections 2 and 3. To see whether the gravitational $f(\chi)$ theory developed in the previous sections can deal with cosmological data, let us now apply the results obtained in those sections to an isotropic Friedmann-Lemaître-Robertson-Walker (FLRW) universe following the procedures first explored by Carranza et al. [10]. In this case, the interval $\mathrm{d}s$ is given by [21]:

$$\mathrm{d}s^2 = c^2\mathrm{d}t^2 - a^2(t)\left\{\frac{\mathrm{d}r^2}{1-\kappa r^2} + r^2\mathrm{d}\Omega^2\right\}, \tag{48}$$

where $a(t)$ is the scale factor of the universe normalised to unity, i.e. $a_0 = 1$, at the present epoch t_0, and the angular displacement $\mathrm{d}\Omega^2 := \mathrm{d}\theta^2 + \sin^2\theta\,\mathrm{d}\varphi^2$ for the polar $\mathrm{d}\theta$ and azimuthal $\mathrm{d}\varphi$ angular displacements with a comoving distance coordinate r. In what follows we assume a null space curvature $\kappa = 0$ at the present epoch in accordance with observations and deal with the expansion of the universe dictated by the field equations (41), avoiding any form of dark unknown component. Since we are interested on the compatibility of this cosmological model with SNIa observations, in what follows we assume a dust $p = 0$

model for which the covariant divergence of the energy-momentum tensor vanishes, and so as discussed in section 4 the trajectories of test particles are geodesic.

To begin with, let us rewrite the field equations (41) inspired by the approach first introduced by Capozziello & Fang [8] (see also Capozziello & Faraoni [9]) as follows:

$$G_{\mu\nu} = \frac{8\pi G}{c^4} \left\{ \left(1 + \frac{c^4}{8\pi G} F_T \right) \frac{T_{\mu\nu}}{F_R} + T^{\text{curv}}_{\mu\nu} \right\}, \tag{49}$$

where the Einstein tensor is given by its usual form:

$$G_{\mu\nu} := R_{\mu\nu} - \frac{1}{2} R g_{\mu\nu}. \tag{50}$$

and

$$T^{\text{curv}}_{\mu\nu} := \frac{c^4}{8\pi G F_R} \left[\left(\frac{1}{2} \left(F - R F_R \right) - \Delta F_R \right) g_{\mu\nu} + \nabla_\mu \nabla_\nu F_R \right], \tag{51}$$

represents the *"energy-momentum"* curvature tensor. Since $T_{00} = \rho c^2$, then it will be useful the identification $T_{00} := \rho_{\text{curv}} c^2$. With this last definition and using the fact that the Laplace-Beltrami operator applied to a scalar field ψ is given by [see e.g. 19]:

$$\Delta\psi = \frac{1}{\sqrt{-g}} \partial_\mu \left(\sqrt{-g}\, g^{\mu\nu} \partial_\nu \psi \right), \tag{52}$$

then

$$\rho_{\text{curv}} = \frac{c^2}{8\pi G F_R} \left[\frac{1}{2} \left(R F_R - F \right) - \frac{3H}{c^2} \frac{\mathrm{d}F_R}{\mathrm{d}t} \right], \tag{53}$$

where $H := \dot{a}(t)/a(t)$ represents Hubble's constant.

With the above definitions and using the 00 component of the field's equations (49) and the relation [cf. 11]:

$$R = -\frac{6}{c^2} \left[\frac{\ddot{a}}{a} + \left(\frac{\dot{a}}{a} \right)^2 + \frac{\kappa c^2}{a^2} \right], \tag{54}$$

between Ricci's scalar and the derivatives of the scale factor for a FLRW universe, then the dynamical Friedman's-like equation for a dust flat universe is:

$$H^2 = \frac{8\pi G}{3}\left[\left(1+\frac{c^4}{8\pi G}F_T\right)\frac{\rho}{F_R}+\rho_{\text{curv}}\right]. \tag{55}$$

The energy conservation equation is given by the null divergence of the energy-momentum tensor:

$$\left(\frac{8\pi G}{c^4}+F_T\right)(\dot{\rho}+3H\rho) = -\rho\frac{\mathrm{d}F_T}{\mathrm{d}t}. \tag{56}$$

For completeness, we write down the correspondent generalisation of Raychadhuri's equation for a dust flat universe:

$$2\frac{\ddot{a}}{a}+H^2 = -\frac{8\pi G p_{\text{curv}}}{c^2}, \tag{57}$$

where the "curvature-pressure"

$$p_{\text{curv}} := \omega c^2 \rho_{\text{curv}}, \tag{58}$$

and

$$w = \frac{c^2\left(F - RF_R\right)/2 + \mathrm{d}^2F_R/\mathrm{d}t^2 + 3H\mathrm{d}F_R/\mathrm{d}t}{c^2\left(RF_R - F\right)/2 - 3H\mathrm{d}F_R/\mathrm{d}t}. \tag{59}$$

On the other hand, note that the mass M that appears on the length L_M must be the causally connected mass at a certain cosmic time t, since particles beyond Hubble's (or particle) horizon with respect to a given fundamental observer do not have any gravitational influence on him. At any particular cosmic epoch, this Hubble mass satisfies the spherically symmetric condition implicit in equation (39) and so,

$$M = 4\pi\int_0^{r_{\mathrm{H}}}\rho r^2\,\mathrm{d}r = \frac{4}{3}\pi\rho\frac{c^3}{H^3}, \tag{60}$$

where

$$r_{\mathrm{H}} := \frac{c}{H(t)}, \tag{61}$$

is the Hubble radius or the distance of causal contact at a particular cosmic epoch [21]. In this respect the mass M is measured from the point of view of any given fundamental observer at a particular cosmic time t and so, it does not depend on which system of reference (or coordinates) is measured. As such, the mass M represents an invariant scalar quantity. From this last relation it follows that the length (31) is given by:

$$L_M = \zeta \frac{\left(\frac{4}{3}\pi c^3 G\right)^{3/4}}{c\, a_0^{1/4}} \frac{\rho^{3/4}}{H^{9/4}}, \tag{62}$$

and so, by using relation (21) and the standard power-law assumptions:

$$a(t) = a(t_0)\left(\frac{t}{t_0}\right)^{\alpha}, \qquad \rho(t) = \rho_0 \left(\frac{a}{a(t_0)}\right)^{\beta}. \tag{63}$$

for the unknown constant powers α and β, it follows that:

$$\frac{\mathrm{d}F_R}{\mathrm{d}t} = b(b-1)R^{b-1}L_M^{2(b-1)}H\left[\frac{j-q-2}{1-q} + \frac{3}{2}\left(\beta + \frac{3}{\alpha}\right)\right], \tag{64}$$

$$\frac{\mathrm{d}F_T}{\mathrm{d}t} = \frac{3}{2}(b-1)\frac{R^b L_M^{2b-2}}{\rho c^2}, \tag{65}$$

where

$$q(t) := -\frac{1}{a}\frac{\mathrm{d}^2 a}{\mathrm{d}t^2}H^{-2}, \qquad \text{and} \qquad j := \frac{1}{a}\frac{\mathrm{d}^3 a}{\mathrm{d}t^3}H^{-3}, \tag{66}$$

are the deceleration parameter and the jerk respectively.

With these and the value of L_M from equation (62), the curvature density (53) is given by:

$$\rho_{\mathrm{curv}} = \frac{3H^2}{8\pi G}(b-1)\left[(1-q) - \frac{j-q-2}{1-q} - \frac{3}{2}\left(\beta + \frac{3}{\alpha}\right)\right]. \tag{67}$$

Substitution of the previous relations on Friedmann's equation (55) gives:

$$H^2 = \frac{8\pi G\rho}{3\,Z\,F_R}, \tag{68}$$

where

$$Z := 1 + (b-1)\left[\frac{j-q-2}{1-q} - \frac{4\,(1-q)}{b} + \frac{3}{2}\left(\beta + \frac{3}{\alpha}\right)\right]. \tag{69}$$

is a dimensionless function.

An important result can be obtained evaluating equation (68) at the present epoch, yielding:

$$\mathsf{a}_0 = \left[\frac{9}{4}\zeta^4\,(1-q_0)^2\,(bZ_0)^{2/(b-1)}\left(\Omega^{(0)}_{\text{matt}}\right)^{(3b-5)/(b-1)}\right] c\,H_0, \tag{70}$$

where the density parameter $\Omega^{(0)}_{\text{matt}}$ at the present epoch has been defined by it's usual relation:

$$\Omega^{(0)}_{\text{matt}} := \frac{3H^2\rho}{8\pi G}. \tag{71}$$

In other words, the value of Milgrom's acceleration constant a_0 at the current cosmic epoch is such that

$$\mathsf{a}_0 \approx c \times H_0. \tag{72}$$

The numerical coincidence between the value of Milgrom's acceleration constant a_0 and the multiplication of the speed of light c by the current value of Hubble's constant H_0 has been noted since the early development of MOND [see e.g. 13, and references therein]. Note that equation (72) means that this coincidence relation occurs at approximately the present cosmic epoch in complete agreement with the results by Bernal, Capozziello, Cristofano & de Laurentis [4] where it is shown that a_0 shows no cosmological evolution and hence it can be postulated as a fundamental constant of nature.

For the power law (21) and the assumptions made above, it follows that the energy conservation equation (56) is given by:

$$(\dot{\rho} + 3H\rho) + \frac{c^2}{8\pi G}\left(A\frac{\dot{\rho}}{\rho} + B\,H\right) R^b\,L_M^{2(b-1)} = 0, \tag{73}$$

where:

$$A := \frac{9}{4}(b-1)^2,$$
$$B := \frac{9}{2}\frac{b-1}{b} + \frac{27}{4}\frac{(b-1)^2}{\alpha} + \frac{3}{2}\frac{b(b-1)(j-q-2)}{1-q}.$$

Direct substitution of the density power law (63) into relation (73) gives a constraint equation between α, β and b:

$$\beta = \frac{1}{\alpha}\left(\frac{9-5b}{3b-5}\right). \tag{74}$$

Let us now proceed to fix the so far unknown parameters of the theory α, β and b. To do so, we need reliable observational data and as such, we use the redshift-magnitude SNIa data obtained by Riess et al. [34] and the following well known standard cosmological relations [see e.g. 21]:

$$1 + z = a(t_0)/a(t), \tag{75}$$
$$\mu(z) = 5\log_{10}\left[H_0\, d_L(z)\right] - 5\log_{10} h + 42.38, \tag{76}$$
$$d_L(z) = (1+z)\int_0^z \frac{c}{H(z)}\,\mathrm{d}z, \tag{77}$$

for the cosmological redshift z, the distance modulus μ, the luminosity distance d_L and where the normalised Hubble constant h at the present epoch is given by $h := H_0/\left(100\,\mathrm{km\,s^{-1}}/\,\mathrm{Mpc}\right)$. Also, from equation (63) it follows that

$$H(a) = H_0\left(\frac{a}{a(t_0)}\right)^{-1/\alpha} = H_0(1+z)^{1/\alpha}, \tag{78}$$

and the substitution of this into equation (77) gives the distance modulus d_L as a function of the redshift z. This means that the redshift magnitude relation (76) is a function that depends on the values of the current Hubble constant H_0 and the value of α. Figure 3 shows the best fit to the redshift magnitude relation of SNIa observed by Riess et al. [34], yielding $\alpha = 1.359 \pm 0.139$ and $h = 0.64 \pm 0.009$. The best fit presented on the figure was obtained using the Marquardt-Levenberg fit provided by gnuplot (http://www.gnuplot.info) for non-linear functions. These values do not provide the whole description of the problem, since β and b are still unknown. However, according to the constraint equation (74) only one of them is needed in order to know the other once α is known.

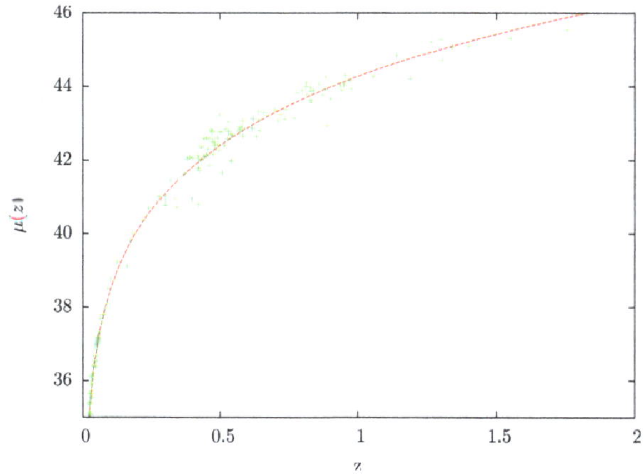

Figure 3. Redshift magnitude plot for SNIa showing the distance modulus μ as a function of the redshift z for SNIa as presented by Riess et al. [34]. The dotted red line shows the best fit to the data with the $f(\chi)$ gravity theory applied to a flat dust FLRW universe (see text) with no dark components. The continuous blue line represents the best fit according to the standard concordance dust ΛCDM model.

The parameter β can be found from conservation of mass arguments, since the total mass of the universe $M_{\text{tot}} = 4\pi \int_0^{R_{\max}} \rho\, r^2\, a^3\, \mathrm{d}r = \text{const.}$, where the upper limit of the integral is the radius of the whole universe. Since $a(t)$ and $\rho(t)$ are time dependent functions, the only way the mass of the universe is conserved is by requiring $a^3 \rho = \text{const.}$ and so, $\beta = 3$. This argument is exactly the one used in standard cosmology when dealing with a dust FLRW universe [see e.g. 21]. Using this value of β and the one already found for α, it follows that $b = 1.57 \pm 0.56$, which is within the expected value of $b = 3/2$ discussed in section 3.

For completeness, we write down a few of the cosmographycal parameters obtained by this $f(\chi)$ gravity applied to the universe:

$$
\begin{aligned}
h = 0.64 \pm 0.009, \quad q_0 = -0.2642 \pm 0.075, \\
j_0 = -0.1246 \pm 0.004.
\end{aligned}
\tag{79}
$$

6. Discussion

As explained by Carranza et al. [10], the obtained value $b \approx 3/2$ is a completely expected result due to the following arguments. As explained in section 2, a gravitational system for which its characteristic size r is such that $x := l_M/r \lesssim 1$ is in the MONDian gravity regime. For the case of the universe, $x \sim$ a few and as such if not totally in the MONDian regime of gravity, then it is far away from the regime of Newtonian gravity. The relativistic version of this means that the universe is close to the regime for which $f(\chi) = \chi^{3/2}$ and so $b = 3/2$. This is a very important result since seen in this way, the accelerated expansion of the universe is due to an extended gravity theory deviating from general relativity. It is quite interesting to note that the function $f(\chi) = \chi^{3/2}$ which at its non-relativistic limit is capable of predicting the correct dynamical behaviour of many astrophysical phenomena, is also able to explain the behaviour of the current accelerated expansion of the universe.

Seen in this way, the behaviour of gravity towards the past (for sufficiently large redshifts z) will differ from $f(\chi) = \chi^{3/2}$ and eventually converge to $f(\chi) = \chi$, i.e. the gravitational regime of gravity is general relativity for sufficiently large redshifts. A very detailed investigation into this needs to be done at different levels in order to be coherent many different cosmological observations [see e.g. 20]. This in turn can serve to calibrate the index p of the transfer function $f(\chi)$ as presented in equation (37), which has a very soft transition when $p = -1$, i.e.,

$$f(\chi) = \frac{\chi^{3/2}}{1+\chi^{3/2}}, \tag{80}$$

and also has a very sharp transition when $p \to \infty$, with the step function:

$$f(\chi) = \begin{cases} \chi^{3/2}, & \text{for } 0 \leq \chi \leq 1, \\ \chi, & \text{for } \chi \geq 1. \end{cases} \tag{81}$$

In this respect, perhaps something close to a sharp transition (81) will be observed since, as mentioned in section 2, at the non-relativistic level different astrophysical observations show a sharp transition from the Newtonian to the MONDian regimes. This sort of decision has to be taken with care and such a full description requires to analyse in full detail the whole Friedmann-like equations:

$$\left(\frac{8\pi G}{c^4} + F_T\right)\left(\dot{\rho} + 3H\rho + \frac{3Hp}{c^2}\right) = \\ -\rho\frac{\mathrm{d}F_T}{\mathrm{d}t} + \frac{1}{c^2}\left(p\frac{\mathrm{d}F_T}{\mathrm{d}t} + F_T\frac{\mathrm{d}p}{\mathrm{d}t}\right), \tag{82}$$

$$H^2 = \frac{8\pi G}{3}\left[\left(1 + \frac{c^4 F_T}{8\pi G}\right)\frac{\rho}{F_R} + \rho_{\text{curv}}\right] - \frac{\kappa c^2}{a^2}, \tag{83}$$

$$2\frac{\ddot{a}}{a} + H^2 + \frac{\kappa c^2}{a^2} = -\frac{8\pi G p}{c^2 F_R} - \frac{2pc^2 F_T}{F_R} - \frac{8\pi G p_{\text{curv}}}{c^2}. \tag{84}$$

These equations are directly obtained from taking the null covariant divergence of the energy momentum tensor, the 00 component of the field equations (49) and the density ρ contains all species of matter and/or radiation. The curvature density ρ_{curv} and the curvature pressure p_{curv} are related to one another by relation (58) with ω given by equation (59).

It is quite remarkable that a metric extended theory of gravity is able to reproduce phenomena from mass and length scales associated to the solar system up to cosmological scales. There are many more astrophysical challenges that this theory needs to address, in particular with respect to lensing at different scales and the dynamics associated to galaxy clusters. These will be addressed elsewhere.

Acknowledgements

This work was supported by a DGAPA-UNAM grant (PAPIIT IN116210-3) and CONACyT 26344. The author acknowledges fruitful discussions at different stages with Tula Bernal, Diego Carranza, Salvatore Capozziello, Rituparno Goswami, Xavier Hernandez, Juan Carlos Hidalgo and Luis Torres.

Author details

Sergio Mendoza

Instituto de Astronomia, Universidad Nacional Autónoma de Mexico, Ciudad Universitaria, Distrito Federal CP 04510, Mexico

References

[1] Bekenstein, J. [2006a]. The modified Newtonian dynamics - MOND and its implications for new physics, *Contemporary Physics* 47: 387–403.

[2] Bekenstein, J. [2006b]. The modified Newtonian dynamics - MOND and its implications for new physics, *Contemporary Physics* 47: 387–403.

[3] Bekenstein, J. D. [2004]. Relativistic gravitation theory for the modified Newtonian dynamics paradigm, *Physical Review D* 70(8): 083509.

[4] Bernal, T., Capozziello, S., Cristofano, G. & de Laurentis, M. [2011]. Mond's Acceleration Scale as a Fundamental Quantity, *Modern Physics Letters A* 26: 2677–2687.

[5] Bernal, T., Capozziello, S., Hidalgo, J. C. & Mendoza, S. [2011]. Recovering MOND from extended metric theories of gravity, *European Physical Journal C* 71: 1794.

[6] Blanchet, L. & Marsat, S. [2012]. Relativistic MOND theory based on the Khronon scalar field, *ArXiv e-prints* .

[7] Capozziello, S., de Laurentis, M. & Faraoni, V. [2010]. A Bird's Eye View of f(R)-Gravity, *The Open Astronomy Journal* 3: 49–72.

[8] Capozziello, S. & Fang, L. Z. [2002]. Curvature Quintessence, *International Journal of Modern Physics D* 11: 483–491.

[9] Capozziello, S. & Faraoni, V. [2010]. *Beyond Einstein Gravity: A Survey of Gravitational Theories for Cosmology and Astrophysics*, Fundamental Theories of Physics, Springer.

[10] Carranza, D. A., Mendoza, S. & Torres, L. A. [2012]. A cosmological dust model with extended $f(\chi)$ gravity, *ArXiv e-prints* .

[11] Dalarsson, M. & Dalarsson, N. [2005]. *Tensor calculus, relativity, and cosmology : a first course.*

[12] Exirifard, Q. [2011]. Lunar system constraints on the modified theories of gravity, *ArXiv e-prints* .

[13] Famaey, B. & McGaugh, S. [2011]. Modified Newtonian Dynamics (MOND): Observational Phenomenology and Relativistic Extensions, *ArXiv e-prints* .

[14] Harko, T., Lobo, F. S. N., Nojiri, S. & Odintsov, S. D. [2011]. f(R,T) gravity, *Physical Review D* 84(2): 024020.

[15] Hernandez, X. & Jiménez, M. A. [2012]. The Outskirts of Globular Clusters as Modified Gravity Probes, *Astrophysical Journal* 750: 9.

[16] Hernandez, X., Jimenez, M. A. & Allen, C. [2012a]. Flattened velocity dispersion profiles in Globular Clusters: Newtonian tides or modified gravity?, *ArXiv e-prints* .

[17] Hernandez, X., Jiménez, M. A. & Allen, C. [2012b]. Wide binaries as a critical test of classical gravity, *European Physical Journal C* 72: 1884.

[18] Hernandez, X., Mendoza, S., Suarez, T. & Bernal, T. [2010]. Understanding local dwarf spheroidals and their scaling relations under MOdified Newtonian Dynamics, *Astronomy and Astrophysics* 514: A101.

[19] Landau, L. & Lifshitz, E. [1975]. *The classical theory of fields*, Course of theoretical physics, Butterworth Heinemann.

[20] Longair, M. [2011]. The Frontiers of Observational Cosmology and the Confrontation with Theory, *Journal of Physics Conference Series* 314(1): 012011.

[21] Longair, M. S. [2008]. *Galaxy Formation.*

[22] Mendoza, S., Bernal, T., Hernandez, X., Hidalgo, J. C. & Torres, L. A. [2012]. Gravitational lensing with $f(\chi) = \chi^{3/2}$ gravity in accordance with astrophysical observations, *ArXiv e-prints* .

[23] Mendoza, S., Hernandez, X., Hidalgo, J. C. & Bernal, T. [2011]. A natural approach to extended Newtonian gravity: tests and predictions across astrophysical scales, *MNRAS* 411: 226–234.

[24] Mendoza, S. & Rosas-Guevara, Y. M. [2007]. Gravitational waves and lensing of the metric theory proposed by Sobouti, *Astronomy and Astrophysics* 472: 367–371.

[25] Meyer, H., Lohrmann, E., Schubert, S., Bartel, W., Glazov, A., Loehr, B., Niebuhr, C., Wuensch, E., Joensson, L. & Kempf, G. [2011]. Test of the Law of Gravitation at small Accelerations, *ArXiv e-prints* .

[26] Milgrom, M. [1983]. A modification of the Newtonian dynamics - Implications for galaxies, *Astrophysical Journal* 270: 371–389.

[27] Milgrom, M. [2008]. The MOND paradigm, *arXiv:0801.3133* .

[28] Milgrom, M. [2009]. New Physics at Low Accelerations (MOND): an Alternative to Dark Matter, *arXiv:0912.2678* .

[29] Milgrom, M. [2010]. New Physics at Low Accelerations (MOND): an Alternative to Dark Matter, *in* J.-M. Alimi & A. Fuözfa (ed.), *American Institute of Physics Conference Series*, Vol. 1241 of *American Institute of Physics Conference Series*, pp. 139–153.

[30] Mishra, P. & Singh, T. P. [2012]. Galaxy rotation curves from a fourth order gravity, *ArXiv e-prints* .

[31] Misner, C. W., Thorne, K. S. & Wheeler, J. A. [1973]. *Gravitation*, San Francisco: W.H. Freeman and Co., 1973.

[32] Nojiri, S. & Odintsov, S. D. [2011]. Unified cosmic history in modified gravity: From F(R) theory to Lorentz non-invariant models, *Physics Reports* 505: 59–144.

[33] Puech, M., Hammer, F., Flores, H., Delgado-Serrano, R., Rodrigues, M. & Yang, Y. [2010]. The baryonic content and Tully-Fisher relation at z ˜ 0.6, *Astronomy and Astrophysics* 510: A68+.

[34] Riess, A. G., Strolger, L.-G., Tonry, J., Casertano, S., Ferguson, H. C., Mobasher, B., Challis, P., Filippenko, A. V., Jha, S., Li, W., Chornock, R., Kirshner, R. P., Leibundgut, B., Dickinson, M., Livio, M., Giavalisco, M., Steidel, C. C., Benítez, T. & Tsvetanov, Z. [2004]. Type Ia Supernova Discoveries at $z > 1$ from the Hubble Space Telescope: Evidence for Past Deceleration and Constraints on Dark Energy Evolution, *Astrophysical Journal* 607: 665–687.

[35] Sedov, L. I. [1959]. *Similarity and Dimensional Methods in Mechanics*, Academic Press.

[36] Sobouti, Y. [2007]. An f(R) gravitation for galactic environments, *Astronomy and Astrophysics* 464: 921–925.

[37] Sotiriou, T. P. & Faraoni, V. [2010]. f(R) theories of gravity, *Reviews of Modern Physics* 82: 451–497.

Chapter 7

Inter-Universal Entanglement

Salvador J. Robles Pérez

Additional information is available at the end of the chapter

1. Introduction

The concept of the multiverse changes many of the preconceptions made in the physics and cosmology of the last century, providing us with a new paradigm that has inevitably influence on major philosophical ideas. The creation of the universe stops being a singled out event to become part of a more general and mediocre process, what can be thought of as a new "Copernican turn" in the natural philosophy of the XXI century. The multiverse also opens the door to new approaches for traditional questions in quantum cosmology. The origin of the universe, the problem of the cosmological constant and the arrow of time, which would eventually depend on the boundary conditions that are imposed on the state of the whole multiverse, challenge us to adopt new and open-minded attitudes for facing up these problems.

It would mean a crucial step for the multiverse proposals if a particular theory could make observable and distinguishable predictions about the current properties of our universe. That would bring the multiverse into the category of a physical theory at the same footing as any other. Then, once the concept of the multiverse has reached a wider acceptance in theoretical cosmology, it is now imperiously needed to develop a precise characterization of the concept of a physical multiverse: one for which the theory could be not only falseable but also indirectly tested, at least in principle. Some claims have been made to that respect [16, 17, 31], although we are far from being able to state the observability of any kind of multiverse nowadays.

In order to see the effects of other universes in the properties of our own universe, it seems to be essential considering any kind of interaction or correlation among the universes of the multiverse. Classical correlations in the state of the multiverse would be induced by the existence of wormholes that would crop up and connect different regions of two or more universes [15, 28, 44, 84]. Quantum correlations in the form of entanglement among the universal states provide us with a another interaction paradigm in the context of the quantum multiverse and it opens the door to a completely new and wider vision of the multiverse.

On the one hand, together with the classical laws of thermodynamics, we can also consider the novel laws of entanglement thermodynamics. This adds a new tool for studying the properties of both the universe and the multiverse. Furthermore, we would expect that the classical and the quantum thermodynamical laws were complementary provided that the quantum theory is a more general framework from which the classical one is recovered as a particular limiting case. Then, local entropic processes of a single universe could be related to the thermodynamical properties of entanglement among universes [60].

On the other hand, the quantum effects of the space-time are customary restricted to the obscure region of the Planck scale or to the neighbourhood of space-time singularities (both local and cosmological). However, cosmic entanglement among different universes of the multiverse could avoid such restriction and still be present along the whole history of a large parent universe [31, 62]. Thus, the effects of inter-universal entanglement on a single universe, and even the boundary conditions of the whole multiverse from which such entanglement would be consequence of, could in principle be tested in a large parent universe like ours. This adds a completely novel feature to the quantum theory of the universe.

The chapter is outlined as follows. In Sec. 2, we shall describe the customary picture in which the universes are spontaneously created from the gravitational vacuum or *space-time foam*. The universes are quantum mechanically described by a wave function that can represent, in the semiclassical regime, either an expanding or a contracting universe. Then, it will be introduced the so-called 'third quantization formalism', where creation and annihilation operators of universes can be defined and it can be given a wave function that represent the quantum state of the multiverse. Afterwards, it will be shown that an appropriate boundary condition of the multiverse allows us to interpret it as made up of entangled pairs of universes.

In Sec. 3, we shall briefly summarize the main features of quantum entanglement in quantum optics, making special emphasis in the characteristics that completely departure from the classical description of light. In Sec. 4, we shall address the question of whether quantum entanglement in the multiverse may induce observable effects in the properties of a single universe. We shall pose a pair of entangled universes and compute the thermodynamical properties of entanglement for each single universe of the entangled pair. It will be shown that the entropy of entanglement can be considered as an arrow of time for single universes and that the vacuum energy of entanglement might allow us to test the whole multiverse proposal. Finally, in Sec. 4, we shall draw some tentative conclusions.

2. Quantum multiverse

2.1. Introduction

A many-world interpretation of Nature can be dated back to the very ancient Greek philosophy[1] or, in a more recent epoch, to the many-world interpretation that Giordano Bruno derived from the heliocentric theory of Copernicus [68], in the XV century, and to the Kant's idea of 'island-universes', term coined by the Prussian naturalist Alenxander von Humboldt in the XIX century [33]. In any case, it was always a very controversial proposal

[1] The interpretation was posed, of course, in a radically different cultural context. However, it is curious reading some of the pieces that have survived from Greek philosophers like Anaximander, Heraclitus or Democritus, in relation to a 'many-world' interpretation on Nature.

perhaps because the mediocre perception that it entails for our world and for the human being itself.

As it happened historically, the controversy disappears when it is properly defined what it is meant by the word 'world'. If Bruno meant by the word 'world' what is now known as a solar system, von Humbolt meant by 'island universes' what we currently know as galaxies. We now uncontroversially know that there exist many solar systems in billions of different galaxies. Maybe, the controversy of the current multiverse proposals could partially be unravelled by first defining precisely what we mean by the word 'universe', in the physics of the XXI century.

Since the advent of the theory of relativity, in the early XX century, we can understand by the word 'universe' a particular geometrical configuration of the space-time as a whole that, following Einstein's equations, is determined by a given distribution of energy-matter in the universe. Furthermore, the geometrical description of the space-time encapsulates the causal relation between material points and, thus, the universe entails everything that may have a causal connection with a particular observer. In other words, the universe is everything we can observe.

Being this true, it does not close the door for the observation of the quantum effects that other universes might have in the properties of our own universe and, thus, it does not prevent us to consider a multiverse scenario. For instance, let us consider a spatially flat space-time endorsed with a cosmological constant. It is well-known that, for a given observer, there is an event horizon beyond which no classical information can be transmitted or received. Thus, two far distant observers are surrounded by their respective event horizons becoming then causally disconnected from each other. These causal enclosures may be interpreted as different universes within the whole space-time manifold[2]. However, cosmic fields are defined upon the whole space-time and, then, some quantum correlations might be present in the state of the field for two distant regions of the space-time, in the same way as non-local correlations appear in an EPR state of light in quantum optics. Therefore, being two observers classically disconnected, they may share common cosmological quantum fields allowing us, in principle, to study the quantum influence that other regions of the space-time may have in the properties of their isolated patches.

This is an example of a more general kind of multiverse proposals for which it can be defined a common space-time to the universes. It includes the multiverse that comes out in the scenario of eternal inflation [45, 46]. There are other proposals[3] in which there is no common space-time among the universes, being the most notable example the landscape of the string theories [9, 71]. In such multidimensional theories, the dimensional reduction that gives rise to our four dimensional universe may contain up to 10^{500} different vacua that can be populated with inflationary universes [29]. Two universes belonging to different vacua share no common space-time. However, it might well be that relic quantum correlations may appear between their quantum states, and even some kind of interaction has been proposed to be observable [31, 48], in principle.

Therefore, even if we have not been exhaustive in the justification of a multiverse scenario, it can easily be envisaged that the multiverse is a plausible cosmological scenario within the framework of the quantum theory provided that this has to be applied to the space-time as a whole. That is the basic assumption of the present chapter.

---

[2] This is the so-called Level I multiverse in Refs. [72, 73].

[3] A more exhaustive classification of multiverses and their properties can be found in Refs. [49, 72, 73].

2.2. Classical universes

In next sections, we shall describe the quantum state of a multiverse made up of homogeneous and isotropic universes. Then, it is worth first noting that homogeneity and isotropy are assumable conditions as far as we deal with large parent universes, where by *large* we mean universes with a length scale which is much greater than the Planck scale even though it can be rather small compared to macroscopic scales. At the Planck length the quantum fluctuations of the metric become of the same order of the metric and the assumptions of homogeneity and isotropy are meaningless. However, except for its very early phase the universe can properly be modeled by a homogeneous and isotropic metric, at least as a first approximation.

We will also consider homogeneous and isotropic scalar fields. This can be more objectionable. It can be considered a good approximation after the inflationary expansion of the universe has rapidly smoothed out the large inhomogeneities of the distribution of matter in the universe, and it clearly is an appropriate assumption for the large scale of the current universe. However, we should keep in mind that the study of inhomogeneities is a keystone for the observational tests of the inflationary scenario. Similarly, they might encode valuable information for testing the properties of inter-universal entanglement. However, as a first approach to the problem, we shall mainly be concerned with a multiverse made up of fully homogeneous and isotropic universes and matter fields.

Therefore, let us consider a space-time described by a closed Friedmann-Robertson-Walker (FRW) metric,

$$ds^2 = -\mathcal{N}^2 dt^2 + a^2(t) d\Omega_3^2, \tag{1}$$

where $\mathcal{N}$ is the lapse function that parameterizes the different foliations of the space-time into space and time, $a(t)$ is the scale factor, and $d\Omega_3^2$ is the usual line element on S^3 [39, 50, 80]. The degrees of freedom of the minisuperspace being considered are then the lapse function, $\mathcal{N}$, the scale factor, a, and n scalar fields, $\vec{\varphi} = (\varphi_1, \ldots, \varphi_n)$, that represent the matter content of the universe. The total action of the space-time minimally coupled to the scalar fields can conveniently be written as [39]

$$S = \int dt L = \int dt \mathcal{N} \left(\frac{1}{2} \frac{G_{AB}}{\mathcal{N}^2} \frac{dq^A}{dt} \frac{dq^B}{dt} - \mathcal{V}(q^I) \right), \tag{2}$$

for $I, A, B = 0, \ldots, n$, where $G_{AB} \equiv G_{AB}(q^I)$, is the minisupermetric of the $n+1$ dimensional minisuperspace, with $\{q^I\} \equiv \{a, \vec{\varphi}\}$, and the summation over repeated indices is implicitly understood in Eq. (2). The minisupermetric G_{AB} is given by [39], $G_{AB} = \text{diag}(-a, a^3, \ldots, a^3)$, and the potential $\mathcal{V}(q^I)$ by

$$\mathcal{V}(q^I) \equiv \mathcal{V}(a, \vec{\varphi}) = a^3 \left(V_1(\varphi_1), \ldots, V_n(\varphi_n) \right) - a, \tag{3}$$

where $V_i(\varphi_i)$ is the potential that corresponds to the field φ_i. The classical equations of motion are obtained by variation of the action (2). Let us for simplicity consider only one scalar field, φ. Variation of the action with respect to the lapse function, fixing afterwards the value $\mathcal{N} = 1$, gives the Friedmann equation

$$\left(\frac{da}{dt}\right)^2 = -1 + a^2\sigma^2 \left(\frac{1}{2}\left(\frac{d\varphi}{dt}\right)^2 + V(\varphi)\right) \equiv -1 + a^2\sigma^2\rho_\varphi, \tag{4}$$

where ρ_φ is the energy density of the scalar field, and [45] $\sigma^2 = \frac{8\pi}{3M_P^2}$, with $M_P \sim 10^{19}$GeV being the Planck mass. Variation of Eq. (2) with respect to the scalar field yields

$$\frac{d^2\varphi}{dt^2} + \frac{3}{a}\frac{da}{dt}\frac{d\varphi}{dt} + \frac{\partial V(\varphi)}{\partial\varphi} = 0. \tag{5}$$

Let us focus on a slow-varying scalar field, which constitutes a particularly interesting case that can model the inflationary stage of the universe. In that case [45, 47], $\frac{d^2\varphi}{dt^2} \ll \frac{3}{a}\frac{da}{dt}\frac{d\varphi}{dt}$ and $(\frac{d\varphi}{dt})^2 \ll V(\varphi)$, and $V(\varphi) \approx V(\varphi_0)$ represents the nearly constant energy density of the scalar field, i.e. $\rho_\varphi \approx V(\varphi_0)$.

A limiting case is that of a constant value of the field, $\dot{\varphi} = 0$ and $\rho_\varphi = V(\varphi_0) \equiv \Lambda$. It effectively describes a de-Sitter space-time with a value Λ of the cosmological constant. Then, the Friedmann equation (4) can be written as

$$\frac{da}{dt} = \sqrt{a^2H^2 - 1}, \tag{6}$$

where, $H^2 \equiv \sigma^2\Lambda$. It can be distinguished two regimes. For values, $a \geq \frac{1}{H}$, the real solution

$$a(t) = \frac{1}{H}\cosh Ht, \tag{7}$$

represents a universe that starts out from a value $a_0 = \frac{1}{H}$ at $t = 0$, and eventually follows an exponential expansion. It corresponds to the Lorentzian regime of the universe. On the other hand, there is no real solution of Eq. (6) for values $a < \frac{1}{H}$. However, we can perform a Wick rotation to Euclidean time, $\tau = it$, by mean of which Eq. (6) transforms into

$$\frac{da_E}{d\tau} = \sqrt{1 - a_E^2H^2}, \tag{8}$$

whose solution,

$$a_E(\tau) = \frac{1}{H}\cos H\tau, \tag{9}$$

is the analytic continuation to Euclidean time of the Lorentzian solution (7). The solution given by Eq. (9) represents an Euclidean space-time that originates at $a_E = 0$ (for $\tau = -\frac{\pi}{2H}$), and expands to the value $a_E = \frac{1}{H}$ at $\tau = 0$. The transition from the Euclidean region to the Lorentzian region occurs at the boundary hypersurface $\Sigma_0 \equiv \Sigma(a_0)$, at $t = 0 = \tau$. This transition should not be seen as a process happening *in time* because the Euclidean time is not actual time (it is *imaginary time*). On the contrary, it precisely corresponds to the appearance of time [39] and to the appearance of the (real) universe, actually.

This is, briefly sketched, the classical picture for the nucleation of a universe from *nothing* [26, 39, 77], depicted in Fig. 1, where by *nothing* we should understand a state of the universe where it does not exist space, time and matter, in the customary sense[4]. Within that picture, the quantum fluctuations of the gravitational vacuum provide it with a *foam structure* [13, 19, 25, 85] where tiny black holes, wormholes and baby universes [70] are virtually created and annihilated (see, Fig. 2). Some of the baby universes may branch off from the parent space-time and become isolated universes that, subsequently, may undergo an inflationary stage and develop into a large parent universe like ours.

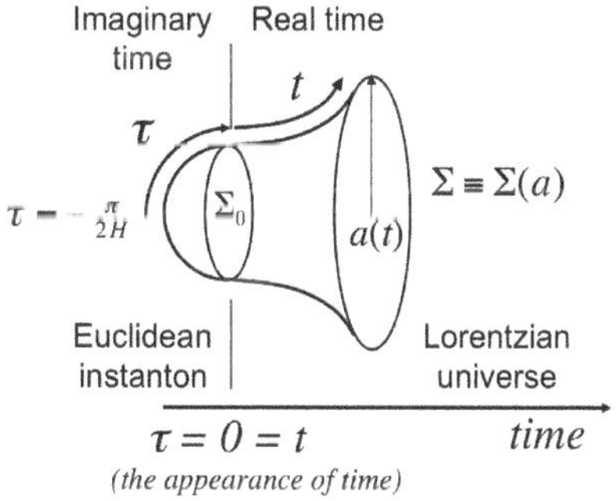

Figure 1. The creation of a De–Sitter universe from a De–Sitter instanton.

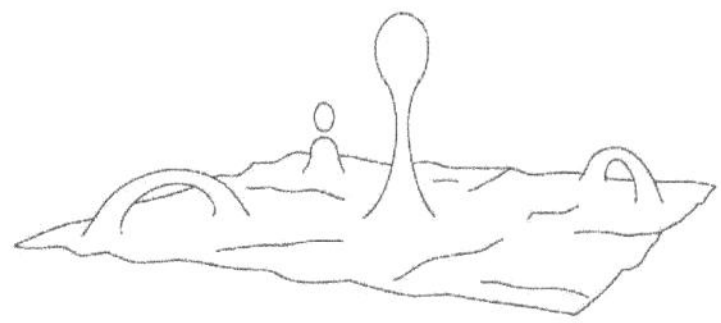

Figure 2. Space-time foam: some of the baby universes may branch off from the parent space-time.

2.3. Quantum state of the multiverse

Following the canonical quantization formalism, the momenta conjugated to the configuration variables q^I are given by $p_I \equiv \frac{\delta L}{\delta(\frac{dq^I}{dt})}$, where $L \equiv L(q^I, \frac{dq^I}{dt})$ is the Lagrangian of

[4] However, it does not correspond to the absolute meaning of 'nothing', in a similar way as the vacuum of a quantum field theory is not 'empty' (see, Ref. [77]).

Eq. (2). The Hamiltonian then reads

$$H \equiv \frac{dq^I}{dt} p_I - L = \mathcal{N}\mathcal{H} \equiv \mathcal{N}\left(G^{AB} p_A p_B + \mathcal{V}(q^I)\right). \quad (10)$$

The invariance of general relativity under time reparametrizations implies that the variation of the Hamiltonian (10) with respect to the lapse function vanishes. We obtain thus the classical Hamiltonian constraint, $\mathcal{H} = 0$, which gives rise to the Friedmann equation (4). The wave function of the universe, ϕ, can then be obtained by performing a canonical quantization of the momenta, $p_I \rightarrow \hat{p}_I \equiv -i\hbar\frac{\partial}{\partial q^I}$, and applying the quantum version of the Hamiltonian constraint to the wave function ϕ, i.e. $\hat{\mathcal{H}}\phi = 0$. With an appropriate choice of factor ordering, it can be written as [39]

$$\left\{-\frac{\hbar^2}{\sqrt{-G}}\frac{\partial}{\partial q^A}\left(\sqrt{-G}\,G^{AB}\frac{\partial}{\partial q^B}\right) + \mathcal{V}(q^I)\right\}\phi(q^I) = 0, \quad (11)$$

where G^{AB} is the inverse of the minisupermetric G_{AB}, with $G^{AB}G_{BC} = \delta^A_C$, and G is the determinant of G_{AB}. For a homogeneous and isotropic universe with a slow-varying field the Wheeler-De Witt equation (11) explicitly yields

$$\hbar^2\frac{\partial^2\phi}{\partial a^2} + \frac{\hbar^2}{a}\frac{\partial\phi}{\partial a} + (a^4V(\varphi) - a^2)\phi = 0, \quad (12)$$

where, $\phi \equiv \phi(a,\varphi)$. Let us note that if we replace $V(\varphi)$ by Λ, the wave function $\phi \equiv \phi_\Lambda(a)$ represents the quantum state of a de-Sitter universe. For later convenience, let us write Eq. (12) as

$$\ddot{\phi} + \frac{\dot{\mathcal{M}}}{\mathcal{M}}\dot{\phi} + \omega^2\phi = 0, \quad (13)$$

where, $\dot{\phi} \equiv \frac{\partial\phi}{\partial a}$ and $\dot{\mathcal{M}} \equiv \frac{\partial\mathcal{M}}{\partial a}$, with $\mathcal{M} \equiv \mathcal{M}(a) = a$, and, $\omega \equiv \omega(a,\varphi) = \frac{a}{\hbar}\sqrt{a^2V(\varphi) - 1}$. It will be useful later on to recall the formal resemblance of Eq. (13) to the equation of motion of a harmonic oscillator. The WKB solutions of Eq. (13) can be written, in the Lorentzian region, as

$$\phi^{\pm}_{WKB}(a,\varphi) = \frac{N(\varphi)}{\sqrt{\mathcal{M}(a)\omega(a,\varphi)}}e^{\pm iS(a,\varphi)}, \quad (14)$$

where $N(\varphi)$ is a normalization factor, and

$$S(a,\varphi) = \int da\,\omega(a,\varphi) = \frac{1}{\hbar}\frac{(a^2V(\varphi) - 1)^{\frac{3}{2}}}{3V(\varphi)}. \quad (15)$$

The positive and negative signs of $\phi^{\pm}_{WKB}$ correspond to the contracting and expanding branches of the universe, respectively. This can be seen by noticing that, for sufficiently

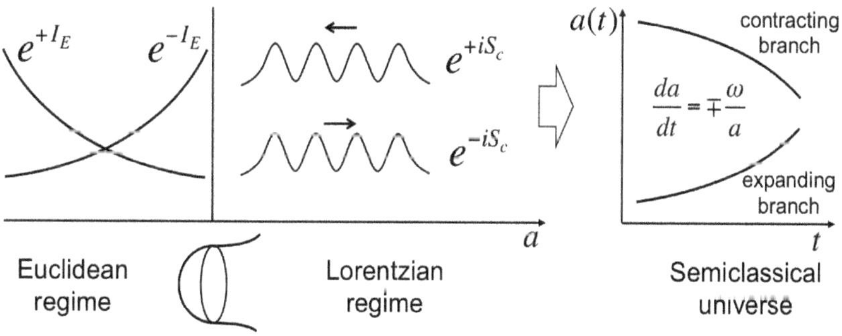

Figure 3. Boundary conditions of the universe.

large values of the scale factor, the Fourier transform of $\phi^{\pm}_{WKB}(a,\varphi)$ is highly peaked around the value of the classical momentum p^c_a [20], i.e. $\tilde{\phi}^{\pm}_{WKB}(p_a,\varphi) \approx \delta(p_a - p^c_a)$. The classical momentum reads, $p^c_a = -a\frac{\partial a}{\partial t}$, and quantum mechanically, for large values of the scale factor, $\hat{p}_a\phi = -i\hbar\dot{\phi} \approx \pm\omega\phi$, where the positive and negative signs correspond to the signs of $\phi^{\pm}_{WKB}$. Then, $\frac{\partial a}{\partial t} \approx \mp\frac{\omega}{a}$, where the negative sign describes a contracting universe and the positive sign an expanding universe. Thus, the solutions $\phi^{\pm}_{WKB}$ of the Wheeler-de Witt equation (13) describe the contracting and expanding branches of the universe, respectively.

In order to fix the state of the universe, a boundary condition has to be imposed on the wave function ϕ_{WKB}. The tunneling boundary condition [78, 79] states that the only modes that survive the Euclidean barrier are the outgoing modes of the minisuperspace that correspond, in the Lorentzian region, to the expanding branches of the universe (see, Fig. 3). Then, the wave function of the universe reads

$$\phi^T(a,\varphi) \approx \frac{N(\varphi)}{\sqrt{\mathcal{M}(a)\omega(a,\varphi)}} e^{-iS(a,\varphi)}, \tag{16}$$

with [39, 79], $N(\varphi) = e^{-\frac{1}{3V(\varphi)}}$. By using the matching conditions, the wave function (16) turns out to be given in the Euclidean region by

$$\phi^T_E(a,\varphi) \approx \frac{e^{-\frac{1}{3V(\varphi)}}}{\sqrt{\mathcal{M}(a)\omega(a,\varphi)}} \left(e^{+I(a,\varphi)} + e^{-I(a,\varphi)}\right), \tag{17}$$

where, $I = iS$, is the Euclidean action. The first term in Eq. (17) may diverge as the scale factor degenerates. However, this is not a problem in terms of Vilenkin's reasoning [78] because the tunneling boundary condition is mainly intended for fixing the state of the wave function on the Lorentzian region where the current probability is defined, cf. [78]. The philosophy of the 'no-boundary' proposal of Hartle and Hawking [24] is quite the contrary. For these authors, the actual quantum description of the universe is given by a path integral performed over all compact Euclidean metrics. The no-boundary condition is then imposed on the Euclidean sector of the wave function. In the case being considered, it is equivalent to impose regularity conditions [22], and thus

$$\phi_E^{NB}(a,\varphi) \approx \frac{N(\varphi)}{\sqrt{\mathcal{M}(a)\omega(a,\varphi)}} e^{-I(a,\varphi)}, \tag{18}$$

with $N(\varphi) = e^{+\frac{1}{3V(\varphi)}}$ [22, 39]. In the Lorentzian sector, the wave function turns out to be given by a linear combination of expanding and contracting branches of the universe [27], i.e.

$$\phi^T(a,\varphi) \approx \frac{e^{+\frac{1}{3V(\varphi)}}}{\sqrt{\mathcal{M}(a)\omega(a,\varphi)}} \cos S \propto \frac{e^{+\frac{1}{3V(\varphi)}}}{\sqrt{\mathcal{M}(a)\omega(a,\varphi)}} \left(e^{+iS(a,\varphi)} + e^{-iS(a,\varphi)}\right). \tag{19}$$

Both expanding and contracting branches suffer subsequently a very effective decoherence process [21, 37] becoming quantum mechanically independent. Thus, observers inhabiting a branch of the universe cannot perceive any effect of the quantum superposition.

2.4. Third quantization formalism

Let us now introduce the so-called 'third quantization' formalism [70], where the creation and the annihilation of universes is naturally incorporated in a parallel way as the creation and annihilation of particles is naturally formulated in a quantum field theory. The third quantization formalism consists of considering the wave function of the universe, $\phi(a,\vec{\varphi})$, as a field defined upon the minisuperspace of variables $(a,\vec{\varphi})$. The minisupermetric of the minisuperspace, $G_{AB} = \mathrm{diag}(-a, a^3, \ldots, a^3)$, where $G_{aa} = -a$, has a Lorentzian signature and it allows us to formally interpret the scale factor as an intrinsic time variable of the minisuperspace. This has not to be confused with a time variable in terms of 'clocks and rods' measured by any observer. The consideration of the scale factor as a time variable within a single universe is a tricky task (see Refs. [23, 30, 34, 38–40, 78]) that will partially be addressed on subsequent sections.

We already noticed the formal analogy between the Wheeler-de Witt equation (13) and the equation of motion of a harmonic oscillator. Taking further the analogy, we can find a (third quantized) action for which the variational principle gives rise to Eq. (13), given by

$$^{(3)}S = \frac{1}{2}\int da\ ^{(3)}L = \frac{1}{2}\int da \left(\mathcal{M}\dot{\phi}^2 - \mathcal{M}\omega^2\phi^2\right). \tag{20}$$

The third quantized momentum is defined as, $^{(3)}P_\phi \equiv \frac{\delta\, ^{(3)}L}{\delta\dot{\phi}} = \mathcal{M}\dot{\phi}$, where $^{(3)}L$ is the Lagrangian of the action (20), and the third quantized Hamiltonian then reads

$$^{(3)}H = \frac{1}{2\mathcal{M}} P_\phi^2 + \frac{\mathcal{M}\omega^2}{2}\phi^2, \tag{21}$$

where $\mathcal{M} \equiv \mathcal{M}(a)$ and $\omega \equiv \omega(a,\varphi)$ are defined after Eq. (13). The configuration variable of the third quantization formalism is the wave function of the universe, ϕ, and the quantum state of the multiverse is thus given by another wave function, $\Psi \equiv \Psi(\phi, a)$, which is the solution of the (third quantized) Schrödinger equation [61, 70]

$$^{(3)}\hat{H}(\phi, -i\hbar\frac{\partial}{\partial\phi}, a)\Psi(\phi, a) = i\hbar\frac{\partial\Psi(\phi, a)}{\partial a}. \tag{22}$$

The customary interpretation of the wave function Ψ is the following [70]: let us expand the quantum state of the multiverse, $|\Psi\rangle$, in an orthonormal basis of number states, $|N\rangle$, i.e.

$$|\Psi\rangle = \sum_N \Psi_N(\phi, a)|N\rangle, \tag{23}$$

then, $|\Psi_N(\phi, a_0)|^2$ gives the probability to find in the multiverse N universes with a value a_0 of the scale factor. We can consider different types of universes having different energy-matter contents represented by the fields $\vec{\varphi}^{(i)}$ of the i-universe. The wave function of the whole multiverse is given then by a linear superposition of wave functions of the form [61, 62]

$$\Psi_{\vec{N}}(\vec{\phi}, a) = \Psi_{N_1}(\phi_1, a)\Psi_{N_2}(\phi_2, a)\cdots\Psi_{N_n}(\phi_n, a), \tag{24}$$

where, $\vec{\phi} \equiv (\phi_1, \phi_2, \ldots, \phi_n)$ and $\vec{N} \equiv (N_1, N_2, \ldots, N_n)$, with N_i being the number of universes of type i, represented by the wave function $\phi_i \equiv \phi(\vec{\varphi}^{(i)}, a)$. Following the canonical interpretation of the wave function in quantum mechanics, $|\Psi_{\vec{N}}(\vec{\phi}, a_0)|^2$ gives the probability to find $\vec{N}$ universes in the multiverse with a value of the scale factor and the scalar fields given by, $a = a_0$ and $\vec{\varphi}^{(i)} = \vec{\varphi}_0^{(i)}$, for the i-universe.

Let us just consider one type, i, of universes. The wave function ϕ_i can be promoted to an operator $\hat{\phi}_i$ that can be written as

$$\hat{\phi}_i(a, \varphi) = A_i(a, \varphi)\hat{b}_{0,i}^\dagger + A_i^*(a, \varphi)\hat{b}_{0,i}, \tag{25}$$

where the probability amplitudes $A_i(a, \varphi)$ and $A_i^*(a, \varphi)$ satisfy the Wheeler-de Witt equation (13), and

$$\hat{b}_{0,i} \equiv \sqrt{\frac{\mathcal{M}_0\omega_0}{\hbar}}\left(\hat{\phi}_i + \frac{i}{\mathcal{M}_0\omega_0}\hat{P}_{\phi_i}\right), \tag{26}$$

$$\hat{b}_{0,i}^\dagger \equiv \sqrt{\frac{\mathcal{M}_0\omega_0}{\hbar}}\left(\hat{\phi}_i - \frac{i}{\mathcal{M}_0\omega_0}\hat{P}_{\phi_i}\right), \tag{27}$$

are the customary creation and annihilation operators of the harmonic oscillator, with $\mathcal{M}_0$ and ω_0 being the mass and frequency terms, $\mathcal{M}(a)$ and $\omega(a, \varphi)$, respectively, evaluated on the boundary hypersurface Σ_0 for which, $a = a_0$ and $\varphi = \varphi_0$. The operators $\hat{b}_{0,i}$ and $\hat{b}_{0,i}^\dagger$ can then be interpreted as the annihilation and creation operators of universes with a value of the scale factor a_0 and an energy density given by $\rho_\varphi \approx V(\varphi_0)$, for the case of a slow-varying field. The kind of universes created and annihilated by $\hat{b}^{\dagger}_{0,i}$ and $\hat{b}_{0,i}$, respectively, also depend on the boundary conditions imposed on the probability amplitudes $A_i(a,\varphi)$ and $A^*_i(a,\varphi)$. Recalling the previous discussion on the boundary conditions of the universe, if the tunneling boundary condition is imposed, then, $\hat{b}^{\dagger}_{0,i}$ ($\hat{b}_{0,i}$) creates (annihilates) expanding branches of the universe. If otherwise the 'no-boundary' proposal is chosen, $\hat{b}^{\dagger}_{0,i}$ ($\hat{b}_{0,i}$) creates (annihilates) linear combinations of expanding and contracting branches.

Therefore, at least for universes with high order of symmetry, the third quantization formalism parallels that of a quantum field theory in a curved space-time, i.e. it can formally be seen as a quantum field theory defined on the curved minisuperspace described by the minisupermetric G_{AB}. The scale factor formally plays the role of the time variable and the matter fields $\vec{\varphi}$ the role of the spatial coordinates. Creation and annihilation operators of universes can properly be defined in the curved minisuperspace. However, as it happens in a quantum field theory, different representations can be chosen to describe the quantum state of the universes. The meaning of such representations needs of a further analysis in terms of the boundary condition that has to be imposed on the quantum state of the whole multiverse.

2.5. Boundary conditions of the multiverse

For a given representation, $\hat{b}^{\dagger}_i$ and $\hat{b}_i$, the eigenvalues of the number operator $\hat{N}_i \equiv \hat{b}^{\dagger}_i \hat{b}_i$ might be interpreted in the third quantization formalism as the number of i-universes in the multiverse, where the index i labels the different kinds of universes considered in the model. However, in terms of the constant operators $\hat{b}_{0,i}$ and $\hat{b}^{\dagger}_{0,i}$ defined in Eqs. (26-27), the number of universes of the multiverse is not conserved because $\hat{N}_{0,i} \equiv \hat{b}^{\dagger}_{0,i}\hat{b}_{0,i}$ is not an invariant operator, i.e.

$$\frac{d\hat{N}_{0,i}}{da} \equiv \frac{i}{\hbar}[{}^{(3)}\hat{H}_i, \hat{N}_{0,i}] + \frac{\partial \hat{N}_{0,i}}{\partial a} = \frac{i}{\hbar}[{}^{(3)}\hat{H}_i, \hat{N}_{0,i}] \neq 0. \tag{28}$$

For a large parent universe, i.e. for values $a \gg 1$, the creation and annihilation operators can asymptotically be taken to be the usual creation and annihilation operators of the harmonic oscillator (21) with the proper frequency ω of the Hamiltonian, i.e.

$$\hat{b}_{\omega,i} \equiv \sqrt{\frac{\mathcal{M}(a)\omega_i(a,\varphi)}{\hbar}} \left(\hat{\phi}_i + \frac{i}{\mathcal{M}(a)\omega_i(a,\varphi)}\hat{P}_{\phi_i}\right), \tag{29}$$

$$\hat{b}^{\dagger}_{\omega,i} \equiv \sqrt{\frac{\mathcal{M}(a)\omega_i(a,\varphi)}{\hbar}} \left(\hat{\phi}_i - \frac{i}{\mathcal{M}(a)\omega_i(a,\varphi)}\hat{P}_{\phi_i}\right), \tag{30}$$

for a given type of i-universes. However, in terms of the asymptotic representation (29-30) the number operator, $\hat{N}_{\omega,i} \equiv \hat{b}^{\dagger}_{\omega,i}\hat{b}_{\omega,i}$, is neither an invariant operator because

$$\frac{d\hat{N}_{\omega,i}}{da} \equiv \frac{i}{\hbar}[{}^{(3)}\hat{H}_i, \hat{N}_{\omega,i}] + \frac{\partial \hat{N}_{\omega,i}}{\partial a} = \frac{\partial \hat{N}_{\omega,i}}{\partial a} \neq 0. \tag{31}$$

It would be expected that the number of universes in the multiverse would be a property of the multiverse independent of any internal property of a particular single universe.

Therefore, it seems appropriate to impose the following boundary condition on the multiverse:

> The number of universes of the multiverse does not depend on the value of the scale factor of a particular single universe.

This boundary condition imposes the restriction that the number operator $\hat{N}_i$ for a particular type of i-universes has to be an invariant operator[5]. We can then follow the theory of invariants developed by Lewis [43] and others [11, 41, 55, 57, 67, 69, 76], and find a Hermitian invariant operator, $\hat{I}_i = \hbar(\hat{b}_i^\dagger \hat{b}_i + \frac{1}{2})$, where [43]

$$\hat{b}_i(a) \equiv \sqrt{\frac{1}{2\hbar}} \left(\frac{1}{R_i} \hat{\phi}_i + i(R_i \hat{P}_{\phi_i} - \dot{R}_i \hat{\phi}_i) \right), \tag{32}$$

$$\hat{b}_i^\dagger(a) \equiv \sqrt{\frac{1}{2\hbar}} \left(\frac{1}{R_i} \hat{\phi}_i - i(R_i \hat{P}_{\phi_i} - \dot{R}_i \hat{\phi}_i) \right), \tag{33}$$

with, $R_i \equiv R_i(a, \varphi)$, that can be written as $R = \sqrt{\phi_{1,i}^2 + \phi_{2,i}^2}$, being $\phi_{1,i}$ and $\phi_{2,i}$ two independent solutions of the Wheeler-de Witt equation (13). In the semiclassical regime, we can use independent combinations of the solutions ϕ_+^{WKB} and ϕ_-^{WKB} so that

$$R_i(a, \varphi) \approx \frac{e^{\pm \frac{1}{3V_i(\varphi)}}}{\sqrt{\mathcal{M}(a)\omega_i(a, \varphi)}}, \tag{34}$$

where the positive sign corresponding to the choice of the no-boundary proposal and the negative sign to the tunneling boundary condition. The number operator for a particular kind of i-universes in the representation given by Eqs. (32-33), $\hat{N}_i \equiv \hat{b}_i^\dagger \hat{b}_i$, is then an invariant operator fulfilling the boundary condition of the multiverse and, thus, the eigenvalues N_i, with $\hat{N}_i |N_i, a\rangle = N_i |N_i, a\rangle$ and $N_i \neq N_i(a)$, can properly be interpreted as the number of i-universes of the multiverse.

In terms of the invariant representation, the Hamiltonian (21) takes the form

$${}^{(3)}\hat{H} = \hbar \left(\beta_+ \, (\hat{b}^\dagger)^2 + \beta_- \, \hat{b}^2 + \beta_0 \, (\hat{b}^\dagger \hat{b} + \frac{1}{2}) \right), \tag{35}$$

where,

$$\beta_+^* = \beta_- = \frac{1}{4} \left\{ \left(R - \frac{i}{R} \right)^2 + \omega^2 R^2 \right\}, \tag{36}$$

$$\beta_0 = \frac{1}{2} \left(\dot{R}^2 + \frac{1}{R^2} + \omega^2 R^2 \right). \tag{37}$$

---

[5] We are not considering transitions from one kind of universes to another.

The Hamiltonian (35) is formally the same Hamiltonian of a degenerated parametric amplifier used in quantum optics [66, 82] (see also, Sec. 3). The quadratic terms are interpreted therein as the creation and annihilation operators of pairs of entangled photons. Similarly, we can interpret the quadratic terms in $\hat{b}^{\dagger}$ and $\hat{b}$ of Eq. (35) as operators that create and annihilate, respectively, pairs of entangled universes. In the case that the universes were distinguishable, the Hamiltonian (35) would take the form of a non-degenerated parametric amplifier [82]

$$^{(3)}\hat{H} = \hbar\left(\beta_{+}\,\hat{b}_1^{\dagger}\hat{b}_2^{\dagger} + \beta_{-}\,\hat{b}_1\hat{b}_2 + \frac{\beta_0}{2}\,(\hat{b}_1^{\dagger}\hat{b}_1 + \hat{b}_2^{\dagger}\hat{b}_2 + 1)\right), \tag{38}$$

where the indices 1 and 2 label the two universes of the entangled pair. The distinguishability of universes is certainly a tricky task. However, observers may exist in the two universes of an entangled pair because the universes share similar properties and, then, the plausible (classical and quantum) communications between these observers would make the universes be distinguishable. Classical communications between the observers of different universes can be conceivable by the presence of wormholes connecting the universes and quantum communications could then be implemented by using quantum correlated fields shared by the two observers. Therefore, it is at least plausible to pose a model of the multiverse made up of entangled pairs of distinguishable universes.

The general quantum state of a multiverse formed by entangled pairs of de-Sitter universes would be given by linear combinations of terms like [61, 62] (see Eq. (24))

$$\Psi_{\vec{N}}(\vec{\phi}, a) = \Psi_{N_1}^{\Lambda_1}(a, \phi_1)\Psi_{N_2}^{\Lambda_2}(a, \phi_2)\cdots\Psi_{N_n}^{\Lambda_n}(a, \phi_n), \tag{39}$$

where, $\vec{\phi} \equiv (\phi_1, \phi_2, \ldots, \phi_n)$, and $\vec{N} \equiv (2N_1, 2N_2, \ldots, 2N_n)$, with N_i being the number of pairs of universes of type i, represented by the wave function $\phi_i \equiv \phi_{\Lambda_i}(a)$ that corresponds to the value Λ_i of the cosmological constant. The wave functions, $\Psi_{N_i}^{\Lambda_i}(\phi_i, a)$, in Eq. (39) are the solutions of the third quantized Schrödinger equation

$$i\hbar\frac{\partial}{\partial a}\Psi_{N_i}^{\Lambda_i}(\phi_i, a) = \hat{H}_i(\phi, p_{\phi}, a)\Psi_{N_i}^{\Lambda_i}(\phi_i, a), \tag{40}$$

with

$$\hat{H}_i = \hbar\left\{\beta_{-}^{(i)}\hat{b}_1^{(i)}\hat{b}_2^{(i)} + \beta_{+}^{(i)}(\hat{b}_1^{(i)})^{\dagger}(\hat{b}_2^{(i)})^{\dagger} + \frac{1}{2}\beta_0^{(i)}\left((\hat{b}_1^{(i)})^{\dagger}\hat{b}_1^{(i)} + (\hat{b}_2^{(i)})^{\dagger}\hat{b}_2^{(i)} + 1\right)\right\}, \tag{41}$$

for each kind of i-universes in the multiverse [62].

3. Quantum entanglement

3.1. Introduction

Back to the early years of the quantum development, in 1935, Schrödinger [64, 65] coined the word 'entanglement' to describe a puzzling feature of the quantum theory that was formerly posed by Einstein, Podolski and Rosen in a famous gedanken experiment [12]. Schrödinger also realized that entanglement is precisely *the characteristic trait of quantum mechanics, the one that enforces its entire departure from classical lines of thought* [64]. Let us briefly show it by following the example given in Ref. [66] (see also Ref. [82]). Let us consider the photo-disintegration of a Hg_2 molecule formed by two atoms of Hg with spin $\frac{1}{2}$. Before the disintegration, the molecule is taken to be in a state of zero angular momentum so that the composite state is given by

$$|\mathrm{Hg}_2\rangle = \frac{1}{\sqrt{2}}\left(|\uparrow_1\downarrow_2\rangle - |\downarrow_1\uparrow_2\rangle\right), \tag{42}$$

where 1 and 2 refer to the atoms of Hg and $|\uparrow(\downarrow)\rangle$ refers to the value $+\frac{1}{2}(-\frac{1}{2})$ of the projection of their spin along the z-axis. After the photo-disintegration, performed with no disturbance of the angular momentum, the two atoms separate each other in opposite directions so we can make independent measurements on them. Before doing any measurement we do not know the particular value of the spin of each atom. However, we do anticipatedly know that if a measurement of the spin projection is performed on the atom 1 yielding a value $+\frac{1}{2}(-\frac{1}{2})$, then, the spin projection of the atom 2 is to be $-\frac{1}{2}(+\frac{1}{2})$. Furthermore, if it is performed a different measurement of the projection of the spin of the particle 1 along, say, the x-axis, we are determining the value of the spin projection of the particle 2 along the same axis, too. This non-local feature of the quantum theory is known as *entanglement* and the state (42) is called an *entangled state*.

In 1964, Bell derived certain inequalities [7, 8] that should be satisfied by any reasonable realistic[6] theory of local variables. The experiments of Aspect [3] and others [2, 63, 74, 83] have shown that the entangled states of the quantum theory violate such inequalities. Furthermore, these states have not only provided us with an experimental test of the quantum postulates but they have also given rise to the development of a completely new branch of physics, the so-called quantum information theory [32, 35, 75], which includes interesting subjects like quantum computation, quantum cryptography, and quantum teleportation, which are currently under a promising state of development.

It is finally worth noticing that the kinematical non-locality of the quantum theory is also the feature that forces us to consider a wave function of the universe. As it is pointed out in Ref. [39], *if gravity is quantized, the kinematical non-separability of quantum theory demands that the whole universe must be described in quantum terms* (cf. p. 4). Every space-time region is entangled to its environment, which is entangled to another environment and so forth, ending up in a quantum description of the whole universe.

3.2. Squeezed and entangled states of light

Squeezed states of light [81] can be seen as a generalization of the coherent states. Let us define the quadrature operators

$$\hat{X}_1 \equiv \hat{a} + \hat{a}^\dagger \ , \ \hat{X}_2 = i(\hat{a}^\dagger - \hat{a}), \tag{43}$$

[6] By a realistic theory we mean a theory that presupposes that the elements of the theory represent elements of physical reality (see Ref. [12]).

where $\hat{a}^\dagger$ and $\hat{a}$ are the usual creation and annihilation operators of the harmonic oscillator. The operators $\hat{X}_1$ and $\hat{X}_2$ are essentially dimensionless position and momentum operators. The uncertainty relation for ΔX_1 and ΔX_2 reads

$$\Delta X_1 \Delta X_2 \geq 1, \tag{44}$$

where, for a coherent state, $\Delta X_1 = \Delta X_2 = 1$. A squeezed state is defined as the quantum state for which one of the quadratures satisfies[7]

$$(\Delta X_i)^2 < 1 \ \ (i = 1 \text{ or } 2). \tag{45}$$

Therefore, for a squeezed state the uncertainty of one of the quadratures is reduced below the limit of the Heisenberg principle at the expense of the increased fluctuations of the other quadrature.

Unlike the generation of coherent states, which is associated with linear terms of the creation and annihilation operators in the Hamiltonian, the generation of squeezed states is associated with quadratic terms of such operators. For instance, let us consider the Hamiltonian that represents in quantum optics a degenerated parametric amplifier [66, 82]

$$\hat{H} = i\hbar\frac{\chi}{2}\left((\hat{a}^\dagger)^2 - \hat{a}^2\right), \tag{46}$$

where χ is a coupling constant. Then, the time evolution of the vacuum state,

$$|s(t)\rangle = \hat{S}(\chi)|0\rangle = e^{\frac{\chi}{2}((\hat{a}^\dagger)^2 - \hat{a}^2)t}|0\rangle, \tag{47}$$

yields a squeezed (vacuum) state, $|s(t)\rangle$, with $\hat{S}_\chi$ being the squeezing operator which satisfies, $\hat{S}^\dagger(\chi) = \hat{S}^{-1}(\chi) = \hat{S}(-\chi)$. It is therefore a unitary operator. The Heisenberg equations of motion for the quadrature amplitudes turn out to be then

$$\frac{d\hat{X}_1}{dt} = \chi\hat{X}_1 \ , \ \frac{d\hat{X}_2}{dt} = -\chi\hat{X}_2, \tag{48}$$

with solutions given by

$$\hat{X}_1(t) = e^{\chi t}\hat{X}_1(0) \ , \ \hat{X}_2(t) = e^{-\chi t}\hat{X}_2(0). \tag{49}$$

Then, for an initial vacuum state, for which $\Delta X_i(0) = 1$, the variances of the quadratures read

$$\Delta X_1(t) = e^{2\chi t} \ , \ \Delta X_2(t) = e^{-2\chi t}. \tag{50}$$

[7] An ideal squeezed state also satisfies $\Delta X_1 \Delta X_2 = 1$

It can clearly be seen that one of the variances (ΔX_2) decreases in time at the expense of the increase of the other (ΔX_1), with $\Delta X_1(t)\Delta X_2(t) = 1$. The squeezed vacuum state is therefore an ideal squeezed state (see footnote 7).

The Hamiltonian given by Eq. (46) is associated with the generation of entangled pairs of photons of equal frequency. For that reason, squeezed states are usually dubbed two photon coherent states [87, 88]. The non-degenerate amplifier is a generalization of the Hamiltonian (46) which generates entangled pairs of distinguishable photons of frequency ω_1 and ω_2, respectively. In that case, the Hamiltonian reads

$$\hat{H} = i\hbar\chi(\hat{a}_1^\dagger\hat{a}_2^\dagger - \hat{a}_1\hat{a}_2), \tag{51}$$

where $\hat{a}_1^\dagger$, $\hat{a}_1$ and $\hat{a}_2^\dagger$, $\hat{a}_2$ are the creation and annihilation operators of modes with frequency ω_1 and ω_2, respectively. The solutions of the Heisenberg equations read [82]

$$\hat{a}_1(t) = \hat{a}_1(0)\cosh\chi t + \hat{a}_2^\dagger(0)\sinh\chi t, \tag{52}$$

$$\hat{a}_2(t) = \hat{a}_2(0)\cosh\chi t + \hat{a}_1^\dagger(0)\sinh\chi t, \tag{53}$$

and the evolution of the two-mode vacuum state is now given by

$$|s_2\rangle = \hat{S}_2(\chi)|0_1 0_2\rangle = e^{(\hat{a}_1^\dagger\hat{a}_2^\dagger - \hat{a}_1\hat{a}_2)\chi t}|0_1 0_2\rangle, \tag{54}$$

where $\hat{S}_2(\chi)$ is the two mode squeeze operator.

Squeezed and entangled states are usually dubbed non-classical states [59] because they may violate some inequalities that should be satisfied in the classical description of light. For instance, in Fig. 4 it is depicted the typical experimental setup to test the violation of the classical inequality $g^{(2)}(0) \geq 1$ (photon bunching [59, 82]), where $g^{(2)}(\tau)$ is the second order correlation function that measures the correlation between the state of the field at two different times t and $t+\tau$. Classically, a beam of light with an initial intensity I_A is split into two beams of equal intensities, $I_{A1} = I_{A2} \equiv I$. If the averaged intensity is defined by

$$\langle I\rangle = \int P(I) I\, dI, \tag{55}$$

for a given positive distribution $P(I)$, then, $g^{(2)}(0)$ can be written as

$$g^{(2)}(0) = \frac{\langle I_{A1} I_{A2}\rangle}{\langle I_{A1}\rangle\langle I_{A2}\rangle} = \frac{\langle I^2\rangle}{\langle I\rangle^2} - 1 + \frac{1}{\langle I\rangle^2}\int dI\, P(I)(I - \langle I\rangle)^2 \geq 1. \tag{56}$$

Quantum mechanically, however, the second order correlation function is defined, for a single mode, as [59, 82]

$$g^{(2)}(0) = \frac{\langle (a^\dagger)^2 a^2 \rangle}{\langle a^\dagger a \rangle^2} \geq 1 - \frac{1}{\langle a^\dagger a \rangle}. \tag{57}$$

There is then room for a quantum violation of the classical inequality $g^{(2)}(0) \geq 1$. For a large number of photons the quantum inequality (57) becomes the classical constraint $g^{(2)}(0) \geq 1$, and light can be described classically.

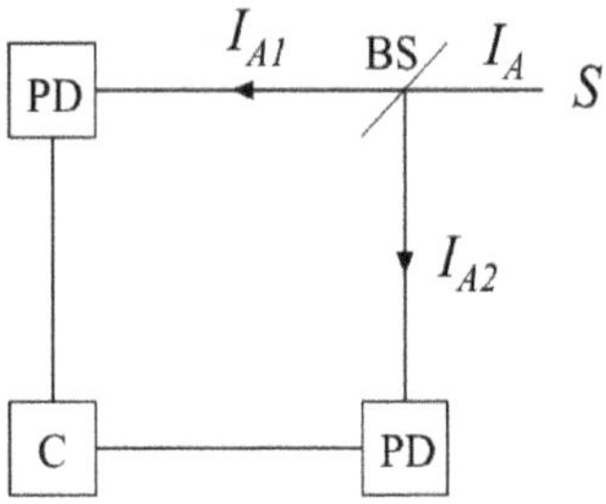

Figure 4. Experimental setup for testing photon antibunching [59]: S, source of light; BS, beam splitter; PD, photodetector; and, C, correlator.

It is worth noticing that what it is violated in an experimental setup involving squeezed and entangled states are some classical assumptions. For instance, in the experimental setup depicted in Fig. 4, the photon is not split into two photons by the beam splitter but it takes either the path that reaches the photo-counter 1 or the path that reaches the photo-counter 2. The fact that the photon is not divided into two photons, as it would happen to an electromagnetic wave, supports the consideration of the photon as a real and individual entity. Moreover, the corpuscular nature of the photon is the postulate that Einstein assumed in order to properly describe the photoelectric effect and it can be considered the germ of quantum mechanics, actually.

However, such a conclusion does not imply that we can interpret the photon as a classical particle. The double-slit experiment clearly shows that the concept of photon as a localized particle is generally meaningless. The quantum concept of particle has rather to be understood as a global property of the field. Their localization and the space-time independence of different particles depend on the *separability* of their states. Furthermore, the violation of the classical inequalities is associated with negative values of the probability distributions. This can clearly be seen from Eq. (56), where a negative value of $P(I)$ is needed to obtain a value $g^{(2)}(0) < 1$. It plainly shows that there are quantum states of light that cannot be described classically [59].

Another test for the non-classicality of some quantum states is given by the violation of the Bell's inequalities. This is achieved, for a two mode state of light, whenever it is satisfied [59]

$$C \equiv \frac{\langle a_1^\dagger a_1 a_2^\dagger a_2 \rangle}{\langle a_1^\dagger a_1 a_2^\dagger a_2 \rangle + \langle (a_1^\dagger)^2 a_1^2 \rangle} \geq \frac{\sqrt{2}}{2}. \tag{58}$$

For the two mode squeezed operators (52-53), it can be checked that

$$\langle a_1^\dagger a_1 a_2^\dagger a_2 \rangle = N^2(6x^4 + 6x^2 + 1) + N(6x^4 + 4x^2) + x^2(2x^2 + 1), \quad (59)$$

$$\langle (a_1^\dagger)^2 a_1^2 \rangle = N^2(6x^4 + 6x^2 + 1) + N(6x^4 + 2x^2 - 1) + 2x^4, \quad (60)$$

where, $x \equiv x(t) = \sinh \chi t$, and the mean value has been computed for initial number states, with $N_1 = N_2 \equiv N$. For an initial vacuum state, $x(0) = 0$ and $N = 0$, then $C = 1 > 0.7$, which implies a maximum violation of Bell's inequalities[8]. This result is expected because the quantum vacuum state is a highly non-local state. For a pair of entangled photons ($N = 1$), it is obtained

$$C = \frac{14x^4 + 11x^2 + 1}{28x^4 + 19x^2 + 1}, \quad (61)$$

which implies a violation of Bell's inequalities for a value, $0.31 > \sinh \chi t > 0$. At later times, the effective number of photons, $\langle N_{eff} \rangle = \sinh^2 \chi t$, produced by the parametric amplifier grows and the quantum correlations are destroyed. The radiation effectively becomes classical, then. However, at shorter times, the two mode squeezed states violate the Bell's inequalities showing their non-classical behaviour.

Therefore, entangled and squeezed states can essentially be seen as non-classical states, which is fundamentally related to the complementary principle of quantum mechanics. Generally speaking, the classical description of light in terms of waves and particles, separately, does not hold: i) the photon has to be considered as an individual entity (particle description), and ii) we have to complementary consider interference as well as non-local effects between the states of two distant photons (wave description).

3.3. Thermodynamics of entanglement

For a physical system whose quantum state is represented by a density matrix[9], $\hat{\rho}(t)$, whose evolution is determined by a Hamiltonian, $\hat{H} \equiv \hat{H}(t)$, we can define the following thermodynamical quantities [1, 14]

$$E(t) = \mathrm{Tr}(\hat{\rho}(t)\hat{H}(t)), \quad (62)$$

$$Q(t) = \int^t \mathrm{Tr}\left(\frac{d\hat{\rho}(t')}{dt'}\hat{H}(t')\right) dt', \quad (63)$$

$$W(t) = \int^t \mathrm{Tr}\left(\hat{\rho}(t')\frac{d\hat{H}(t')}{dt'}\right) dt', \quad (64)$$

where $\mathrm{Tr}(\hat{O})$ denotes the trace of the operator $\hat{O}$. The quantities $E(t)$, $Q(t)$, and $W(t)$, are the quantum informational analogue to the energy, heat and work, respectively. The first principle of thermodynamics,

[8] Let us notice that for a pure entangled state like $|\psi\rangle = \frac{1}{\sqrt{2}}(|00\rangle + |11\rangle)$, $\langle\psi|(a_1^\dagger)^2 a_1^2|\psi\rangle = 0$ and thus $C = 1$, too.

[9] In this section, it turns out to be convenient to use the density matrix formalism. This can generally be found in the bibliography (see, for instance, Refs. [14, 32, 35, 54, 75]). Let us just briefly note that for a pure state $|\psi\rangle$, the density matrix is given by $\hat{\rho} = |\psi\rangle\langle\psi|$, and for a mixed state, $\hat{\rho} = \sum_i \lambda_i |i\rangle\langle i|$, where $\lambda_i < 1$ are the eigenvalues of the density matrix, with $\sum_i \lambda_i = 1$, and the vectors $\{|i\rangle\}$ form an orthonormal basis.

$$dE = \delta W + \delta Q, \quad (65)$$

is then directly satisfied. The quantum informational analogue to the entropy is defined through the von Neumann formulae [14, 35, 54, 75]

$$S(\rho) = -\text{Tr}\left(\hat{\rho}(t)\ln\hat{\rho}(t)\right) = -\Sigma_i \lambda_i(t)\ln\lambda_i(t), \quad (66)$$

where $\lambda_i(t)$ are the eigenvalues of the density matrix, and $0 \ln 0 \equiv 0$. For a pure state, $\hat{\rho}^n = \hat{\rho}$ and $\lambda_i = \delta_{ij}$ for some value j. Then, the entropy vanishes. For a mixed state, $S > 0$. It can be distinguished two terms [1] in the variation of entropy,

$$dS = \frac{\delta Q}{T} + \sigma. \quad (67)$$

The first term corresponds to the variation of the entropy due to the change of heat. The second term in Eq. (67) is called [1] *entropy production*, and it accounts for the variation of entropy due to any adiabatic process. The second principle of thermodynamics states that the change of entropy has to be non-negative for any adiabatic process, i.e. $\sigma \geq 0$.

Let us now analyze the thermodynamical properties of a two mode squeezed state, Eq. (54), represented by the density matrix

$$\hat{\rho} = |s_2\rangle\langle s_2| = \hat{S}_2(r)|0_1 0_2\rangle\langle 0_1 0_2|\hat{S}_2^\dagger(r), \quad (68)$$

where the squeezing operator is given by, $\hat{S}_2(r) \equiv e^{(\hat{a}_1^\dagger \hat{a}_2^\dagger - \hat{a}_1 \hat{a}_2) r(t)}$, with $r(t) = \chi t$, and $|0_1 0_2\rangle \equiv |0_1\rangle|0_2\rangle$, with $|0_1\rangle$ and $|0_2\rangle$ being the initial ground states of each single mode, respectively. The reduced density matrix that represents the quantum state of each single mode can be obtained by tracing out the degrees of freedom of the partner mode, i.e.

$$\hat{\rho}_1 \equiv \text{Tr}_2\hat{\rho} = \sum_{N_2=0}^{\infty} \langle N_2|\hat{\rho}\, N_2\rangle, \quad (69)$$

and similarly for $\hat{\rho}_2$ by replacing the indices 2 and 1. By making use of the disentangling theorem [10, 86], the squeezing operator $\hat{S}_2(r)$ can be written as

$$\hat{S}_2(r) = e^{\Gamma(t)\hat{a}_1^\dagger \hat{a}_2^\dagger} e^{-g(t)(\hat{a}_1^\dagger \hat{a}_1 + \hat{a}_2^\dagger \hat{a}_2 + 1)} e^{-\Gamma(t)\hat{a}_1 \hat{a}_2}, \quad (70)$$

where

$$\Gamma(t) \equiv \tanh r(t) \quad , \quad g(t) \equiv \ln\cosh r(t), \quad (71)$$

with, $r(t) = \chi t$. We can thus compute the reduced density matrix (69), yielding

$$\hat{\rho}_1(t) = e^{-2g(t)} \sum_{N_1=0}^{\infty} e^{2N_1 \ln \Gamma(t)} |N_1\rangle\langle N_1| = \frac{1}{\cosh^2 r(t)} \sum_{N_1=0}^{\infty} \left(\tanh^2 r(t)\right)^{N_1} |N_1\rangle\langle N_1|. \quad (72)$$

It turns out to be that $\hat{\rho}_1$ describes a thermal state

$$\hat{\rho}_1(t) = \frac{1}{Z(t)} \sum_{N_1=0}^{\infty} e^{-\frac{\omega_1}{T(t)}(N_1+\frac{1}{2})} |N_1\rangle\langle N_1|, \quad (73)$$

where, $Z^{-1} = 2\sinh\frac{\omega_1}{2T(t)}$, with a time dependent temperature of entanglement given by

$$T(t) = \frac{\omega_1}{2\ln\frac{1}{\Gamma(t)}}, \quad (74)$$

with ω_1 being the frequency of the mode. It is worth mentioning that the thermal state (73) is indistinguishable from a classical mixture [36, 56]. In that sense, it can be seen as a classical state. However, it has been obtained from the partial trace of a composite entangled state which is, as it has previously been shown, a quantum state having no classical analogue.

We can now compute the thermodynamical quantities given by Eqs. (62-64) and Eq. (66) for the thermal state (73). The entropy of entanglement, i.e. the quantum entropy that corresponds to the reduced density matrix $\hat{\rho}_1$, reads

$$S_{ent}(t) = -\mathrm{Tr}(\hat{\rho}_1 \ln \hat{\rho}_1) = \cosh^2 r(t) \ln \cosh^2 r(t) - \sinh^2 r(t) \ln \sinh^2 r(t). \quad (75)$$

The total energy $E_1 \equiv E(\rho_1)$ yields

$$E_1(t) = \omega_1(\sinh^2 r(t) + \frac{1}{2}) \equiv \omega_1(\langle N(t)\rangle + \frac{1}{2}), \quad (76)$$

where $\langle N(t)\rangle$ is an effective mean number of photons due to the squeezing effect. For a mode of constant frequency ω_1, the variation of work vanishes because

$$\delta W_1 = \frac{d\omega_1}{dt}(\sinh^2 r(t) + \frac{1}{2}) = 0. \quad (77)$$

The variation of heat is however different from zero. It reads

$$\delta Q = \omega_1 \sinh 2r(t) \frac{\partial r(t)}{\partial t} dt. \quad (78)$$

It can also be checked that

$$\sigma \equiv \frac{dS_{ent}}{dt} - \frac{1}{T(t)} \frac{\delta Q}{\delta t} = 0 \ , \ \forall t. \tag{79}$$

Therefore, the second principle of thermodynamics provides us with no arrow of time because the entropy production σ identically vanishes at any time. In a non-reversible process, however, the constraint $\sigma > 0$ would give rise to the *entanglement thermodynamical arrow of time* [36, 56].

4. Quantum entanglement in the multiverse

4.1. Creation of entangled pairs of universes

First, we shall present a plausible scenario for the nucleation of a pair of entangled universes. The Wheeler-de Witt equation (12) for a de-Sitter universe with a massless scalar field reads

$$\hbar^2 \ddot{\phi} + \frac{\hbar^2}{a} \dot{\phi} - \frac{\hbar^2}{a^2} \phi'' + (\Lambda a^4 - a^2)\phi = 0, \tag{80}$$

where, $\phi \equiv \phi(a, \varphi)$ is the wave function of the universe with, $\dot{\phi} \equiv \frac{\partial \phi}{\partial a}$ and $\phi' \equiv \frac{\partial \phi}{\partial \varphi}$, and Λ is the cosmological constant. As it was already pointed out in Sec. 2, in the third quantization formalism the wave function ϕ is promoted to an operator $\hat{\phi}$ that, in the case now being considered, can be decomposed in normal modes as

$$\hat{\phi}(a, \varphi) = \int dk \left(e^{ik\varphi} A_k(a) \hat{b}_k^\dagger + e^{-ik\varphi} A_k^*(a) \hat{b}_k \right), \tag{81}$$

where, $\hat{b}_k \equiv \hat{b}_k(a_0)$ and $\hat{b}_k^\dagger \equiv \hat{b}_k^\dagger(a_0)$, are the constant operators defined in Eqs. (26-27), now with the mode-dependent frequency,

$$\omega_k(a) = \frac{1}{\hbar} \sqrt{\Lambda a^4 - a^2 + \frac{\hbar^2 k^2}{a^2}}, \tag{82}$$

evaluated at a_0. The probability amplitudes $A_k(a)$ and $A_k^*(a)$ satisfy the equation of the damped harmonic oscillator,

$$\ddot{A}_k(a) + \frac{\dot{\mathcal{M}}}{\mathcal{M}} \dot{A}_k(a) + \omega_k^2 A_k(a) = 0, \tag{83}$$

with, $\mathcal{M} \equiv \mathcal{M}(a) = a$, and $\omega_k \equiv \omega_k(a)$. Let us recall that the real values of the frequency (82) define the oscillatory regime of the wave function of the universe in the Lorentzian region, and the complex values define the exponential regime of the Euclidean region. Let us first consider the zero mode of the wave function, i.e. $k = 0$. Then, the wave function

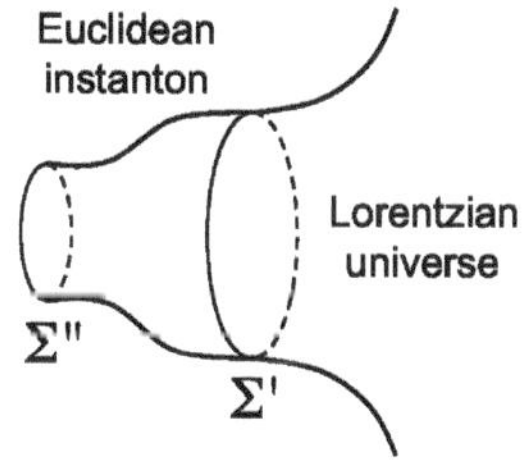

Figure 5. Before reaching the collapse, the instanton finds the transition hypersurface Σ''.

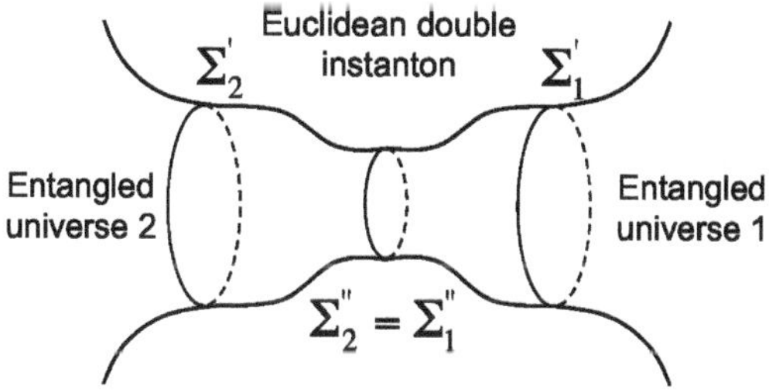

Figure 6. Creation of a pair of entangled universes from a pair of instantons.

$\phi_\Lambda(a)$ quantum mechanically describes the nucleation of a de-Sitter universe from a de-Sitter instanton [26, 39, 77, 78] depicted in Sec. 2, with a transition hypersurface $\Sigma_0 \equiv \Sigma(a_0)$ located at $a_0 = \frac{1}{\sqrt{\Lambda}}$ (see, Fig 1).

For values of k different from zero, the quantum correction term given in Eq. (82) introduces a novelty. For the value, $k_m > k > 0$, where $k_m^2 \equiv \frac{4}{27\hbar^2\Lambda^2}$, there are two transition hypersurfaces from the Euclidean to the Lorentzian region, $\Sigma' \equiv \Sigma(a_+)$ and $\Sigma'' \equiv \Sigma(a_-)$, respectively, located at [62]

$$a_+ \equiv \frac{1}{\sqrt{3\Lambda}}\sqrt{1+2\cos\left(\frac{\theta_k}{3}\right)}, \tag{84}$$

$$a_- \equiv \frac{1}{\sqrt{3\Lambda}}\sqrt{1-2\cos\left(\frac{\theta_k+\pi}{3}\right)}, \tag{85}$$

where, in units for which $\hbar = 1$,

$$\theta_k \equiv \arctan\frac{2k\sqrt{k_m^2-k^2}}{k_m^2-2k^2}. \tag{86}$$

The picture is then rather different from the one depicted in Fig. 1. First, at the transition hypersurface Σ' the universe finds the Euclidean region (let us notice that for $k \to 0$, $a_+ \to a_0$ and $a_- \to 0$). However, before reaching the collapse, the Euclidean instanton finds a new transition hypersurface Σ'' (see Fig 5). Then, following a mechanism that parallels that proposed by Barvinsky and Kamenshchik in Refs. [4–6], two instantons can be matched by identifying their hypersurfaces Σ'' (see Fig. 6). The instantons can thus be created in pairs which would eventually give rise to an entangled pair of universes. Let us notice that this is a quantum effect having no classical analog because the quantum correction term in Eq. (82) does not appear in the classical theory.

The matching hypersurface $\Sigma'' \equiv \Sigma''(a_-)$, where $a_- \equiv a_-(\theta_k)$ is given by Eq. (85) with Eq. (86), depends on the value k of the mode. Therefore, the matched instantons can only be joined for an equal value of the mode of their respective scalar fields. The universes created from such a double instanton are then entangled, with a composite quantum state given by

$$\phi_{I,II} = \int dk \left(e^{ik(\varphi_I+\varphi_{II})} A_{I,k}(a) A_{II,k}(a)\, \hat{b}^\dagger_{I,k} \hat{b}^\dagger_{II,k} + e^{-ik(\varphi_I+\varphi_{II})} A^*_{I,k}(a) A^*_{II,k}(a)\, \hat{b}_{I,k}\, \hat{b}_{II,k} \right), \quad (87)$$

where $\varphi_{I,II}$ are the values of the scalar field of each single universe, labelled by I and II, respectively. The cross terms like $A_{I,k}A^*_{II,k}$ cannot be present in the state of the pair of universes because the orthonormality relations between the modes [62]. Then, the composite quantum state must necessarily be the entangled state represented by Eq. (87).

It is also worth mentioning that, in the model being considered, there is no Euclidean regime for values $k \geq k_m$ and, therefore, no universes are created from the space-time foam with such values of the mode. Then, k_m can be considered the natural cut-off of the model. Let us also note that a similar behavior of the modes of the universe would be obtained for a non-massless scalar field provided that the potential of the scalar field, $V(\varphi)$, satisfies the boundary condition [38, 39], $V(\varphi) \to 0$ for $a \to 0$.

4.2. Entangled and squeezed states in the multiverse

Entangled states, like those found in the preceding section or those appearing in the phantom multiverse [18, 62], can generally be posed in the quantum multiverse. Furthermore, the canonical representations of the harmonic oscillator that represent the quantum state of the multiverse, in the model described in Sec. 2, are related by squeezed transformations [41]. Thus, squeezed states may generally be considered in the quantum multiverse.

As we saw in Sec. 3, entangled and squeezed states are usually dubbed 'non-classical' states because they are related to the violation of classical inequalities. Such violation is fundamentally associated to the complementary principle of quantum mechanics. In the multiverse, squeezed and entangled states may also violate the classical inequalities [62]. However, the conceptual meaning of such violation can be quite different from that given in quantum optics. For instance, if the existence of entangled and squeezed states would imply a violation of Bell's inequalities, then, it could not be interpreted in terms of locality or non-locality because these concepts are only well-defined inside a universe, where space and time are meaningful. In the quantum multiverse, there is generally no common space-time among the universes and, therefore, the violation of Bell's inequalities would be rather related to the interdependence of the quantum states that represent different universes of the multiverse.

Like in quantum optics [75], the violation of the classical inequalities in the multiverse depends on the representation which is chosen to describe the quantum states of the universes [62]. Unlike quantum optics, we do not have an experimental device to measure other universes rather than our own universe[10]. However, the extension of the complementary principle to the quantum description of the multiverse entails two main consequences. On the one hand, if the wave function of the universe has to be described in terms of 'particles', it means that in some appropriate representation we can formally distinguish the universal states as individual entities, giving rise therefore to the multiverse scenario. On the other hand, if it has to be complementary described in terms of waves, then, interference between the quantum states of two or more universes can generally be considered as well.

4.3. Thermodynamical properties of entangled universes

Let us consider a multiverse made up of homogeneous and isotropic universes with a slow-varying scalar field φ, recalling that in the case for which $\dot{\varphi} = 0$ and $V(\varphi_0) \equiv \Lambda$, the model effectively represents a multiverse formed by de-Sitter universes.

Let us consider one type of universes and describe the quantum state of the multiverse in terms of the annihilation and creation operators $\hat{b}(a)$ and $\hat{b}^\dagger(a)$ given in Eqs. (32-33). The vacuum state of the multiverse, $|\bar{0}\rangle$, is then defined as the eigenstate of the annihilation operator $\hat{b}(a)$ with eigenvalue zero, i.e. $\hat{b}(a)|\bar{0}\rangle \equiv 0$. On the other hand, observers inhabiting a large parent universe would quantum mechanically describe the state of their respective universes in the asymptotic representation given by Eqs. (29-30), with a ground state $|0\rangle$ defined by, $\hat{b}_\omega(a)|0\rangle \equiv 0$.

We can consider therefore two representations: the one derived from a consistent formulation of the boundary condition of the whole multiverse, or *invariant representation*, given by the operators $\hat{b}(a)$ and $\hat{b}^\dagger(a)$, and the asymptotic representation given by the operators $\hat{b}_\omega(a)$ and $\hat{b}^\dagger_\omega(a)$, which might be called the *observer representation*. They both are related by the squeezing transformation

$$\hat{b} = \mu_\omega \hat{b}_\omega + \nu_\omega \hat{b}^\dagger_\omega, \tag{88}$$

$$\hat{b}^\dagger = \mu^*_\omega \hat{b}^\dagger_\omega + \nu^*_\omega \hat{b}_\omega, \tag{89}$$

where, $\mu_\omega \equiv \mu_\omega(a,\varphi)$ and $\nu_\omega \equiv \nu_\omega(a,\varphi)$, are given by

$$\mu_\omega(a,\varphi) = \frac{1}{2\sqrt{\mathcal{M}(a)\omega(a,\varphi)}}\left(\frac{1}{R} + R\mathcal{M}(a)\omega(a,\varphi) - i\dot{R}\right), \tag{90}$$

$$\nu_\omega(a,\varphi) = \frac{1}{2\sqrt{\mathcal{M}(a)\omega(a,\varphi)}}\left(\frac{1}{R} - R\mathcal{M}(a)\omega(a,\varphi) - i\dot{R}\right), \tag{91}$$

with $|\mu_\omega|^2 - |\nu_\omega|^2 = 1$, and $R = R(a,\varphi)$ is given, in the semiclassical regime, by Eq. (34),

$$R \approx \frac{e^{\pm\frac{1}{3V(\varphi)}}}{\sqrt{\mathcal{M}(a)\omega(a,\varphi)}},$$

[10] In some sense, we are the 'measuring device' of our universe.

where the positive sign corresponds to the choice of the no-boundary condition and the negative sign to the tunneling boundary condition. Let us further assume that the multiverse is in the invariant vacuum state $|\bar{0}\rangle$. The density matrix that represents the quantum state of the multiverse turns out to be then

$$\hat{\rho}(a,\varphi) \equiv |\bar{0}\rangle\langle\bar{0}| = \hat{\mathcal{U}}_S^{\dagger}|0_1 0_2\rangle\langle 0_1 0_2|\hat{\mathcal{U}}_S, \tag{92}$$

where $|0_1 0_2\rangle \equiv |0_1\rangle|0_2\rangle$, with $|0_1\rangle$ and $|0_2\rangle$ being the ground states of a pair of entangled universes in their respective observer representations. Similarly to Eq. (54), the squeezing operator $\hat{\mathcal{U}}_S$ is given by [62]

$$\hat{\mathcal{U}}_S(a,\varphi) = e^{r(a,\varphi)\hat{b}_1\hat{b}_2 - r(a,\varphi)\hat{b}_1^{\dagger}\hat{b}_2^{\dagger}}, \tag{93}$$

where the squeezing parameter, $r(a,\varphi)$, reads

$$r(a,\varphi) \equiv \text{arcsinh}|v_{\omega}(a,\varphi)|, \tag{94}$$

with $v_{\omega}(a,\varphi)$ being given by Eq. (91). We can then follow the procedure of Sec. 3.3 to compute the reduced density matrix, $\hat{\rho}_1$, that represents the quantum state of one single universe of the entangled pair. It is given then by the thermal state [62]

$$\hat{\rho}_1(a,\varphi) \equiv \text{Tr}_2\hat{\rho} = \frac{1}{Z}\sum_{N=0}^{\infty} e^{-\frac{\omega(a,\varphi)}{T}(N+\frac{1}{2})}|N\rangle\langle N|, \tag{95}$$

with, $|N\rangle \equiv |N\rangle_2$ and $Z^{-1} = 2\sinh\frac{\omega}{2T}$. The two universes of the entangled pair evolve, in the observer representation of each single universe, in thermal equilibrium with a temperature of entanglement given by

$$T \equiv T(a,\varphi) = \frac{\omega(a,\varphi)}{2\ln\frac{1}{\Gamma(a,\varphi)}}, \tag{96}$$

where, $\Gamma(a,\varphi) \equiv \tanh r(a,\varphi)$. The total energy reads

$$E(a) = \omega(a)(\langle N\rangle + \frac{1}{2}), \tag{97}$$

where, $\langle N\rangle \equiv |v_{\omega}|^2$. The variation of the quantum informational analogues to the work, W, and heat, Q, now read

$$\delta W = \delta\omega\,(\langle N\rangle + \frac{1}{2}) \approx \frac{\partial\omega(a,\varphi)}{\partial a}(\langle N\rangle + \frac{1}{2})\,da, \tag{98}$$

$$\delta Q = \omega\,\delta\langle N\rangle \approx \omega(a,\varphi)\frac{\partial\langle N\rangle}{\partial a}\,da, \tag{99}$$

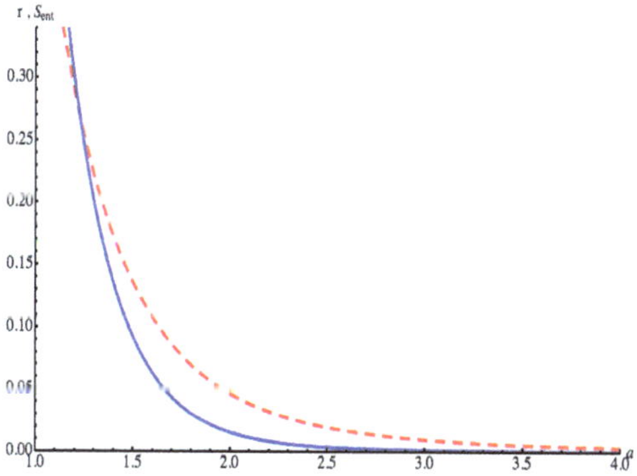

Figure 7. Parameter of squeezing, r (dashed line), and entropy of entanglement, S_{ent} (continuous line), with respect to the value of the scale factor, a.

where in the last equalities it has been taken into account that for a slow-varying field, $\delta\omega \approx \dot{\omega}\, da$ and $\delta\langle N\rangle \approx \langle \dot{N}\rangle\, da$. From Eqs. (97-99) it can be checked that the first principle of thermodynamics, $\delta E = \delta W + \delta Q$, is directly satisfied. The entropy of entanglement, Eq. (75), reads

$$S_{ent}(a,\varphi) = |\mu_\omega(a,\varphi)|^2 \ln |\mu_\omega(a,\varphi)|^2 - |\nu_\omega(a,\varphi)|^2 \ln |\nu_\omega(a,\varphi)|^2, \tag{100}$$

with, $|\mu_\omega(a,\varphi)| = \cosh r(a,\varphi)$ and $|\nu_\omega(a,\varphi)| = \sinh r(a,\varphi)$. Therefore, like in Sec. 3.3, the second principle of thermodynamics is also satisfied because the entropy production vanishes for any values of the scale factor and the scalar field, i.e. $\sigma \equiv \sigma(a,\varphi) = 0$.

4.3.1. *Entropy of entanglement as an arrow of time for single universes*

Let us summarize the general picture described so far. The multiverse stays in a squeezed vacuum state which is the product state of the wave functions that correspond to the state of pairs of entangled i-universes (see Eq. (39)), where the index i labels all the species of universes considered in the multiverse. The multiverse stays therefore in a highly non-classical state. Furthermore, the quantum entropy of a pure state is zero and, therefore, there is no thermodynamical arrow of time in the multiverse. Let us recall that, in the third quantization formalism, the scale factor was just taken as a formal time-like variable given by the Lorentzian structure of the minisupermetric. However, the minisuperspace is not space-time and, therefore, the scale factor has no meaning of a physical (i.e. a measurable) time, a priori, in the multiverse. It might well be said that (physical) time and (physical) evolution are concepts that really make sense within a single universe.

For an observer inside a universe, this is described by a thermal state which is indistinguishable from a classical mixture (see Eq. (73), and the comments thereafter), i.e. it is seen as a classical universe. The entropy of entanglement for a single universe is a monotonic function of the scale factor. However, the entropy production identically vanishes for any increasing or decreasing rate of the scale factor so that the customary formulation of the second principle of thermodynamics does not impose any arrow of time in the universe within the present approach. Although the universe can be seen as a classical mixture by an observer inside the universe, its quantum state has been obtained from a highly non-classical state. Thus, it would not be expected that the classical constraint $\sigma \geq 0$ would impose any arrow of time in the model.

The second principle of entanglement thermodynamics [58] does provide us with an arrow of time for single universes. In the quantum multiverse, it can be reparaphrased as follows: by local operations and classical communications alone, the amount of entanglement between the universes cannot increase. Let us recall that by *local* operations we mean in the multiverse anything that happens within a single universe, i.e. everything we can observe. Therefore, the growth of cosmic structures, particle interactions and even the presence of life in the universe cannot increase the amount of entanglement between a pair of entangled universes provided that all these features are due to local interactions. They should decrease the rate of entanglement in a non-reversible universe with dissipative processes, actually.

The amount of entanglement between the pair of universes only decreases for growing values of the scale factor (see Fig. 7). Thus, the second law of the entanglement thermodynamics implies that the universe has to expand once it is created in an entangled pair, as seen by an observer inside the universe. Furthermore, if the classical thermodynamics and the thermodynamics of entanglement were related, it could be followed that the negative change of entropy would be balanced by the creation of cosmic structures and other local processes that increase the local (classical) entropy. The decrease of the entropy of entanglement is larger for a small value of the scale factor. Then, the growth of local structures in the universe would be favored in the earliest phases of the universe, as it is expected.

4.3.2. *Energy of entanglement and the vacuum energy of the universe*

In the model being considered, $\sigma \equiv dS - \frac{\delta Q}{T}$ =0, and thus, the variation of the entropy of entanglement is related to the quantum informational heat, Q, by

$$dS = \frac{\delta Q}{T}. \tag{101}$$

Eq. (101) can be compared with the equation that is customary used to define the energy of entanglement [42, 51–53], $dE_{ent} = TdS$. Then, in the case being considered, we can identify the energy of entanglement, E_{ent}, with the informational heat, Q, and interpret it as a vacuum energy for each single universe of an entangled pair. It is given by the integral of Eq. (99), with

$$\langle N \rangle \approx \frac{9\, e^{\pm \frac{2}{3V(\varphi)}}}{16V(\varphi)} \frac{1}{a^8} + \sinh^2 \frac{1}{3V(\varphi)}, \tag{102}$$

where it has been used that, $\mathcal{M}(a) = a$ and $\omega = a\sqrt{a^2 V(\varphi) - 1} \approx a^2 \sqrt{V(\varphi)}$, in units for which $\hbar = 1$. For a slow-varying field, $\varphi \approx \varphi_0$ and $\delta \langle N \rangle \approx \frac{\partial \langle N \rangle}{\partial a} da$, and therefore

$$\delta Q \approx \omega \frac{\partial \langle N \rangle}{\partial a} da = -\frac{9\, e^{\pm \frac{2}{3V(\varphi_0)}}}{2\sqrt{V(\varphi_0)}} a^{-7} da, \tag{103}$$

whose integration yields

$$E_{ent} = Q(a, \varphi_0) = \frac{3}{4} \frac{e^{\pm \frac{2}{3V(\varphi_0)}}}{\sqrt{V(\varphi_0)}} a^{-6}. \tag{104}$$

The energy of entanglement (104) provides us with a curve that might be compared with the evolution of the vacuum energy of the universe. From Eq. (104), it can be seen that the vacuum energy would follow a different curve depending on whether the tunneling condition or the no-boundary condition is imposed on the state of a single universe. The boundary condition imposed on a single universe might therefore be discriminated from observational data, at least in principle. However, the model being considered is unrealistic for at least two reasons. First, after the inflationary stage the universe becomes hot [46, 47] and the slow-roll approximation is no longer valid. Secondly, if the energy of entanglement is to be considered as a vacuum energy, it should have been considered as a variable of the model from the beginning. More realistic matter fields and the backreaction should be taken into account to make a first serious attempt to observational fitting. However, the important thing that is worth noticing is that the vacuum energy of entanglement might thus be tested as well as the whole multiverse proposal. Furthermore, different boundary conditions would provide us with different curves for the energy of entanglement along the entire evolution of the universe. Therefore, the boundary conditions of the whole multiverse might be tested as well by direct observation, which is a completely novel feature in quantum cosmology.

5. Conclusions: The physical multiverse

In this chapter, we have presented a quantum mechanical description of a multiverse made up of large and disconnected regions of the space-time, called universes, with a high degree of symmetry. We have obtained, within the framework of a third quantization formalism, a wave function that quantum mechanically represents the state of the whole multiverse, and an appropriate boundary condition for the state of the multiverse has allowed us to interpret it as formed by entangled pairs of universes.

If universes were entangled to each other, then, the violation of classical inequalities like the Bell's inequalities could no longer be associated to the concepts of locality or non-locality because there is not generally a common space-time among the universes of the quantum multiverse. It would rather be related to the independence or interdependence of the quantum states that represent different universes. Furthermore, the complementary principle of quantum mechanics, being applied to the space-time as a whole, enhances us to: i) look for an appropriate boundary condition for which universes should be described as individual entities forcing us to consider a multiverse; and, ii) take into account as well interference effects between the quantum states of two or more universes.

For a pair of entangled universes, the quantum thermodynamical properties of each single universe have been computed. In the scenario of a multiverse made up of entangled pairs of universes, the picture is the following: the multiverse may state in the pure state that corresponds to the product state of the ground states derived from the boundary condition imposed on the multiverse, for each type of single universes. Then, the entropy of the whole multiverse vanishes and there is thus no physical arrow of time in the multiverse. For single universes, however, it appears an arrow of time derived from the entropy of entanglement with their partner universes.

The entropy of entanglement decreases for an increasing value of the scale factor. The second principle of thermodynamics is however satisfied because the process is non-adiabatic, in the quantum informational sense, and the entropy production is zero. In fact, the entropy production is zero for any increasing or decreasing rate of the scale factor, imposing therefore no correlation between the cosmic arrow of time, which is given by expansion or contraction rate of a single universe, and the customary formulation of the second principle of thermodynamics. This is in contrast to what it happens inside a single universe, where there is a correlation between the cosmic arrow of time and the entropy of matter fields [27, 40]. Let us recall that the entropy of entanglement is a quantum feature having no classical analogue and, thus, it is not expected that it imposes an arrow of time through the customary, i.e. classical, formulation of the second principle of thermodynamics.

The second principle of entanglement thermodynamics, which states [58] that the entropy of entanglement cannot be increased by any *local* operation and any classical communication alone, does impose an arrow of time on single universes [60]. It should be noticed that by *local* we mean in the multiverse anything that happens in a single universe. Therefore, everything that we observe, i.e. the creation of particles, the growth of cosmic structures, and even life, cannot make the inter-universal entanglement to grow provided that all these processes are internal to a single universe. In an actual and non-reversible universe they should induce a decreasing of the entropy of inter-universal entanglement, enhancing therefore the expansion of the universe that would induce a correlation between the growth of cosmic structures and the entanglement arrow of time.

In the model presented in this chapter, the energy of entanglement between a pair of entangled universes provides us with a vacuum energy for each single universes. The energy of entanglement of the universe is high in the early stage of the universes becoming very small at later times. That behavior might be compatible with an initial inflationary universe, for which a high value of the vacuum energy is assumed, that would eventually evolve to a state with a very small value of the cosmological constant, like the current state of the universe. However, it is not expected that such a simple model of the universe would fit with actual observational tests. A more realistic model of the universes that form the multiverse, in which genuine matter fields were considered, is needed to make a serious attempt of observational fitting. However, the fact that inter-universal entanglement provides us with testable properties of our universe opens the door to future developments that would make falseable the multiverse proposal giving an observational support to the quantum multiverse.

In conclusion, the question of whether the multiverse is a physical theory or just a mathematical construction, derived however from the general laws of physics, holds on whether the existence of other universes may affect the properties of the observable universe. Inter-universal entanglement is a novel feature that supplies us with new explanations for unexpected cosmic phenomena and it might allow us to test the whole multiverse proposal. It will be the future theoretical developments and the improved observational tests what will make us to decide whether to adopt or deny such a cosmological scenario.

Acknowledgements

In memory of Prof. González Díaz, who always encouraged us to adopt a fearless attitude as well in science as in everyday life. The author wishes to thank I. Garay and P. V. Moniz for the revision of the text and their valuable comments.

Author details

Salvador J. Robles Pérez

Grupo de Relatividad y Cosmología, Instituto de Física Fundamental - Consejo Superior de Investigaciones Científicas (IFF-CSIC), Madrid, Spain, and Estación Ecológica de Biocosmología (EEBM), Medellín, Spain

References

[1] R. Alicki et al. Thermodynamics of quantum informational systems - Hamiltonian description. *Open Syst. Inf. Dyn.*, 11:205–217, 2004.

[2] M. Ansmann et al. Violation of Bell's inequality in Josephson phase qubits. *Nature*, 461:504–506, 2009.

[3] A. Aspect, J. Dalibard, and G. Roger. Experimental test of Bell's inequalities using time-varying analyzers. *Phys. Rev. Lett.*, 49:1804–1807, 1982.

[4] A. O. Barvinsky. Why there is something rather than nothing (out of everything)? *Phys. Rev. Lett.*, 99:071301, 2007.

[5] A. O. Barvinsky and A. Y. Kamenshchik. Cosmological landscape from nothing: Some like it hot. *JCAP*, 0609:014, 2006.

[6] A. O. Barvinsky and A. Y. Kamenshchik. Cosmological landscape and Euclidean quantum gravity. *J. Phys. A*, 40:7043–7048, 2007.

[7] J. S. Bell. On the Einstein Podolsky Rosen paradox. *Physics*, 1:195, 1964.

[8] J. S. Bell. *Speakable and unspeakable in quantum mechanics.* Cambridge University Press, Cambridge, UK, 1987.

[9] R. Bousso and J. Polchinski. Quantization of four-form fluxes and dynamical neutralization of the cosmological constant. *JHEP*, 0006:006, 2000.

[10] V. Buzek. Jaynes-Cummings model with intensity-dependent coupling interacting with squeezed vacuum. *Phys. Lett. A*, 139:231–235, 1989.

[11] C. M. A. Dantas, I. A. Pedrosa, and B. Baseia. Harmonic oscillator with time-dependent mass and frequency and a perturbative potential. *Phys. Rev. A*, 45:1320, 1992.

[12] A. Einstein, B. Podolsky, and N. Rosen. Can quantum-mechanical description of physical reality be considered complete? *Phys. Rev.*, 47:777–780, 1935.

[13] L. J. Garay. Thermal properties of space-time foam. *Phys. Rev. D*, 58:124015, 1998.

[14] J. Gemmer et al. *Quantum thermodynamics.* Springer-Verlag, Berlin, Germany, 2009.

[15] P. F. González-Díaz. Wormholes and ringholes in a dark-energy universe. *Phys. Rev. D*, 68:084016, 2003.

[16] P. F. González-Díaz. Thermal properties of time machines. *Phys. Rev. D*, 85(105026), 2012.

[17] P. F. González-Díaz and A. Alonso-Serrano. Observing another universe through ringholes and Klein-bottle holes. *Phys. Rev. D*, 84:023008, 2011.

[18] P. F. González-Díaz and S. Robles-Pérez. Quantum theory of an accelerating universe. *Int. J. Mod. Phys. D*, 17:1213–1228, 2008.

[19] J. R. III Gott and Li-Xin Li. Can the universe create itself? *Phys. Rev. D*, 58:023501, 1998.

[20] J. J. Halliwell. Correlations in the wave function of the universe. *Phys. Rev. D*, 36:3626–3640, 1987.

[21] J. J. Halliwell. Decoherence in quantum cosmology. *Phys. Rev. D*, 39:2912–2923, 1989.

[22] J. J. Halliwell. Introductory lectures on quantum cosmology. In S. Coleman, J. B. Hartle, T. Piran, and S. Weinberg, editors, *Quantum Cosmology and Baby Universes*, volume 7. World Scientific, London, UK, 1990.

[23] J. B. Hartle. Spacetime quantum mechanics and the quantum mechanics of spacetime. In B. Julia and J. Zinn-Justin, editors, *Gravitation and Quantization: Proceedings of the 1992 Les Houches Summer School*. North Holland, Amsterdam, 1995.

[24] J. B. Hartle and S. W. Hawking. Wave function of the universe. *Phys. Rev. D*, 28:2960, 1983.

[25] S. W. Hawking. Spacetime foam. *Nucl. Phys. B*, 144:349–362, 1978.

[26] S. W. Hawking. Quantum cosmology. In B. S. De Witt and R. Stora, editors, *Relativity, groups and topology II, Les Houches, Session XL, 1983*. Elsevier Science Publishers B. V., 1984.

[27] S. W. Hawking. Arrow of time in cosmology. *Phys. Rev. D*, 32:2489–2495, 1985.

[28] S. W. Hawking. Wormholes in spacetime. *Phys. Rev. D*, 37:904–910, 1988.

[29] S. W. Hawking and T. Hertog. Populating the landscape: A top-down approach. *Phys. Rev. D*, 73:123527, 2006.

[30] S. W. Hawking, R. Laflamme, and G. W. Lyons. Origin of time asymmetry. *Phys. Rev. D*, pages 1546–1550, 1992.

[31] R. Holman, L. Mersini-Houghton, and T. Takahashi. Cosmological avatars of the landscape I, II. *Phys. Rev. D*, 77:063510,063511, 2008.

[32] R. Horodecki et al. Quantum entanglement. *Rev. Mod. Phys.*, 81:865–942, 2009.

[33] A. von Humboldt. *Cosmos*. Harper and Brothers, New York, 1845.

[34] C. J. Isham. Canonical quantum gravity and the problem of time. 1992.

[35] G. Jaeger. *Quantum information.* Springer, Berlin, Germany, 2007.

[36] D. Jennings and T. Rudolph. Entanglement and the thermodynamic arrow of time. *Phys. Rev. E*, 81:061130, 2010.

[37] C. Kiefer. Decoherence in quantum electrodynamics and quantum gravity. *Phys. Rev. D*, 46:1658–1670, 1992.

[38] C. Kiefer. Quantum cosmology and the arrow of time. *Braz. J. Phys.*, 35:296–299, 2005.

[39] C. Kiefer. *Quantum gravity.* Oxford University Press, Oxford, UK, 2007.

[40] C. Kiefer and H. D. Zeh. Arrow of time in a recollapsing quantum universe. *Phys. Rev. D*, 51:4145–4153, 1995.

[41] S. P. Kim and D. N. Page. Classical and quantum action-phase variables for time-dependent oscillators. *Phys. Rev. A*, 64:012104, 2001.

[42] J-W Lee et al. Dark energy from vacuum entanglement. *JCAP*, 0708:005, 2007.

[43] H. R. Lewis and W. B. Riesenfeld. An exact quantum theory of the time-dependent harmonic oscillator and of a charged particle in a time-dependent electromagnetic field. *J. Math. Phys.*, 10:1458–1473, 1969.

[44] Li-Xin Li. Two open universes connected by a wormhole: exact solutions. *J. Geom. Phys.*, 40:154–160, 2001.

[45] A. Linde. Eternally existing self-reproducing chaotic inflationary universe. *Phys. Lett. B*, 175(4):395–400, 1986.

[46] A. Linde. The inflationary multiverse. In B. Carr, editor, *Universe or Multiverse*, chapter 8. Cambridge University Press, Cambridge, UK, 2007.

[47] A. Linde. Inflationary cosmology. *Lect. Notes Phys.*, 738:1–54, 2008.

[48] L. Mersini-Houghton. The void: Imprint of another universe? *New Scientist*, pages 11–24, 2007.

[49] L. Mersini-Houghton. Thoughts on defining the multiverse. 2008.

[50] C. W. Misner, K. S. Thorne, and J. A. Wheeler. *Gravitation.* W. H. Freeman and Company, San Francisco, 1970.

[51] S. Mukohyama. Comments on entanglement entropy. *Phys. Rev. D*, 58:104023, 1998.

[52] S. Mukohyama, M. Seriu, and H. Kodama. Can the entanglement entropy be the origin of black-hole entropy? *Phys. Rev. D*, 55:7666–7679, 1997.

[53] R. Müller and C. O. Lousto. Entanglement entropy in curved spacetimes with event horizons. *Phys. Rev. D*, 52:4512–4517, 1995.

[54] J. v Neumann. *Mathematical Foundations of Quantum Mechanics*. Princeton University Press, New Jersey, US, 1996.

[55] T. J. Park. Canonical transformations for time-dependent harmonic oscillators. *Bull. Korean Chem. Soc.*, 25(285-288), 2004.

[56] M. H. Partovi. Entanglement versus stosszahlansatz: Disappearance of the thermodynamic arrow in a high-correlation environment. *Phys. Rev. E*, 77:021110, 2008.

[57] I. A. Pedrosa. Comment on "coherent states for the time-dependent harmonic oscillator". *Phys. Rev. D*, 36:1279, 1987.

[58] M. B. Plenio and V. Vedral. Teleportation, entanglement and thermodynamics in the quantum world. *Comtemp. Phys.*, 39:431–446, 1998.

[59] M. D. Reid and D. F. Walls. Violations of classical inequalities in quantum optics. *Phys. Rev. A*, 34:1260–1276, 1986.

[60] S. Robles-Pérez. Entanglement arrow of time in the multiverse (submitted). 2012.

[61] S. Robles-Pérez and P. F. González-Díaz. Quantum state of the multiverse. *Phys. Rev. D*, 81:083529, 2010.

[62] S. Robles-Pérez and P. F. González-Díaz. Quantum entanglement in the multiverse (submitted). 2011.

[63] M. A. Rowe et al. Experimental violation of a Bell's inequality with efficient detection. *Nature*, 409:791–794, 2001.

[64] E. Schrödinger. Discussion of probability relations between separated systems. *Mathematical proceedings of the Cambridge Philosophical Society*, 31:555–563, 1936.

[65] E. Schrödinger. Probability relations between separated systems. *Mathematical proceedings of the Cambridge Philosophical Society*, 32:446–452, 1936.

[66] M. O. Scully and M. S. Zubairy. *Quantum optics*. Cambridge University Press, Cambridge, UK, 1997.

[67] D. Sheng et al. Quantum harmonic oscillator with time-dependent mass and frequency. *Int. J. Theor. Phys.*, 34:355–368, 1995.

[68] D. W. Singer. *Giordano Bruno, His Life and Thought, With Annotated translation of His Work - On the infinite Universe and Worlds (1584)*. Schuman, NY, USA, 1950.

[69] D-Y. Song. Unitary relation between a harmonic oscillator of time-dependent frequency and a simple harmonic oscillator with and without an inverse-square potential. *Phys. Rev. A*, 62:014103, 2000.

[70] A. Strominger. Baby universes. In S. Coleman, J. B. Hartle, T. Piran, and S. Weinberg, editors, *Quantum Cosmology and Baby Universes*, volume 7. World Scientific, London, UK, 1990.

[71] L. Susskind. The anthropic landscape of string theory. 2003.

[72] M. Tegmark. Parallel universes. *Scientific American*, 288(5), May 2003.

[73] M. Tegmark. The multiverse hierarchy. In B. Carr, editor, *Universe or Multiverse*, chapter 7. Cambridge University Press, Cambridge, UK, 2007.

[74] W. Tittel et al. Experimental demonstration of quantum correlations over more than 10 km. *Phys. Rev. A*, 57:3229, 1998.

[75] V. Vedral. *Introduction to quantum information science*. Oxford University Press, Oxford, UK, 2006.

[76] D. G. Vergel and J. S. Villaseñor. The time-dependent quantum harmonic oscillator revisited: Applications to quantum field theory. *Ann. Phys.*, 324:1360, 2009.

[77] A. Vilenkin. Creation of universes from nothing. *Phys. Lett. B*, 117:25–28, 1982.

[78] A. Vilenkin. Boundary conditions in quantum cosmology. *Phys. Rev. D*, 33:3560–3569, 1986.

[79] A. Vilenkin. Quantum cosmology and the initial state of the universe. *Phys. Rev. D*, 37:888–897, 1988.

[80] R. M. Wald. *General Relavity*. The University of Chicago Press, Chicago, USA, 1984.

[81] D. F. Walls. Squeezed states of light. *Nature*, 306:141–146, 1986.

[82] D. F. Walls and G. J. Milburn. *Quantum optics*. Springer-Verlag, Berlin, Germany, 2008.

[83] G. Weihs et al. Violation of Bell's inequality under strict Einstein locality conditions. *Phys. Rev. Lett.*, 81:5039–5043, 1998.

[84] J. A. Wheeler. Geons. *Phys. Rev.*, 97:511–536, 1955.

[85] J. A. Wheeler. On the nature of quantum geometrodynamics. *Ann. Phys.*, 2:604–614, 1957.

[86] K. Wodkiewicz and J. H. Eberly. Coherent states, squeezed fluctuations, and the su(2) and su(1) groups in quantum-optics applications. *J. Opt. Soc. Am. B*, 2:458–466, 1985.

[87] H. P. Yuen. Generalized coherent states and the statistics of two-photon lasers. *Phys. Lett. A*, 51:1–2, 1975.

[88] H. P. Yuen. Two-photon coherent states of the radiation field. *Phys. Rev. A*, 13:2226–2243, 1976.

Chapter 8

Introduction to Palatini Theories of Gravity and Nonsingular Cosmologies

Gonzalo J. Olmo

Additional information is available at the end of the chapter

1. Introduction

The impact of Einstein's fundamental idea of gravitation as a curved space-time phenomenon on our current understanding of the Universe has been enormously successful. A key aspect of his celebrated theory of General Relativity (GR) is that the spatial sections of four dimensional space-time need not be Euclidean. The Minkowskian description is just an approximation valid on (relatively) local portions of space-time. On larger scales, however, one must consider deformations induced by the matter on the geometry, which must be dictated by some set of field equations. In this respect, the predictions of GR are in agreement with experiments in scales that range from millimeters to astronomical units, scales in which weak and strong field phenomena can be observed [39]. The theory is so successful in those regimes and scales that it is generally accepted that it should work also at larger and shorter scales, and at weaker and stronger regimes. The validity of these assumptions, obviously, is not guaranteed *a priori* regardless of how beautiful and elegant the theory might appear. Therefore, not only must we keep confronting the predictions of the theory with experiments and/or observations at new scales, but also we have to demand theoretical consistency with the other physical interactions and, in particular, in the quantum regime.

For the above reasons, we believe that scrutinizing the implicit assumptions and mathematical structures behind the classical formulation of GR could help better understand the starting point of some current approaches that go beyond our standard model of gravitational physics. At the same time, this could provide new insights useful to address from a different perspective some current open questions, such as the existence of black hole and big bang singularities or the cosmic speedup problem. In this sense, Einstein himself stated that *"the question whether the structure of [the spacetime] continuum is Euclidean, or in accordance with Riemann's general scheme, or otherwise, is . . . a physical question which must be answered by experience, and not a question of a mere convention to be selected on practical grounds"* [10]. From these words it follows that questioning the regime of applicability of the Riemannian nature of the geometry associated with the gravitational field and considering more general frameworks are legitimate questions that should be explored by all available means (theoretical and experimental). These are some of the basic points to be addressed in this work.

In this chapter we explore in some detail the implications of relaxing the Riemannian condition on the geometry by allowing the connection to be determined from first principles, not by choice or convention. This approach, known as metric-affine or Palatini formalism [24], assumes that metric and connection are equally fundamental and independent geometrical entities. In consequence, any geometrical theory of gravity formulated in this approach must provide enough equations to determine the form of the metric and the connection (within the unavoidable indeterminacy imposed by the underlying gauge freedom). We derive and discuss the field equations of a rather general family of Palatini theories and then focus on two particular subfamilies which have attracted special attention in recent years, namely, $f(R)$ and $f(R,Q)$ theories. The interest in studying these particular theories lies in their ability to avoid (or soften in some cases) big bang and black hole singularities and their relation with recent approaches to quantum gravity. Here we will focus on the early-time cosmology of such theories.

The content is organized as follows. We begin by briefly reviewing in section 2 the basics of differentiable manifolds with affine and metric structures, to emphasize that metric and connection are equally fundamental and independent geometrical objects. In section 3 a derivation of the field equations for a generic action depending on the metric and the Riemann tensor is presented taking into account also the presence of torsion. In section 4 we discuss a particular family of Lagrangians of the form $f(R, R_{\mu\nu}R^{\mu\nu})$ in combination with perfect fluid matter, and prepare the notation and field equations needed to study the dynamics of those theories. We then focus on the early-time characteristics of isotropic and anisotropic homogeneous cosmologies 5 and show that nonsingular bouncing solutions exist for $f(R)$ and $f(R,Q)$ models (subsections 5.5 and 5.6, respectively). We conclude with a discussion of the results presented and point out some open questions that should be addressed in the future.

2. Differentiable manifolds, affine connections, and the metric

In this section we quickly review some of the mathematical structures needed to construct a geometric theory of the gravitational interactions. The goal is to put forward that metric and connection are equally fundamental and independent geometrical entities, an aspect usually overlooked in the construction of phenomenological extensions of GR. We will thus be more sketchy than mathematically accurate. For a more exhaustive and precise discussion of these topics see your favorite book on differentiable manifolds (or, for instance, [20]).

In the geometric description of gravitational theories, one begins by identifying physical events with points on an n-dimensional manifold $\mathcal{M}$. The next natural step is to provide this manifold with a differentiable structure. One then labels the points $p \in \mathcal{M}$ with a set of charts $(\mathcal{U}_i, \varphi_i)$, where the $\mathcal{U}_i$ are subsets of $\mathcal{M}$ and φ_i are maps from $\mathcal{U}_i$ to $\mathbb{R}^n$ (or an open subset of $\mathbb{R}^n$) such that every $p \in \mathcal{M}$ lies in at least one of the charts $(\mathcal{U}_i, \varphi_i)$. If for any two charts $(\mathcal{U}_i, \varphi_i)$ and $(\mathcal{U}_j, \varphi_j)$ that overlap at some nonzero subset of points the map $\varphi_i \circ \varphi_j{}^{-1}$ is not just continuous but differentiable, then we say that $\mathcal{M}$ is a differentiable manifold. Since the Euclidean view of vectors as arrows connecting two points of the manifold is not valid in general, to get a consistent definition we need to introduce first the concept of curve and tangent vector to a curve at a point. We thus say that a smooth curve $\gamma(t)$ in $\mathcal{M}$ is a differentiable map that to each point of a segment associates a point in $\mathcal{M}$, $\gamma(t)$:

$t \in [0,1] \to \mathcal{M}$. In a chart $(\mathcal{U}, \varphi)$, the points of the curve have the following coordinate representation: $x = \varphi(p_t) = \varphi \circ \gamma(t)$. If we consider now a function f on $\mathcal{M}$, where f is a map that to every $p \in \mathcal{M}$ assigns a real number ($f : \mathcal{M} \to \mathbb{R}$), the rate of change of f along the curve $\gamma(t)$ using the coordinates of the chart $(\mathcal{U}, \varphi)$ is given by

$$\frac{df(\varphi \circ \gamma(t))}{dt} = \frac{df(x(t))}{dt} = \frac{\partial f}{\partial x^\mu}\frac{dx^\mu(t)}{dt} \equiv X^\mu(t)\frac{\partial f}{\partial x^\mu} , \quad (1)$$

where we have defined the components of the tangent vector to the curve in this chart as $X^\mu \equiv dx^\mu(t)/dt$. Vectors can thus be seen as differential operators $X = X^\mu \partial_\mu$ whose action on functions is of the form $X[f] = X^\mu \partial_\mu f$, thus providing a natural notion of directional derivative for functions. The set $\{e_\mu \equiv \partial_\mu\}$ defines a (coordinate) basis of the tangent space of vectors at the point p, which we denote $T_p\mathcal{M}$. Obviously, vectors exist without specifying the coordinates. Under changes of coordinates, we have $V = V^\mu e_\mu = \tilde{V}^\alpha \tilde{e}_\alpha = \tilde{V}^\alpha \frac{\partial x^\mu}{\partial \tilde{x}^\alpha} e_\mu$, which implies the well-known transformation law $V^\mu = \tilde{V}^\alpha \frac{\partial x^\mu}{\partial \tilde{x}^\alpha}$ for the vector components. When a vector is assigned smoothly to each point of $\mathcal{M}$, it is called a vector field over $\mathcal{M}$. Each component of a vector field is thus a smooth function from $\mathcal{M} \to \mathbb{R}$. Given a vector field X, an integral curve of X is defined as the curve whose tangent vector coincides with X. For infinitesimal displacements of magnitude ϵ in the direction of X, a given point p of coordinates x^μ becomes $\sigma^\mu_\epsilon(x) = x^\mu + \epsilon X^\mu(x)$. This transformation also induces a correspondence between vectors of the tangent spaces $T_x\mathcal{M}$ and $T_{\sigma_\epsilon(x)}\mathcal{M}$. The effect of these transformations on a vector field $Y(x)$ leads to the concept of Lie derivative, whose action on vector fields is defined as

$$\mathcal{L}_X Y = [X^\nu \partial_\nu Y^\mu(x) - Y^\nu \partial_\nu X^\mu(x)]\, e_\mu \equiv [X, Y] . \quad (2)$$

This derivative operator is independent of the choice of coordinates and follows naturally from the differential structure of the manifold. It satisfies a number of useful properties such as bilinearity in its two arguments, $\mathcal{L}_X(Y+Z) = \mathcal{L}_X Y + \mathcal{L}_X Z$, $\mathcal{L}_{X+Y} Z = \mathcal{L}_X Z + \mathcal{L}_Y Z$, and the chain rule $\mathcal{L}_X fY = (\mathcal{L}_X f)Y + f\mathcal{L}_X Y$, with $\mathcal{L}_X f = X[f]$. Though the Lie derivative provides a natural *directional derivative* for functions, it does not work in the same way for vectors and tensors of higher rank. In fact, since the partial derivatives of the vector X appear explicitly in $\mathcal{L}_X Y$, two vectors whose components at a given point have the same values but whose partial derivatives at the point differ do not yield a vector that points in the same direction, i.e., they are not proportional. Therefore, in order to introduce a proper notion of *directional derivative* for vectors and tensors, we need to introduce a new structure called **connection** which specifies how vectors (and tensors in general) are transported along a curve.

Manifolds with a connection. We are thus going to introduce a derivative operator, which we denote by ∇, such that given two vector fields X and Y we obtain a new vector field Z defined by $Z \equiv \nabla_X Y$. This derivative operator must be bilinear in its two arguments, $\nabla_X(Y + Z) = \nabla_X Y + \nabla_X Z$, $\nabla_{X+Y} Z = \nabla_X Z + \nabla_Y Z$, must satisfy the chain rule $\nabla_X(fY) = (\nabla_X f)Y + f\nabla_X Y$, with $\nabla_X f = X[f]$, and must also behave as a natural directional derivative in the sense that $\nabla_{fX} Y = f\nabla_X Y$ to guarantee that any two proportional vectors yield a result that points in the same direction. In a given coordinate basis, we have $\nabla_X Y = X^\mu \nabla_{e_\mu}(Y^\nu e_\nu) = X^\mu \left(e_\mu[Y^\nu]e_\nu + Y^\nu \nabla_{e_\mu} e_\nu\right)$. If our manifold is $m-$dimensional, defining m^3 functions called connection coefficients $\Gamma^{\lambda}_{\mu\nu}$ by $\nabla_{e_\mu} e_\nu \equiv \Gamma^{\lambda}_{\mu\nu} e_\lambda$ we find that the last requirement, $\nabla_{fX} Y = f \nabla_X Y$, is naturally satisfied. We thus find that

$$\nabla_X Y = X^\mu \left[\frac{\partial Y^\lambda}{\partial x^\mu} + \Gamma^{\lambda}_{\mu\nu} Y^\nu \right] e_\lambda \,. \tag{3}$$

The connection coefficients specify how the basis vectors change from point to point and, in principle, can be arbitrarily defined. Under changes of coordinates, these coefficients transform as follows:

$$\nabla_{e_\mu} e_\nu \equiv \Gamma^{\lambda}_{\mu\nu} e_\lambda = \frac{\partial \tilde{x}^\alpha}{\partial x^\mu} \nabla_{\tilde{e}_\alpha} \left(\frac{\partial \tilde{x}^\beta}{\partial x^\nu} \tilde{e}_\beta \right) = \frac{\partial \tilde{x}^\alpha}{\partial x^\mu} \left[\frac{\partial x^\lambda}{\partial \tilde{x}^\nu} \frac{\partial^2 \tilde{x}^\gamma}{\partial x^\lambda \partial x^\nu} + \frac{\partial \tilde{x}^\beta}{\partial x^\nu} \tilde{\Gamma}^{\gamma}_{\alpha\beta} \right] \tilde{e}_\gamma = \Gamma^{\lambda}_{\mu\nu} \frac{\partial \tilde{x}^\gamma}{\partial x^\lambda} \tilde{e}_\gamma \,, \tag{4}$$

which implies

$$\Gamma^{\lambda}_{\mu\nu} = \frac{\partial x^\lambda}{\partial \tilde{x}^\gamma} \frac{\partial \tilde{x}^\alpha}{\partial x^\mu} \frac{\partial \tilde{x}^\beta}{\partial x^\nu} \tilde{\Gamma}^{\gamma}_{\alpha\beta} + \frac{\partial x^\lambda}{\partial \tilde{x}^\gamma} \frac{\partial^2 \tilde{x}^\gamma}{\partial x^\mu \partial x^\nu} \,. \tag{5}$$

This transformation law indicates that the connection coefficients do not transform as tensorial quantities. Therefore, the connection cannot have an intrinsic geometrical meaning as a measure of how much a manifold is curved. As intrinsic geometric objects, we can define the torsion tensor

$$T(X,Y) = \nabla_X Y - \nabla_Y X - [X,Y] \,, \tag{6}$$

and the Riemann curvature tensor

$$R(X,Y)Z = \nabla_X \nabla_Y Z - \nabla_Y \nabla_X Z - \nabla_{[X,Y]} Z \,. \tag{7}$$

In a coordinate basis, these tensors have the following components:

$$T(e_\mu, e_\nu) = \left(\Gamma^{\lambda}_{\mu\nu} - \Gamma^{\lambda}_{\nu\mu} \right) e_\lambda \,, \tag{8}$$

$$R(e_\mu, e_\nu) e_\lambda = \left[\partial_\mu \Gamma^{\beta}_{\nu\lambda} - \partial_\nu \Gamma^{\beta}_{\mu\lambda} + \Gamma^{\kappa}_{\nu\lambda} \Gamma^{\gamma}_{\mu\kappa} - \Gamma^{\kappa}_{\mu\lambda} \Gamma^{\gamma}_{\nu\kappa} \right] e_\gamma \,. \tag{9}$$

With the introduction of the connection, one can define the notion of parallel transport. Given a curve $\gamma(t)$ such that its tangent vector in a given chart has coordinates $X^\mu = dx^\mu(t)/dt$, we say that a vector Y is parallel transported along $\gamma(t)$ if $\nabla_X Y = 0$. In components, this equation reads $\frac{dY^\mu}{dt} + \Gamma^{\mu}_{\alpha\beta} \frac{dx^\alpha(t)}{dt} Y^\beta = 0$, where $d/dt \equiv X^\mu \partial_\mu$. Geodesics are defined as those curves which are parallel transported along themselves, namely, $\nabla_X X = 0$ or $\frac{dX^\mu}{dt} + \Gamma^{\mu}_{\alpha\beta} X^\alpha X^\beta = 0$.

Manifolds with a metric. So far we have been able to construct a number of geometrical objects such as directional derivatives of vectors and tensors in general, the torsion and Riemann tensors, geodesic curves, ... without the need to introduce a metric tensor. A metric tensor provides a notion of distance between nearby points and allows, among other things, to determine lengths, angles, areas, and volumes of objects which are locally defined in space-time. Formally, a (pseudo-Riemannian) metric tensor is a symmetric bilinear form that at each $p \in \mathcal{M}$ satisfies $g_p(U,V) = g_p(V,U)$ for any two vectors $U, V \in T_p\mathcal{M}$ and $g_p(U,V) = 0$ for any $U \in T_p\mathcal{M}$ iff $V = 0$. The metric tensor allows to define an inner product between vectors and also gives rise to an isomorphism between $T_p\mathcal{M}$ and the dual space of one-forms $T_p^*\mathcal{M}$. In a coordinate basis, it can be represented by $g = g_{\mu\nu}dx^\mu \otimes dx^\nu$, where the differentials dx^μ form a basis of $T_p^*\mathcal{M}$. In manifolds with a metric, one can impose a particular relation between the metric and the connection by demanding that the scalar product of any two vectors which are parallel transported along any curve remains covariantly constant. This condition can be translated into[1] $\nabla_\mu g_{\alpha\beta} = 0$, which implies that (recall that $T^\rho_{\beta\sigma} \equiv \Gamma^\rho_{\beta\sigma} - \Gamma^\rho_{\sigma\beta}$)

$$2\Gamma^\lambda_{(\mu\nu)} + \left(T^\rho_{\nu\sigma}g_{\rho\mu} + T^\rho_{\mu\sigma}g_{\rho\nu}\right)g^{\sigma\lambda} = g^{\lambda\rho}\left[\partial_\mu g_{\rho\nu} + \partial_\nu g_{\rho\mu} - \partial_\rho g_{\mu\nu}\right] . \tag{10}$$

From the right-hand side of this equation, one defines the Levi-Civita connection as

$$L^\lambda_{\mu\nu} \equiv \frac{g^{\lambda\rho}}{2}\left[\partial_\mu g_{\rho\nu} + \partial_\nu g_{\rho\mu} - \partial_\rho g_{\mu\nu}\right] . \tag{11}$$

From this definition it follows that when the torsion vanishes, the connection is symmetric and coincides with the Levi-Civita connection. In that case, when $\Gamma^\lambda_{\mu\nu} = L^\lambda_{\mu\nu}$, we say that the associated geometry is Riemannian. It should be noted that though connections are not tensors, the difference between any two connections is a tensor. This, in particular, allowed us to construct the torsion tensor. With more generality, when the manifold is provided with a metric, any connection $\Gamma^\lambda_{\mu\nu}$ can be expressed as

$$\Gamma^\lambda_{\mu\nu} = L^\lambda_{\mu\nu} + A^\lambda_{\mu\nu} , \tag{12}$$

where $A^\lambda_{\mu\nu}$ is a tensor (which needs not be symmetric in its lower indices). Therefore, Palatini theories of gravity, in which metric and connection are regarded as independent fields, can be seen as theories in which an additional rank-three tensor field $A^\lambda_{\mu\nu}$ has been added to the gravitational Lagrangian.

3. Dynamics of Palatini theories

From the above quick review of the properties of differentiable manifolds with metric and affine structures, it is clear that metric and connection are equally fundamental and independent geometrical entities. In the construction of theories of gravity based on geometry, we will thus assume this independence and will require those theories to yield equations that allow to determine both the metric and the connection and the possible relations between them. For simplicity, we will assume that the matter is only coupled to the

---

[1] From now on we use the more standard notation $\nabla_\mu \equiv \nabla_{e_\mu}$

metric (which is consistent with the experimental tests of the equivalence principle [39]) but will allow an independent connection to appear in the gravitational sector of the theory. As pointed out above, this is equivalent to having, besides the metric, a rank-three gravitational tensor field. From a geometric perspective, this possibility seems much more natural and fundamental than considering, for instance, scalar fields in the gravitational sector, though scalar-tensor theories have traditionally received much more attention in the literature.

We begin by deriving the field equations of Palatini theories in a very general case and then consider some simplifications to make contact with the literature. For a generic Palatini theory in which the connection appears through the Riemann tensor or contractions of it, the action can be written as follows [25]

$$S = \frac{1}{2\kappa^2}\int d^4x\sqrt{-g}f(g_{\mu\nu}, R^{\alpha}{}_{\beta\mu\nu}) + S_m[g_{\mu\nu}, \psi] , \tag{13}$$

where S_m is the matter action, ψ represents collectively the matter fields, κ^2 is a constant with suitable dimensions (if $f = R$, then $\kappa^2 = 8\pi G$), and

$$R^{\alpha}{}_{\beta\mu\nu} = \partial_\mu\Gamma^{\alpha}_{\nu\beta} - \partial_\nu\Gamma^{\alpha}_{\mu\beta} + \Gamma^{\alpha}_{\mu\lambda}\Gamma^{\lambda}_{\nu\beta} - \Gamma^{\alpha}_{\nu\lambda}\Gamma^{\lambda}_{\mu\beta} \tag{14}$$

represents the components of the Riemann tensor, the field strength of the connection $\Gamma^{\alpha}_{\mu\beta}$. Note that since the connection is determined dynamically, i.e., we assume independence between the metric and affine structures of the theory, we cannot assume any *a priori* symmetry in its lower indices. This means that in the variation of the action to obtain the field equations we must bear in mind that $\Gamma^{\alpha}_{\beta\gamma} \neq \Gamma^{\alpha}_{\gamma\beta}$, i.e., we admit the possibility of nonvanishing torsion. It should be noted that in GR energy and momentum are the sources of curvature, while torsion is sourced by the spin of particles [14]. The fact that torsion is usually not considered in introductory courses on gravitation may be rooted in the educational tradition of this subject and the fact that the spin of particles was discovered many years after the original formulation of GR by Einstein. Another reason may be that the effects of torsion are very weak in general, except at very high densities, where the role of torsion becomes dominant and may even avoid the formation of singularities (see [30] for a recent discussion and earlier literature on the topic). For these reasons, and to motivate and facilitate the exploration of the effects of torsion in extensions of GR, our derivation of the field equations will be as general as possible (within reasonable limits). We will assume a symmetric metric tensor $g_{\mu\nu} = g_{\nu\mu}$ and the usual definitions for the Ricci tensor $R_{\mu\nu} \equiv R^{\rho}{}_{\mu\rho\nu}$ and the Ricci scalar $R \equiv g^{\mu\nu}R_{\mu\nu}$. The variation of the action (13) with respect to the metric and the connection can be expressed as

$$\delta S = \frac{1}{2\kappa^2}\int d^4x\sqrt{-g}\left[\left(\frac{\partial f}{\partial g^{\mu\nu}} - \frac{f}{2}g_{\mu\nu}\right)\delta g^{\mu\nu} + \frac{\partial f}{\partial R^{\alpha}{}_{\beta\mu\nu}}\delta R^{\alpha}{}_{\beta\mu\nu}\right] + \delta S_m . \tag{15}$$

Straightforward manipulations show that $\delta R^{\alpha}{}_{\beta\mu\nu}$ can be written as

$$\delta R^{\alpha}{}_{\beta\mu\nu} = \nabla_\mu\left(\delta\Gamma^{\alpha}_{\nu\beta}\right) - \nabla_\nu\left(\delta\Gamma^{\alpha}_{\mu\beta}\right) + 2S^{\lambda}_{\mu\nu}\delta\Gamma^{\alpha}_{\lambda\beta} , \tag{16}$$

where $S^{\lambda}_{\mu\nu} \equiv (\Gamma^{\lambda}_{\mu\nu} - \Gamma^{\lambda}_{\nu\mu})/2$ now represents the torsion tensor (note the additional $\frac{1}{2}$ factor as compared to our initial definition in Eq.(8)) From now on we will use the notation $P_{\alpha}{}^{\beta\mu\nu} \equiv \frac{\partial f}{\partial R^{\alpha}{}_{\beta\mu\nu}}$. In order to put the $\delta R^{\alpha}{}_{\beta\mu\nu}$ term in (15) in suitable form, we need to note that

$$I_{\Gamma} = \int d^4x \sqrt{-g} P_{\alpha}{}^{\beta\mu\nu} \nabla_{\mu} \delta\Gamma^{\alpha}_{\nu\beta} = \int d^4x \left[\nabla_{\mu}(\sqrt{-g} J^{\mu}) - \delta\Gamma^{\alpha}_{\nu\beta} \nabla_{\mu} \left(\sqrt{-g} P_{\alpha}{}^{\beta\mu\nu} \right) \right] , \tag{17}$$

where $J^{\mu} \equiv P_{\alpha}{}^{\beta\mu\nu} \delta\Gamma^{\alpha}_{\nu\beta}$. Since, in general, $\nabla_{\mu}(\sqrt{-g}J^{\mu}) = \partial_{\mu}(\sqrt{-g}J^{\mu}) + 2S^{\sigma}_{\sigma\mu}\sqrt{-g}J^{\mu}$, we find that (17) can be written as

$$I_{\Gamma} = \int d^4x \left[\partial_{\mu}(\sqrt{-g}J^{\mu}) - \delta\Gamma^{\alpha}_{\nu\beta} \left\{ \nabla_{\mu} \left(\sqrt{-g} P_{\alpha}{}^{\beta\mu\nu} \right) - 2S^{\sigma}_{\sigma\mu} \sqrt{-g} P_{\alpha}{}^{\beta\mu\nu} \right\} \right] . \tag{18}$$

Using this result, (15) becomes

$$\begin{aligned} \delta S = \frac{1}{2\kappa^2} \int d^4x & \left[\sqrt{-g} \left(\frac{\partial f}{\partial g^{\mu\nu}} - \frac{f}{2} g_{\mu\nu} \right) \delta g^{\mu\nu} + \partial_{\mu} \left(\sqrt{-g} J^{\mu} \right) \right. \\ & + \left. \left\{ -\frac{1}{\sqrt{-g}} \nabla_{\mu} \left(\sqrt{-g} P_{\alpha}{}^{\beta[\mu\nu]} \right) + S^{\nu}_{\sigma\rho} P_{\alpha}{}^{\beta\sigma\rho} + 2S^{\sigma}_{\sigma\mu} P_{\alpha}{}^{\beta[\mu\nu]} \right\} 2\sqrt{-g}\delta\Gamma^{\alpha}_{\nu\beta} \right] + \delta S_m . \end{aligned} \tag{19}$$

We thus find that the field equations can be written as follows

$$\kappa^2 T_{\mu\nu} = \frac{\partial f}{\partial g^{(\mu\nu)}} - \frac{f}{2} g_{\mu\nu} \tag{20}$$

$$\kappa^2 H_{\alpha}{}^{\nu\beta} = -\frac{1}{\sqrt{-g}} \nabla_{\mu} \left(\sqrt{-g} P_{\alpha}{}^{\beta[\mu\nu]} \right) + S^{\nu}_{\sigma\rho} P_{\alpha}{}^{\beta\sigma\rho} + 2S^{\sigma}_{\sigma\mu} P_{\alpha}{}^{\beta[\mu\nu]} , \tag{21}$$

where $P_{\alpha}{}^{\beta[\mu\nu]} = (P_{\alpha}{}^{\beta\mu\nu} - P_{\alpha}{}^{\beta\nu\mu})/2$, $T_{\mu\nu} = -\frac{2}{\sqrt{-g}} \frac{\delta S_m}{\delta g^{\mu\nu}}$ is the energy-momentum tensor of the matter, and $H_{\alpha}{}^{\nu\beta} = -\frac{1}{\sqrt{-g}} \frac{\delta S_m}{\delta \Gamma^{\alpha}_{\nu\beta}}$ represents the coupling of matter to the connection. For simplicity, from now on we will assume that $H_{\alpha}{}^{\nu\beta} = 0$. Eq. (21) can be put in a more convenient form if the connection is decomposed into its symmetric and antisymmetric (torsion) parts, $\Gamma^{\alpha}_{\mu\nu} = C^{\alpha}_{\mu\nu} + S^{\alpha}_{\mu\nu}$, such that $\nabla_{\mu} A_{\nu} = \partial_{\mu} A_{\nu} - C^{\alpha}_{\mu\nu} A_{\alpha} - S^{\alpha}_{\mu\nu} A_{\alpha} = \nabla^{C}_{\mu} A_{\nu} - S^{\alpha}_{\mu\nu} A_{\alpha}$ and $\nabla_{\mu}\sqrt{-g} = \nabla^{C}_{\mu}\sqrt{-g} - S^{\alpha}_{\mu\alpha}\sqrt{-g}$. By doing this, (21) turns into

$$\kappa^2 H_{\alpha}{}^{\nu\beta} = -\frac{1}{\sqrt{-g}} \nabla^{C}_{\mu} \left(\sqrt{-g} P_{\alpha}{}^{\beta[\mu\nu]} \right) + S^{\lambda}_{\mu\alpha} P_{\lambda}{}^{\beta[\mu\nu]} - S^{\beta}_{\mu\lambda} P_{\alpha}{}^{\lambda[\mu\nu]} . \tag{22}$$

3.1. Example: f(R,Q) theories

Eqs. (20) and (22) can be used to write the field equations for the metric and the connection for specific choices of the Lagrangian $f(g_{\mu\nu}, R^{\alpha}{}_{\beta\mu\nu})$. To make contact with the literature [26],

[5] ,[23], we now focus on the case $f(R,Q) = f(g^{\mu\nu}R_{\mu\nu}, g^{\mu\nu}g^{\alpha\beta}R_{\mu\alpha}R_{\nu\beta})$. For this family of Lagrangians, we obtain

$$P_{\alpha}{}^{\beta\mu\nu} = \delta_{\alpha}{}^{\mu}M^{\beta\nu} = \delta_{\alpha}{}^{\mu}\left(f_R g^{\beta\nu} + 2f_Q R^{\beta\nu}\right) , \tag{23}$$

where $f_X = \partial_X f$. Inserting this expression in (22) and tracing over α and ν, we find that $\nabla^C_\lambda[\sqrt{-g}M^{\beta\lambda}] = (2\sqrt{-g}/3)[S^{\sigma}_{\lambda\sigma}M^{\beta\lambda} + (3/2)S^{\beta}_{\lambda\mu}M^{\lambda\mu}]$. Using this result, the connection equation can be put as follows

$$\frac{1}{\sqrt{-g}}\nabla^C_\alpha\left[\sqrt{-g}M^{\beta\nu}\right] = S^{\nu}_{\alpha\lambda}M^{\beta\lambda} - S^{\nu}_{\beta\lambda}M^{\lambda\nu} - S^{\lambda}_{\alpha\lambda}M^{\beta\nu} + \frac{2}{3}\delta^{\nu}_{\alpha}S^{\sigma}_{\lambda\sigma}M^{\beta\lambda} \tag{24}$$

The symmetric and antisymmetric combinations of this equation lead, respectively, to

$$\frac{1}{\sqrt{-g}}\nabla^C_\alpha\left[\sqrt{-g}M^{(\beta\nu)}\right] = S^{\nu}_{\alpha\lambda}M^{[\beta\lambda]} - S^{\beta}_{\alpha\lambda}M^{[\nu\lambda]} - S^{\lambda}_{\alpha\lambda}M^{(\beta\nu)} + \frac{S^{\sigma}_{\lambda\sigma}}{3}\left(\delta^{\nu}_{\alpha}M^{\beta\lambda} + \delta^{\beta}_{\alpha}M^{\nu\lambda}\right) \tag{25}$$

$$\frac{1}{\sqrt{-g}}\nabla^C_\alpha\left[\sqrt{-g}M^{[\beta\nu]}\right] = S^{\nu}_{\alpha\lambda}M^{(\beta\lambda)} - S^{\beta}_{\alpha\lambda}M^{(\nu\lambda)} - S^{\lambda}_{\alpha\lambda}M^{[\beta\nu]} + \frac{S^{\sigma}_{\lambda\sigma}}{3}\left(\delta^{\nu}_{\alpha}M^{\beta\lambda} - \delta^{\beta}_{\alpha}M^{\nu\lambda}\right) . \tag{26}$$

Important simplifications can be achieved considering the new variables

$$\tilde{\Gamma}^{\lambda}_{\mu\nu} = \Gamma^{\lambda}_{\mu\nu} + \alpha\delta^{\lambda}_{\nu}S^{\sigma}_{\sigma\mu} , \tag{27}$$

and taking the parameter $\alpha = 2/3$, which implies that $\tilde{S}^{\lambda}_{\mu\nu} \equiv \tilde{\Gamma}^{\lambda}_{[\mu\nu]}$ is such that $\tilde{S}^{\sigma}_{\sigma\nu} = 0$. The symmetric and antisymmetric parts of $\tilde{\Gamma}^{\lambda}_{\mu\nu}$ are related to those of $\Gamma^{\lambda}_{\mu\nu}$ by

$$\tilde{C}^{\lambda}_{\mu\nu} = C^{\lambda}_{\mu\nu} + \frac{1}{3}\left(\delta^{\lambda}_{\nu}S^{\sigma}_{\sigma\mu} + \delta^{\lambda}_{\mu}S^{\sigma}_{\sigma\nu}\right) \tag{28}$$

$$\tilde{S}^{\lambda}_{\mu\nu} = S^{\lambda}_{\mu\nu} + \frac{1}{3}\left(\delta^{\lambda}_{\nu}S^{\sigma}_{\sigma\mu} - \delta^{\lambda}_{\mu}S^{\sigma}_{\sigma\nu}\right) \tag{29}$$

Using these variables, Eqs. (25) and (26) take the following compact form

$$\frac{1}{\sqrt{-g}}\nabla^{\tilde{C}}_\alpha\left[\sqrt{-g}M^{(\beta\nu)}\right] = \left[\tilde{S}^{\nu}_{\alpha\lambda}g^{\beta\kappa} + \tilde{S}^{\beta}_{\alpha\lambda}g^{\nu\kappa}\right]g^{\lambda\rho}M_{[\kappa\rho]} \tag{30}$$

$$\frac{1}{\sqrt{-g}}\nabla^{\tilde{C}}_\alpha\left[\sqrt{-g}M^{[\beta\nu]}\right] = \left[\tilde{S}^{\nu}_{\alpha\lambda}g^{\beta\kappa} - \tilde{S}^{\beta}_{\alpha\lambda}g^{\nu\kappa}\right]g^{\lambda\rho}M_{(\kappa\rho)} . \tag{31}$$

In these equations, $M^{(\beta\nu)} = f_R g^{\beta\nu} + 2f_Q R^{(\beta\nu)}(\Gamma)$, and $M^{[\beta\nu]} = 2f_Q R^{[\beta\nu]}(\Gamma)$, where $R_{(\beta\nu)}(\Gamma) = R_{(\beta\nu)}(\tilde{\Gamma})$ and $R_{[\beta\nu]}(\Gamma) = R_{[\beta\nu]}(\tilde{\Gamma}) + \frac{2}{3}\left(\partial_\beta S^{\sigma}_{\sigma\nu} - \partial_\nu S^{\sigma}_{\sigma\beta}\right)$.

In the recent literature on Palatini theories, only the torsionless case has been studied in detail. When torsion is considered in $f(R)$ theories, Eqs. (30) and (31) recover the results presented in [24]. In general, those equations put forward that when the traceless torsion tensor $\tilde{S}^{\nu}_{\alpha\lambda}$ vanishes, the symmetric and antisymmetric parts of $M^{\beta\nu}$ decouple. The dynamics of these theories, therefore, can be studied in different levels of complexity. The simplest case will be studied here and consists on setting $S^{\nu}_{\alpha\lambda}$ and $R_{[\mu\nu]}$ to zero. A more detailed discussion of the other cases can be found in [28].

3.2. Volume-invariant and torsionless $f(R,Q)$

When the torsion is set to zero, it can be shown [32], [12] that the vanishing of $R_{[\mu\nu]}$ guarantees the existence of a volume element that is covariantly conserved by $\Gamma^{\alpha}_{\mu\nu}$. The rank-two tensor that defines that volume element must be a solution of (30), which in this case takes the form

$$\nabla^{\Gamma}_{\alpha}\left[\sqrt{-g}\left(f_R g^{\beta\nu}+2f_Q R^{\beta\nu}(\Gamma)\right)\right]=0\,. \tag{32}$$

Note that here $R^{\beta\nu}(\Gamma)$ is symmetric because we are taking $R_{[\mu\nu]}(\Gamma)=0$. To obtain the solution of (32), we first consider (20) particularized to our theory (with $S^{\nu}_{\alpha\lambda}$ and $R_{[\mu\nu]}$ set to zero),

$$f_R R_{\mu\nu}-\frac{f}{2}g_{\mu\nu}+2f_Q R_{\mu\alpha}R^{\alpha}{}_{\nu}=\kappa^2 T_{\mu\nu}\,, \tag{33}$$

and rewrite it in the following form

$$f_R B_{\mu}{}^{\nu}-\frac{f}{2}\delta_{\mu}{}^{\nu}+2f_Q B_{\mu}{}^{\alpha}B_{\alpha}{}^{\nu}=\kappa^2 T_{\mu}{}^{\nu}\,, \tag{34}$$

where we have defined $B_{\mu}{}^{\nu}\equiv R_{\mu\alpha}g^{\alpha\nu}$. This equation can be seen as a second-order algebraic equation for the matrix $\hat{B}$, whose components are $[\hat{B}]_{\mu}{}^{\nu}\equiv B_{\mu}{}^{\nu}$. The solutions to this equation imply that $\hat{B}$ is an algebraic function of the components of the stress-energy tensor $T_{\mu}{}^{\nu}$, i.e., $\hat{B}=\hat{B}(\hat{T})$. This relation is very important because it allows to express (32) in the form

$$\nabla^{\Gamma}_{\alpha}\left[\sqrt{-g}g^{\beta\lambda}\left(f_R\delta^{\nu}_{\lambda}+2f_Q B_{\lambda}{}^{\nu}\right)\right]=0\,, \tag{35}$$

where now f_R, f_Q and $B_{\alpha}{}^{\nu}$ are functions of the stress-energy tensor of the matter. The connection, therefore, can be obtained by elementary algebraic manipulations [23]. To do it, one defines a rank-two symmetric tensor $h^{\mu\nu}$ such that $\sqrt{-g}g^{\beta\lambda}\left(f_R\delta^{\nu}_{\lambda}+2f_Q B_{\lambda}{}^{\nu}\right)=\sqrt{-h}h^{\beta\nu}$, which turns (35) into the well-known equation $\nabla_{\mu}\left[\sqrt{-h}h^{\beta\nu}\right]=0$, and implies that $\Gamma^{\alpha}_{\beta\nu}$ is given by the Christoffel symbols of the tensor $h_{\mu\nu}$, i.e.,

$$\Gamma^{\alpha}_{\beta\gamma}=\frac{h^{\alpha\rho}}{2}\left(\partial_{\beta}h_{\rho\gamma}+\partial_{\gamma}h_{\rho\beta}-\partial_{\rho}h_{\beta\gamma}\right)\,. \tag{36}$$

From the defining expression of $h_{\mu\nu}$, one finds that the relation between $h_{\mu\nu}$ and $g_{\mu\nu}$ can be expressed as follows

$$h_{\mu\nu} = \sqrt{\det\hat{\Sigma}}[\Sigma^{-1}]_\mu{}^\alpha g_{\alpha\nu} \ , \quad h^{\mu\nu} = \frac{g^{\mu\alpha}\Sigma_\alpha{}^\nu}{\sqrt{\det\hat{\Sigma}}} \ , \tag{37}$$

where we have defined the matrix $\Sigma_\alpha{}^\nu \equiv (f_R\delta_\alpha^\nu + 2f_Q B_\alpha{}^\nu)$. With these relations and definitions, the field equations for the metric $h_{\mu\nu}$ can be written in compact form expressing (34) as $B_\mu{}^\alpha\Sigma_\alpha{}^\nu = \frac{f}{2}\delta_\mu^\nu + \kappa^2 T_\mu{}^\nu$ and using the relation $B_\mu{}^\alpha\Sigma_\alpha{}^\nu = \sqrt{\det\hat{\Sigma}}R_{\mu\alpha}(h)h^{\alpha\nu}$ to obtain [27]

$$R_\mu{}^\nu(h) = \frac{1}{\sqrt{\det\hat{\Sigma}}}\left(\frac{f}{2}\delta_\mu^\nu + \kappa^2 T_\mu{}^\nu\right) \ . \tag{38}$$

In general, it will be more convenient to work with the field equations for the auxiliary metric $h_{\mu\nu}$ because their form is more tractable. Nonetheless, if one insists on writing the field equations using the metric $g_{\mu\nu}$, one must note that the connection (36) is related to the Levi-Civita connection of $g_{\mu\nu}$ by the tensor (recall Eq.(12))

$$A^\alpha_{\beta\gamma} \equiv \Gamma^\alpha_{\beta\gamma} - L^\alpha_{\beta\gamma} = \frac{h^{\alpha\rho}}{2}\left[\nabla^L_\mu h_{\rho\nu} + \nabla^L_\nu h_{\rho\mu} - \nabla^L_\rho h_{\mu\nu}\right] \ . \tag{39}$$

The Riemann tensors of $\Gamma^\alpha_{\beta\gamma}$ and $L^\alpha_{\beta\gamma}$ are thus related as follows

$$R^\alpha{}_{\beta\mu\nu}(\Gamma) = R^\alpha{}_{\beta\mu\nu}(L) + \nabla^L_\mu A^\alpha_{\nu\beta} - \nabla^L_\nu A^\alpha_{\mu\beta} + A^\lambda_{\nu\beta}A^\alpha_{\mu\lambda} - A^\lambda_{\mu\beta}A^\alpha_{\nu\lambda} \ , \tag{40}$$

which allows to express (38) in terms of the Ricci tensor of the metric $g_{\mu\nu}$, the usual covariant derivatives of $L^\alpha_{\beta\gamma}$, and the matter.

4. $f(R,Q)$ theories with a perfect fluid

The explicit form of the matrix $\hat{\Sigma}$ that relates the metrics $h_{\mu\nu}$ and $g_{\mu\nu}$ can only be found once all the sources that make up $T_{\mu\nu}$ have been specified. In our discussion we will just consider a perfect fluid or a sum of non-interacting perfect fluids such that

$$T_{\mu\nu} = (\rho + P)u_\mu u_\nu + P g_{\mu\nu} \tag{41}$$

with $\rho = \sum_i \rho_i$ and $P = \sum_i P_i$. In order to find an expression for $\hat{\Sigma}$, we first rewrite (34) using matrix notation as

$$2f_Q\hat{B}^2 + f_R\hat{B} - \frac{f}{2}\hat{I} = \kappa^2\hat{T} \ . \tag{42}$$

Using (41) this equation can be rewritten as follows

$$2f_Q\left(\hat{B}+\frac{f_R}{4f_Q}\hat{I}\right)^2=\left(\kappa^2P+\frac{f}{2}+\frac{f_R^2}{8f_Q}\right)\hat{I}+\kappa^2(\rho+P)u_\mu u^\mu\,. \tag{43}$$

Denoting $\lambda^2\equiv\left(\kappa^2P+\frac{f}{2}+\frac{f_R^2}{8f_Q}\right)$ and making explicit the matrix representation, (43) becomes

$$2f_Q\left(\hat{B}+\frac{f_R}{4f_Q}\hat{I}\right)^2=\begin{pmatrix}\lambda^2-\kappa^2(\rho+P) & \vec{0}\\ \vec{0} & \lambda^2\hat{I}_{3X3}\end{pmatrix}\,, \tag{44}$$

where $\hat{I}_{3X3}$ denotes 3-dimensional identity matrix. Since the right-hand side of (44) is a diagonal matrix, it is immediate to compute its square root, which leads to

$$\sqrt{2f_Q}\left(\hat{B}+\frac{f_R}{4f_Q}\hat{I}\right)=\begin{pmatrix}s_1\sqrt{\lambda^2-\kappa^2(\rho+P)} & \vec{0}\\ \vec{0} & \lambda\hat{S}_{3X3}\end{pmatrix}\,, \tag{45}$$

where s_1 denotes a sign, which can be positive or negative, and $\hat{S}_{3X3}$ denotes a 3X3 diagonal matrix with elements $\{s_i=\pm1\}$. For consistency of the theory in the limit $f_Q\to0$, we must have $s_1=1$ and $\hat{S}_{3X3}=\hat{I}_{3X3}$. This result allows to express $\hat{\Sigma}$ as follows

$$\hat{\Sigma}=\begin{pmatrix}\sigma_1 & \vec{0}\\ \vec{0} & \sigma_2\hat{I}_{3X3}\end{pmatrix}\,, \tag{46}$$

where σ_1 and σ_2 take the form

$$\begin{aligned}\sigma_1&=\frac{f_R}{2}\pm\sqrt{2f_Q}\sqrt{\lambda^2-\kappa^2(\rho+P)}\\ \sigma_2&=\frac{f_R}{2}+\sqrt{2f_Q}\lambda\,.\end{aligned} \tag{47}$$

Note that we have kept the two signs $\pm$ in σ_1. The reason for this will be understood later, when particular models are considered. The point is that in some cases of physical interest, at high densities one should take the negative sign in front of the square root to guarantee that σ_1 is continuous and differentiable accross the point where the square root vanishes. This technical issue does not arise for σ_2.

4.1. Workable models: $f(R,Q)=\tilde{f}(R)+\alpha Q$

So far we have made progress without specifying the form of the Lagrangian $f(R,Q)$. However, in order to find the explicit dependence of $R=B_\mu{}^\mu$ and $Q=B_\mu{}^\alpha B_\alpha{}^\mu$ with the ρ and P of the fluids, we must choose a Lagrangian explicitly. Restricting the function $f(R,Q)$ to the family $f(R,Q)=\tilde{f}(R)+\alpha Q$, we will see that it is possible to find the generic dependence of

Q with ρ and P, while R is found to depend only on the combination $T = -\rho + 3P$ [23]. The reason for this follows from the trace of (33) with $g^{\mu\nu}$, which for this family of Lagrangians gives the algebraic relation $R\tilde{f}_R - 2\tilde{f} = \kappa^2 T$ and implies that $R = R(T)$ (like in Palatini $f(R)$ theories). For these theories, we have that $f_Q = \alpha$, which is a constant. Therefore, from the trace of (44) we find

$$\sqrt{2f_Q}\left(R + \frac{f_R}{f_Q}\right) = \sqrt{\lambda^2 - \kappa^2(\rho + P)} + 3\lambda\,, \tag{48}$$

which can be cast as

$$\left[\sqrt{2f_Q}\left(R + \frac{f_R}{f_Q}\right) - 3\lambda\right]^2 = \lambda^2 - \kappa^2(\rho + P) \tag{49}$$

After a bit of algebra we find that

$$\lambda = \frac{\sqrt{2f_Q}}{8}\left[3\left(R + \frac{f_R}{f_Q}\right) \pm \sqrt{\left(R + \frac{f_R}{f_Q}\right)^2 - \frac{4\kappa^2(\rho + P)}{f_Q}}\right] \tag{50}$$

From this expression and the definition of λ^2, we find

$$\alpha Q = -\left(\tilde{f} + \frac{\tilde{f}_R^2}{4f_Q} + 2\kappa^2 P\right) + \frac{f_Q}{16}\left[3\left(R + \frac{\tilde{f}_R}{f_Q}\right) \pm \sqrt{\left(R + \frac{\tilde{f}_R}{f_Q}\right)^2 - \frac{4\kappa^2(\rho + P)}{f_Q}}\right]^2\,, \tag{51}$$

where R, $\tilde{f}$, and $\tilde{f}_R$ are functions of $T = -\rho + 3P$.

5. Nonsingular cosmologies in $f(R,Q)$ theories

The difficulties faced by GR to provide a consistent description of singularities and quantum phenomena at high energies (microscopic or Planck scales) is generally seen as an indication that we should go beyond the standard geometric structures to successfully quantize the theory and avoid singularities. This idea has motivated a variety of approaches that range from the consideration of higher-dimensional superstrings and other extended objects [11], to non-commutative geometries or non-perturbative quantization methods [3], [31], [35] , to name just a few well-known cases. Unfortunately, the formidable task of building a satisfactory quantum theory of gravity is not yet complete. Moreover, even if we managed to get such a theory, we would still have to face the challenge of testing its predictions. In this sense, it should be noted that since the quantum gravitational regime is so far from our current and future experimental capabilities, our only hope might be to use the information available in the cosmic microwave background radiation to verify or rule out our theories [1]. How much of the quantum gravitational regime could be contrasted with these yet-to-come theories is not clear. This is due, in part, because the theorized rapid accelerated expansion that took place during the inflationary period may have washed out many of the relevant proper signatures needed to distinguish the predictions of different quantum theories of gravity.

A conservative approach, therefore, consists on exploring the quantum properties and interactions of the matter fields in the very early universe using the well-established methods of quantum field theory in curved space-times [29]. The success of this approach has been confirmed in combination with models of inflation and sheds relevant light on the mechanisms that may have caused the primordial spectra of scalar and tensorial perturbations [38], [19], [9],[18], [15]. The applicability of this approach, however, becomes unreliable at increasing energies as the regime of the classical big bang singularity is approached and the quantum fluctuations of the gravitational field can no longer be neglected. At that stage, a complete quantum theory of gravity seems necessary to provide a consistent description of the ongoing physical processes. Obviously, different quantum theories could lead to completely different quantum gravitational scenarios and, therefore, a generic quantum origin for the universe cannot be guessed *a priori* by any logical means.

In recent years, bouncing cosmological models have attracted much attention [21]. These are scenarios in which the big bang singularity is replaced by a quantum-induced bounce that connects an earlier phase of contraction with the subsequent expanding phase (in which we happen to exist). In such scenarios, aside from the quantum regime, the contracting and expanding phases are expected to asymptote an effective classical geometry whose dynamics, on consistency grounds, should match that of GR at low energies. In this context, and as an intermediate step between the quantum field theory approach in the (singular) curved background provided by GR and a (nonsingular) full theory of quantum gravity, one could consider the case of a smooth effective geometry free from big bang singularities on top of which quantum matter fields could still be treated perturbatively in a consistent way. This view would somehow disentangle the non-perturbative part of the quantum gravitational sector into an effective classical, nonsingular geometry, plus perturbative quantum corrections that propagate on top of the regular effective background. The absence of curvature singularities would make the treatment of quantum fields on the resulting geometry more reliable, and could help shed new light on the effects of the matter-gravity interaction in the very-early universe.

In the literature there exist many interesting examples of (quantum and non-quantum) cosmological models that avoid the big bang singularity by means of a bounce. Roughly, those models can be classified in two large groups, depending on whether they contain a modified gravitational sector or a modified matter sector (see [21] for details and a very complete list of references). Generically, modified gravity theories imply the existence of new dynamical degrees of freedom, such as gravitational scalar fields (like in scalar-tensor theories), higher-derivatives of the metric, extra dimensions, ... The consideration of exotic matter sources may be justified, in some cases, from an effective field theory approach, such as in the case of non-linear theories of electrodynamics, which naturally arise in low-energy limits of string theories. In the remainder of this chapter, we are going to study bouncing cosmological models from the modified gravity perspective provided by the Palatini theories discussed above. This approach is particularly interesting because, despite being a modified-gravity approach, the underlying mechanisms that modify the gravitational dynamics are not associated with new dynamical degrees of freedom or higher-derivative equations. In fact, it is the nontrivial role played by the matter in the determination of the space-time connection that induces nonlinearities in the matter sector that end up changing the dynamics at very high matter-energy densities. In this sense, it should be noted that the gravitational field equations in vacuum exactly recover those of GR (with possibly a cosmological constant, depending on the particular Lagrangian chosen). For this reason, this type of theories can be regarded as a *minimal extension of the standard model of gravitational physics*, because they only appreciably depart from GR in regions that contain sources and when those sources reach the energy-density scales that characterize the correcting terms of the Lagrangian.

5.1. Homogeneous cosmologies in $f(R,Q)$ theories

In this section we introduce the basic definitions and formulas needed to derive the equations for the evolution of the expansion and shear [37] for an arbitrary Palatini $f(R,Q)$ theory of the kind presented in Section 3.2 . These magnitudes will be very useful to extract information about the geometric properties of the space-time and to determine whether cosmic singularities are present or not. We focus on homogeneous cosmologies of the Bianchi I type (a different expansion factor for each spatial direction) because that will allow us to test the rebustness of our results against deviations from the idealized Friedmann-Robertson-Walker spacetimes (same expansion rate in all the spatial directions). We will also particularize our results to the case of $f(R)$ theories, i.e., no dependence on Q.

We consider a Bianchi I spacetime with physical line element of the form

$$ds^2 = g_{\mu\nu}dx^\mu dx^\nu = -dt^2 + \sum_{i=1}^{3} a_i^2(t)(dx^i)^2 \tag{52}$$

In terms of this line element, using the relation between metrics (37) and the expression (46) for the matrix $\hat{\Sigma}$ of a collection of perfect fluids, the nonzero components of the auxiliary metric $h_{\mu\nu}$ become

$$h_{tt} = -\left(\frac{\sigma_2^2}{\sqrt{\sigma_1\sigma_2}}\right) \equiv -S \tag{53}$$

$$h_{ij} = \sqrt{\sigma_1\sigma_2}a_i^2\delta_{ij} \equiv \Omega a_i^2\delta_{ij} \tag{54}$$

The relevant Christoffel symbols associated with $h_{\mu\nu}$ are the following:

$$\Gamma^t_{tt} = \frac{\dot{S}}{2S} \tag{55}$$

$$\Gamma^t_{ij} = \frac{\Omega a_i^2}{2S}\left[\frac{\dot{\Omega}}{\Omega} + \frac{2\dot{a}_i}{a_i}\right]\delta_{ij} \tag{56}$$

$$\Gamma^i_{tj} = \frac{\delta^i_j}{2}\left[\frac{\dot{\Omega}}{\Omega} + \frac{2\dot{a}_i}{a_i}\right] \tag{57}$$

The nonzero components of the corresponding Ricci tensor are

$$R_{tt}(h) = -\sum_i \dot{H}_i - \sum_i H_i^2 - \frac{3}{2}\frac{\ddot{\Omega}}{\Omega} + \frac{3}{4}\frac{\dot{\Omega}}{\Omega}\left(\frac{\dot{S}}{S} + \frac{\dot{\Omega}}{\Omega}\right) + \frac{1}{2}\left(\frac{\dot{S}}{S} - \frac{2\dot{\Omega}}{\Omega}\right)\sum_i H_i \tag{58}$$

$$R_{ij}(h) = \frac{\delta_{ij}a_i^2}{2}\frac{\Omega}{S}\left[2\dot{H}_i + \frac{\ddot{\Omega}}{\Omega} - \left(\frac{\dot{\Omega}}{\Omega}\right)^2 + \frac{\dot{\Omega}}{\Omega}\sum_k H_k + \frac{1}{2}\frac{\dot{\Omega}}{\Omega}\left(\frac{3\dot{\Omega}}{\Omega} - \frac{\dot{S}}{S}\right) + \right.$$
$$\left. + 2H_i\left\{\sum_k H_k + \frac{1}{2}\left(\frac{3\dot{\Omega}}{\Omega} - \frac{\dot{S}}{S}\right)\right\}\right], \tag{59}$$

where $H_k \equiv \dot{a}_k/a_k$. For completeness, we give an expression for the corresponding scalar curvature

$$R(h) = \frac{1}{S}\left[2\sum_k \dot{H}_k + \sum_k H_k^2 + \left(\sum_k H_k\right)^2 + \left(3\frac{\dot{\Omega}}{\Omega} - \left\{\frac{\dot{S}}{S} - \frac{\dot{\Omega}}{\Omega}\right\}\right)\sum_k H_k + 3\frac{\ddot{\Omega}}{\Omega} - \frac{3}{2}\frac{\dot{\Omega}}{\Omega}\frac{\dot{S}}{S}\right] \tag{60}$$

From the above formulas, one can readily find the corresponding ones in the isotropic, flat configuration by just replacing $H_i \to H$. For the spatially nonflat case, the $R_{tt}(h)$ component is the same as in the flat case. The $R_{ij}(h)$ component, however, picks up a new piece, $2K\gamma_{ij}$, where γ_{ij} represents the nonflat spatial metric of $g_{ij} = a_i^2\gamma_{ij}$. The Ricci scalar then becomes $R(h) \to R^{K=0}(h) + \frac{6K}{a^2\Omega}$.

5.2. Shear

From the previous formulas and the field equation (38), we find that the combination $R_i{}^i - R_j{}^j$ (no summation over indices) leads to

$$R_i{}^i - R_j{}^j = \frac{1}{S}\left[\dot{H}_{ij} + H_{ij}\left\{\sum_k H_k + \frac{1}{2}\left(\frac{3\dot{\Omega}}{\Omega} - \frac{\dot{S}}{S}\right)\right\}\right] = 0\,, \tag{61}$$

where we have defined $H_{ij} \equiv H_i - H_j$. Note that the final equality $R_i{}^i - R_j{}^j = 0$, follows from the fact that the right hand sides of $R_i{}^i$ and $R_j{}^j$ as given by (38) are equal. Expressing (61) in the form

$$R_i{}^i - R_j{}^j = \frac{d}{dt}\left[\ln H_{ij} + \ln(a_1a_2a_3) + \ln\Omega^{3/2} - \ln S^{1/2}\right] = 0\,, \tag{62}$$

we see that it can be readily integrated regardless of the number and particular equations of state of the fluids involved. The result is

$$H_{ij} = C_{ij}\frac{S^{\frac{1}{2}}}{\Omega^{\frac{3}{2}}}\frac{C_{ij}}{(a_1 a_2 a_3)} = \frac{C_{ij}}{\sigma_1}\frac{V_0}{V(t)}, \tag{63}$$

where the constants $C_{ij} = -C_{ji}$ satisfy the relation $C_{12} + C_{23} + C_{31} = 0$, V_0 represents a reference volume, and $V(t) = V_0 a_1 a_2 a_3$ represents the volume of the universe. It is worth noting that writing explicitly the three equations (63) and combining them in pairs, one can write the individual Hubble rates as follows

$$\begin{aligned} H_1 &= \theta + \frac{(C_{12} - C_{31})}{3\sigma_1}\left(\frac{V_0}{V(t)}\right) \\ H_2 &= \theta + \frac{(C_{23} - C_{12})}{3\sigma_1}\left(\frac{V_0}{V(t)}\right) \\ H_3 &= \theta + \frac{(C_{31} - C_{23})}{3\sigma_1}\left(\frac{V_0}{V(t)}\right) \end{aligned} \tag{64}$$

where θ is the expansion of a congruence of comoving observers and is defined as $3\theta = \sum_i H_i$. Using these relations, the shear $\sigma^2 = \sum_i (H_i - \theta)^2$ of the congruence takes the form

$$\sigma^2 = \frac{(C_{12}^2 + C_{23}^2 + C_{31}^2)}{9\sigma_1^2}\left(\frac{V_0}{V(t)}\right)^2, \tag{65}$$

where we have used the relation $(C_{12} + C_{23} + C_{31})^2 = 0$.

5.3. Expansion

We now derive an equation for the evolution of the expansion with time and a relation between expansion and shear. From previous results, one finds that

$$G_{tt}(h) \equiv -\frac{1}{2}\sum_k H_k^2 + \frac{1}{2}\left(\sum_k H_k\right)^2 + \frac{\dot{\Omega}}{\Omega}\sum_k H_k + \frac{3}{4}\left(\frac{\dot{\Omega}}{\Omega}\right)^2 \tag{66}$$

In terms of the expansion and shear, this equation becomes

$$G_{tt} \equiv 3\left(\theta + \frac{\dot{\Omega}}{2\Omega}\right)^2 - \frac{\sigma^2}{2}. \tag{67}$$

From the field equation (38), we find that

$$G_{tt} = \frac{f + \kappa^2(\rho + 3P)}{2\sigma_1}, \tag{68}$$

which in combination with (67) yields

$$3\left(\theta+\frac{\dot{\Omega}}{2\Omega}\right)^2=\frac{f+\kappa^2(\rho+3P)}{2\sigma_1}+\frac{\sigma^2}{2}\,. \tag{69}$$

For a set of non-interacting fluids with equations of state $w_i = P_i/\rho_i$, we have that $\Omega = \Omega(\rho_i, w_i)$ and, therefore, $\dot{\Omega} = \sum_i \Omega_{\rho_i}\dot{\rho}_i$, where $\Omega_{\rho_i} \equiv \partial\Omega/\partial\rho_i$. Since for those fluids the conservation equation is $\dot{\rho}_i = -3\theta(1+\omega_i)\rho_i$, we find that $\dot{\Omega} = -3\theta\sum_i(1+\omega_i)\rho_i\Omega_{\rho_i}$. With this result, (69) can be written as

$$3\theta^2\left(1+\frac{3}{2}\Delta_1\right)^2=\frac{f+\kappa^2(\rho+3P)}{2\sigma_1}+\frac{\sigma^2}{2}\,, \tag{70}$$

where we have defined

$$\Delta_1=-\sum_i(1+w_i)\rho_i\frac{\partial_{\rho_i}\Omega}{\Omega}\,. \tag{71}$$

Note that in this last equation $w_i = w_i(\rho_i)$, i.e., they need not be constants. For fluids with constant w_i, the conservation equation implies that their density depends on the volume of the universe according to $\rho_i(t) = \rho_i(t_0)\left(\frac{V_0}{V(t)}\right)^{1+w_i}$. This implies that once a particular Lagrangian is specified, the equations of state $P_i = w_i\rho_i$ are given, and the anisotropy constants C_{ij} are chosen, the right-hand side of Eqs. (65) and (70) can be parametrized in terms of $V(t)$. This, in turn, allows us to parametrize the H_i functions of (64) in terms of $V(t)$ as well. This will be very useful later for our discussion of particular models.

In the isotropic case ($\sigma^2 = 0\,, \theta = \dot{a}/a \equiv \mathcal{H}$) with nonzero spatial curvature, (70) takes the following form:

$$\mathcal{H}^2=\frac{1}{6\sigma_1}\frac{\left[f+\kappa^2(\rho+3P)-\frac{6K\sigma_2}{a^2}\right]}{\left[1+\frac{3}{2}\Delta_1\right]^2} \tag{72}$$

The evolution equation for the expansion can be obtained by noting that the R_{ij} equations, which are of the form $R_{ij} \equiv (\Omega/2S)g_{ij}[\ldots] = (f/2+\kappa^2P)g_{ij}/\sigma_2$, can be summed up to give

$$2(\dot{\theta}+3\theta^2)+\theta\left(\frac{6\dot{\Omega}}{\Omega}-\frac{\dot{S}}{S}\right)+\left\{\frac{\ddot{\Omega}}{\Omega}+\frac{1}{2}\frac{\dot{\Omega}}{\Omega}\left(\frac{\dot{\Omega}}{\Omega}-\frac{\dot{S}}{S}\right)\right\}=\frac{[f+2\kappa^2P]}{\sigma_1}\,. \tag{73}$$

5.4. Limit to $f(R)$

We now consider the limit $f_Q \to 0$, namely, the case in which the Lagrangian only depends on the Ricci scalar R. Doing this we will obtain the corresponding equations for shear and expansion in the $f(R)$ case without the need for extra work. From the definitions of λ^2 (see below eq.(43)), and σ_1 and σ_2 in (47), it is easy to see that in the limit $f_Q \to 0$ we get

$$\sigma_1 \to \sigma_2 \to f_R \tag{74}$$

$$S \to \Omega \to f_R \,. \tag{75}$$

With these rules it is easy to see that $h_{\mu\nu} = f_R g_{\mu\nu}$, which makes (38) boil down to the expected field equations for Palatini $f(R)$ theories, namely, $f_R R_{\mu\nu}(h) - \frac{f}{2} g_{\mu\nu} = \kappa^2 T_{\mu\nu}$. Equation (63) turns into

$$H_{ij} = \frac{C_{ij}}{f_R} \frac{V_0}{V(t)}, \tag{76}$$

from which one can easily obtain expressions for H_1, H_2 and H_3 as in (64). The shear becomes

$$\sigma^2 = \frac{(C_{12}^2 + C_{23}^2 + C_{31}^2)}{9 f_R^2} \left(\frac{V_0}{V(t)} \right)^2 , \tag{77}$$

where $C_{12} + C_{23} + C_{31} = 0$. The relation between expansion and shear for a collection of non-interacting perfect fluids now becomes

$$3\theta^2 \left(1 + \frac{3}{2} \tilde{\Delta}_1 \right)^2 = \frac{f + \kappa^2(\rho + 3P)}{2 f_R} + \frac{\sigma^2}{2} \tag{78}$$

where $\tilde{\Delta}_1$ is given by (71) but with Ω replaced by f_R. In the isotropic case with nonzero K we find

$$\mathcal{H}^2 = \frac{1}{6 f_R} \frac{\left[f + \kappa^2(\rho + 3P) - \frac{6 K f_R}{a^2} \right]}{\left[1 + \frac{3}{2} \tilde{\Delta}_1 \right]^2} . \tag{79}$$

5.5. Bouncing $f(R)$ cosmologies

We now present the cosmological dynamics of simple $f(R)$ models to illustrate how this family of theories modifies the standard Big Bang picture of the early universe. Consider, for instance, the model[2] $f(R) = R + a R^2 / R_P$, where $R_P = l_P^{-2} = c^3 / \hbar G$ is the Planck curvature. From the trace equation $R f_R - 2f = \kappa^2 T$ (see Sec.4.1), we find that this model leads to the same relation between the matter and the scalar curvature as in GR, namely, $R = -\kappa^2 T$. This implies that the theory behaves as GR whenever the energy density is much smaller than the Planck density scale $\rho_P \equiv R_P / \kappa^2$. Since by definition $\theta = \frac{1}{3} \sum_i \frac{\dot{a}_i}{a_i} = \frac{1}{3} \frac{d}{dt} \ln a_1 a_2 a_3 = \frac{1}{3} \frac{\dot{V}}{V}$, where $V = V_0 a_1 a_2 a_3$ represents the volume of the universe (with $V_0 = V(t_0)$), Eq. (78) for this quadratic model with dust and radiation leads to

[2] Note that the constant a could be absorbed into a redefinition of R_P and, therefore, only its sign is relevant.

$$\theta^2 = \frac{1}{9}\left(\frac{\dot{V}}{V}\right)^2 = \frac{\left(\rho_d + \rho_r + \frac{a\rho_d}{2\rho_P}\right)\left(1 + \frac{2a\rho_d}{\rho_P}\right)}{3\left(1 - \frac{a\rho_d}{\rho_P}\right)^2} + \frac{(C_{12}^2 + C_{23}^2 + C_{31}^2)}{54\left(1 - \frac{a\rho_d}{\rho_P}\right)^2}\left(\frac{V_0}{V(t)}\right)^2 , \tag{80}$$

where $\rho_d = \rho_{d,0}\left(\frac{V_0}{V(t)}\right)$ and $\rho_r = \rho_{r,0}\left(\frac{V_0}{V(t)}\right)^{4/3}$.
In general, an homogeneous cosmological model experiences a bounce when the expansion θ vanishes, which implies an extremum (a maximum or a minimum) of the volume of the Universe. If $V(t)$ vanishes at some finite time, then a big bang or big crunch singularity is found, depending on whether $\dot{V} > 0$ or $\dot{V} < 0$ at that time. Focusing for the moment on the isotropic case, $C_{12}^2 + C_{23}^2 + C_{31}^2 = 0$, we find that a bounce occurs if $a < 0$ when ρ_d reaches the value $\rho_d^B = \rho_P/(2|a|)$ [see Fig.1)] . This value of the density implies that $f_R = 1 - 2a\rho_d/\rho_P = 0$. This condition, $f_R = 0$, characterizes the location of the bounce in Palatini $f(R)$ theories with a single fluid with constant equation of state [5]. For our quadratic model, in particular, bouncing solutions exist if the dynamics allows to reach the density $\rho_B = \frac{\rho_P}{2a(3w-1)} > 0$. This means that for $a > 0$ fluids with $w > 1/3$ avoid the initial singularity, whereas for $a < 0$ it takes $w < 1/3$. The case $a = 0$ naturally recovers the equations of GR. It is worth noting that a cosmic bounce may arise even for presureless matter, $w = 0$, if $a < 0$, which implies that exotic sources of matter-energy that violate the energy conditions are not necessary to avoid the big bang singularity in this framework. The reason for this is that at high energies gravitation may become repulsive for matter sources with $w > -1$, whereas it is attractive at low energy-densities for those same sources. Note also that the pure radiation universe, $w = 1/3$, is a peculiar case because it does not produce any modified dynamics in Palatini $f(R)$ theories. On physical grounds, however, it should be noted that due to quantum effects related with the trace anomaly of the electromagnetic field, a gas of photons in a $SU(N)$ gauge theory with N_f fermion flavors has an effective equation of state given by

$$w_{eff}^{rad} = \frac{1}{3} - \frac{5\alpha^2}{18\pi^2}\frac{\left(N_c + \frac{5}{4}N_f\right)\left(\frac{11}{3}N_c - \frac{2}{3}N_f\right)}{2 + \frac{7}{2}\frac{N_c N_f}{N_c^2 - 1}} , \tag{81}$$

where N_c is the color number of the gauge theory (which has $N_c(N_c - 1)$ generators) [13], [17]. Therefore, a universe filled with photons should be able to avoid the singularity if $a > 0$. In physically realistic scenarios, one should consider the co-existence of several fluids and take into account the time dependence of the number of effective degrees of freedom and the transfer of energy among different species [15], which leads to the possibility of having different effective fluids at different stages of the cosmic expansion. In this sense, the "dust plus radiation" model represented by (80) needs not be accurate at all times because dust particles may become relativistic at high energies and contribute to ρ_r rather than to ρ_d. This suggests that the choice/determination of the sign of the parameter a is not a trivial issue and would require a very careful and elaborate analysis (which goes beyond the scope of this introductory work).

When anisotropies are taken into account, one finds that bouncing solutions are still possible as long as the amount of anisotropy is not too large. In Fig.2, we see that increasing the

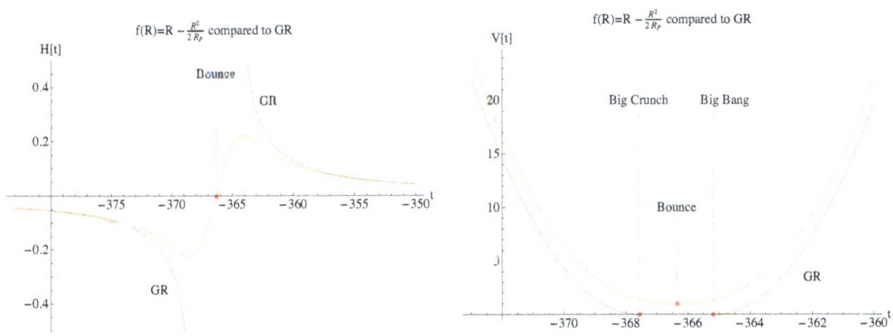

Figure 1. Representation of the Hubble function (left) and volume of the Universe (right) as a function of time for the model $f(R) = R - R^2/2R_P$ in a universe filled with dust and radiation (for the numerical integration $\rho_{d,0} = 10^3\rho_{r,0}$, and $V = 10^5 V_0$). The GR solutions corresponding to a contracting branch, which ends in a big crunch, and an expanding branch, which begins with a big bang, are represented together with the bouncing solution of the Palatini model that interpolates between those singular solutions.

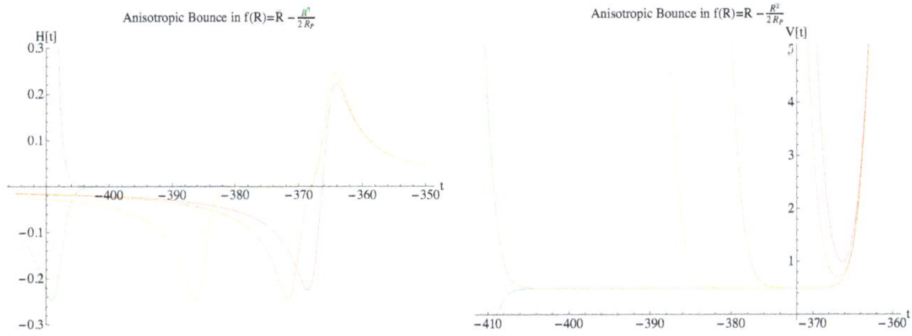

Figure 2. Representation of the expansion (left) and volume of the Universe (right) as a function of time for the model $f(R) = R - R^2/2R_P$ in a universe filled with dust and radiation with anisotropies (for the numerical integration $\rho_{d,0} = 10^3\rho_{r,0}$, and $V = 10^5 V_0$). From right to left, we have plotted the bouncing cases $C^2 = 0, 40, 40.60211073, 40.60211073942454489657$, and the collapsing case with $C^2 = 40.60211073942454489658$. Fine tunning the value of C^2 even more should allow to keep the universe in its minimum for longer periods of time in the past, which eventually should lead to an asymptotically static solution.

value of $C^2 \equiv C_{12}^2 + C_{23}^2 + C_{31}^2$ from zero, the volume of the universe presents a minimum as long as $C^2 < C_c^2$. If $C^2 > C_c^2$, the collapse is unavoidable and $V \to 0$ in a finite time. The critical case $C^2 \to C_c^2$ represents a configuration that is neither a bouncing universe nor a big bang. It corresponds to a state in which the volume of the universe remains constant in the past and expands in the future. Though this solution is clearly unstable and fine-tuned, its existence puts forward the possibility of obtaining static regular solutions corresponding to ultracompact objects, which could shed new light on the internal structure of black holes and/or topological deffects when Planck scale corrections to the gravitational action are taken into account. It should be noted, however, that in order to obtain this asymptotically static solution one must cross from the domain where $f_R > 0$ to the region where $f_R < 0$. Since the shear, as defined in (77) for $f(R)$ theories with perfect fluids, is proportional to $1/f_R^2$, the crossing through $f_R = 0$ implies a divergence in some curvature scalars of the theory. Whether this divergence is a true (or strong) physical singularity in the sense defined in [36], [8], [16] is an open question that will be explored elsewhere. In any case, we remark that the existence of that divergence does not have any effect on the time evolution of the expansion θ, as can be seen in Fig.2.

5.6. Nonsingular universes in $f(R,Q)$ Palatini theories

In the previous section we have seen that Palatini $f(R)$ models are able to avoid the big bang singularity in idealized homogeneous and isotropic scenarios but run into trouble when anisotropies are present. The divergence of the shear is a generic problem for those $f(R)$ theories in which the function f_R vanishes at some point, regardless of the number and equation of state of the fluids involved. Though the nature of this divergence has not been identified yet with that of a strong singularity, which besides the divergence of some components of the Riemann, Ricci, and Weyl tensors also requires the divergence of some of their integrals, its very presence is a disturbing aspect that one would like to overcome within the framework of Palatini theories. In this sense, a natural step is to study the behavior in anisotropic scenarios of some simple generalization of the $f(R)$ family to see if the situation improves. Using Lagrangians of the form presented in (4.1), we will show next that completely regular bouncing solutions exist for both isotropic and anisotropic homogeneous cosmologies.

5.6.1. Isotropic universe

Consider Eq.(72) particularized to the following $f(R,Q)$ Lagrangian

$$f(R,Q) = R + a\frac{R^2}{R_P} + b\frac{Q}{R_P} \tag{82}$$

For this theory, we find that $R = \kappa^2(\rho - 3P)$ and $Q = Q(\rho, P)$ is given by (51) with $\alpha \equiv b/R_P$. From now on we assume that the parameter b of the Lagrangian is positive and has been absorbed into a redefinition of R_P, which is assumed positive. This restriction is necessary (though not sufficient) if one wants the scalar Q to be bounded from above when fluids with $w > -1$ are considered. Stated differently, when $b/R_P > 0$, positivity of the square root of Eq.(51) establishes that there may exist a maximum for the combination $\rho + P$.

In order to have (72) well defined, one must make sure that the choice of sign in front of the square root of σ_1 in (47) is the correct one. In this sense, we find that to recover the $f(R)$ limit and GR at low curvatures, we must take the positive sign, i.e., $\sigma_1 = \sigma_1^+$. However, when considering particular models, which are characterized by the constant a and, for instance, a constant equation of state w, one realizes that the square root may reach a zero at some high density. Beyond that point, we may need to switch from σ_1^+ to σ_1^- to guarantee that σ_1 is a continuous and differentiable function (see Fig.3 for an illustration of this point). Bearing in mind this technical subtlety, one can then proceed to represent the Hubble function for different choices of parameters and fluid combinations to determine whether bouncing solutions exist or not.

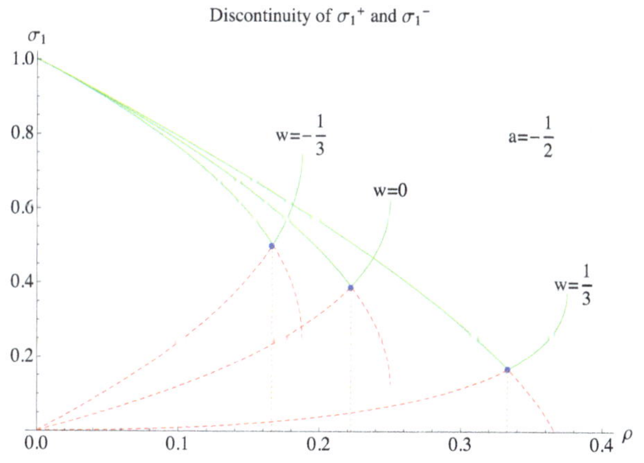

Figure 3. Illustration of the need to combine the two branches of σ_1 to obtain a continuous and differentiable curve. The branch that starts at $\sigma_1 = 1$ has the plus sign in front of the square root (continuous green line). When the square root vanishes (at the blue dot), the function must be continued through the dashed red branch, which corresponds to the negative sign in front of the square root.

The classification of the bouncing solutions of the model (82) with a fluid with constant w was carried out in [5]. It was found that for every value of the parameter a there exist an infinite number of bouncing solutions, which depend on the particular equation of state w. The bouncing solutions can be divided into two large classes:

- **Class I:** $a \geq 0$. The bounce occurs when the scalar Q reaches its maximum value and happens for all equations of state satisfying the condition

$$w > w_{min} = \frac{a}{2+3a} \,. \tag{83}$$

  From this equation it follows that a radiation dominated universe, with $w = 1/3$, always bounces for any $a > 0$.

- **Class II:** $a \leq 0$. This case is more involved because the bounce can occur either at the point where Q reaches its maximum or when σ_1 vanishes. This last case can only happen at high curvatures when we are in the branch defined by $\sigma_1 = \sigma_1^-$. To proceed with the classification, we divide this sector into several intervals:
  - **If** $-1/4 < a \leq 0$. The bounce occurs if

$$-\frac{1}{3}+\frac{1}{3}\sqrt{\frac{1+4a}{1+a}} < w < \infty \tag{84}$$

    We see that when $a = 0$ we find agreement with the discussion of case I. As a approaches the limiting value $-1/4$, the bouncing solutions extend up to $w \to -1/3$.
  - **If** $-1/3 \leq a \leq -1/4$. Numerically one finds that the bouncing solutions cannot be extended below $w < -1$ and occur if $-1 < w < \infty$, where $w = -1$ is excluded.

- **If** $-1 \leq a \leq -1/3$. In this case, one finds numerically that the bouncing solutions are restricted to the interval $-1 < w < \frac{\alpha+\beta a}{(1+3a)^2} > 1$, where $\alpha = 1.1335$ and $\beta = -3.3608$.
- **If** $a \leq -1$. Similarly as the family $a \geq 0$, this set of models also allows for a simple characterization of the bouncing solutions, which correspond to the interval $-1 < w < a/(2+3a)$. In the limiting case $a = -1$ we obtain the condition $-1 < w < 1$ (compare this with the numerical fit above, which gives $-1 < w < 1.12$).

5.7. Anisotropic universe

Using Eqs. (70) and (72), the expansion can be written as follows:

$$\theta^2 = H^2 + \frac{1}{6}\frac{\sigma^2}{(1+\frac{3}{2}\Delta_1)^2}, \tag{85}$$

where H represents the Hubble function in the $K = 0$ isotropic case. To better understand the behavior of θ^2, let us consider when and why H^2 vanishes. Using the results of [5] summarized above, one finds that H^2 vanishes either when the density reaches the value $\rho_{Q_{max}}$ or when the function σ_1 vanishes. These two conditions imply a divergence in the quantity $(1+\frac{3}{2}\Delta_1)^2$, which appears in the denominator of H^2 and, therefore, force the vanishing of H^2 (isotropic bounce). Technically, these two types of divergences can be easily characterized. From the definition of Δ_1 in (71), one can see that $\Delta_1 \sim \partial_\rho\Omega/\Omega$. Since $\Omega \equiv \sqrt{\sigma_1\sigma_2}$, it is clear that Δ_1 diverges when $\sigma_1 = 0$. The divergence due to reaching $\rho_{Q_{max}}$ is a bit more elaborate. One must note that $\partial_\rho\Omega$ contain terms that are finite plus a term of the form $\partial_\rho\lambda$, with λ defined below Eq. (43). In this λ there is a Q term hidden in the function $f(R,Q)$, which implies that $\partial_\rho\lambda \sim \partial_\rho Q/R_P$ plus other finite terms. From the definition of Q it follows that $\partial_\rho Q$ has finite contributions plus the term $\partial_\rho\Phi/\sqrt{\Phi}$, where $\Phi \equiv (1+(1+2a)R/R_P)^2 - 4\kappa^2(\rho+P)/R_P$, which diverges when Φ vanishes. This divergence of $\partial_\rho Q$ indicates that Q cannot be extended beyond the maximum value Q_{max}. Now, since the shear goes like $\sigma^2 \sim 1/(\sigma_1)^2$ [see Eq.(65)], we see that the condition $\sigma_1 = 0$ implies a divergence on σ^2 (though θ^2 remains finite). This is exactly the same type of divergence that we already found in the $f(R)$ models, where $\sigma_1 \to f_R$. Since in the $f(R)$ models the bounce can only occur when $f_R = 0$, there is no way to avoid the divergence of the shear in the anisotropic case within the $f(R)$ setting. On the contrary, since the quadratic $f(R,Q)$ model (82) allows for a second mechanism for the bounce, which takes place at $\rho_{Q_{max}}$, there is a natural way out of the problem with the shear. Summarizing, we conclude that for universes governed by the Lagrangian (82) and containing a single stiff fluid there exist completely regular bouncing solutions in the anisotropic case for $w > \frac{a}{2+3a}$ if $a \geq 0$, for $w_0 < w < \infty$ if $-1/3 \leq a \leq 0$, for $w_0 < w < (\alpha+\beta a)/(1+3a)^2$ if $-1 \leq a \leq -1/3$, and for $-1/3 < w < a/(2+3a)$ if $a \leq -1$, where $w_0 < 0$ is defined as the equation of state for which the (isotropic) bounce occurs when $Q = Q_{max}$ and $\sigma_1 = 0$ simultaneously (see [5] for details). These results imply that for $a < 0$ the interval $0 \leq w \leq 1/3$ is always included in the family of completely regular isotropic and anisotropic bouncing solutions, which contain the dust and radiation cases. For $a \geq 0$, the radiation case is always nonsingular too.

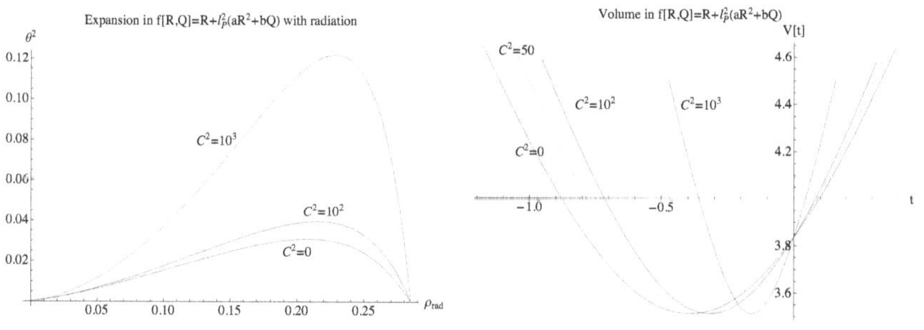

Figure 4. Representation of the expansion squared (left) and volume of the Universe (right) as a function of time for the model $f(R,Q) = R - l_P^2 R^2/2 + l_P^2 Q$ in a radiation universe with anisotropies. We have plotted the bouncing cases $C^2 = 0, 50, 10^2, 10^3$. Note that the bounce always occurs at the same maximum density (minimum volume). Note also that the time spent in the bouncing region decreases as the anisotropy grows. The starting point of the time integration is chosen such that at $t = 0$ the two branches of σ_1 coincide.

5.8. Example: Radiation universe

As an illustrative example, we consider here the particular case of a universe filled with radiation. Besides its obvious physical interest, this case leads to a number of algebraic simplifications that make more transparent the form of some basic definitions

$$Q = \frac{3R_P^2}{8}\left[1 - \frac{8\kappa^2\rho}{3R_P} - \sqrt{1 - \frac{16\kappa^2\rho}{3R_P}}\right] \tag{86}$$

$$\sigma_1^\pm = \frac{1}{2} \pm \frac{1}{2\sqrt{2}}\sqrt{5 - 3\sqrt{1 - \frac{16\kappa^2\rho}{3R_P}} - \frac{24\kappa^2\rho}{R_P}} \tag{87}$$

$$\sigma_2 = \frac{1}{2} + \frac{1}{2\sqrt{2}}\sqrt{5 - 3\sqrt{1 - \frac{16\kappa^2\rho}{3R_P}} - \frac{8\kappa^2\rho}{3R_P}} \tag{88}$$

Note that the coincidence of the two branches of σ_1 occurs at $\kappa^2\rho = R_P/6$, where $\sigma_1^\pm = \frac{1}{2}$. It is easy to see that at low densities (86) leads to $Q \approx 4(\kappa^2\rho)^2/3 + 32(\kappa^2\rho)^3/9R_P + 320(\kappa^2\rho)^4/27R_P^2 + \ldots$, which recovers the expected result for GR, namely, $Q = 3P^2 + \rho^2$. From this formula we also see that the maximum value of Q occurs at $\kappa^2\rho_{max} = 3R_P/16$ and leads to $Q_{max} = 3R_P^2/16$. At this point the shear also takes its maximum allowed value, namely, $\sigma_{max}^2 = \sqrt{3/16}R_P^{3/2}(C_{12}^2 + C_{23}^2 + C_{31}^2)$, which is always finite. At ρ_{max} the expansion vanishes producing a cosmic bounce regardless of the amount of anisotropy [see Fig.4].

6. Conclusions and open questions

In this chapter we have tried to convey the idea that in the construction of extended theories of gravity, one should bear in mind the fact that metric and connection are equally fundamental and independent objects. This observation allows to broaden the spectrum of available possibilities to go beyond the *standard model of gravitation*. In fact, any theory of

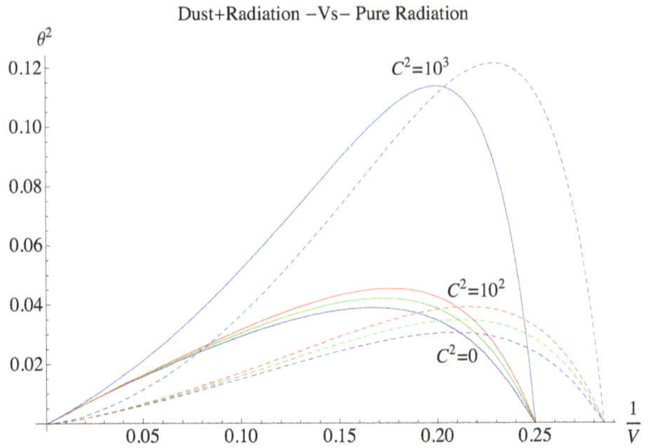

Figure 5. Comparison of the expansion in a universe filled with dust and radiation ($\rho_{0,rad} = 10^{-3}\rho_{0,dust}$) and a radiation dominated universe (dashed lines) for several values of the anisotropy.

gravity based on a geometry in which the connection has been forced to be given by the Christoffel symbols of the metric admits an alternative formulation in which the form of the connection is dictated by the theory itself, i.e., it is not given by convention or selected on practical grounds.

In our exploration of Palatini theories, we have seen that assuming that metric and connection are independent geometrical objects has non-trivial effects on the resulting field equations as compared with the usual *metric* formulation of the same theories. For the particular family of $f(R,Q)$ models studied here, we have seen that the metric is governed by second-order equations that boil down to GR in vacuum. This is in sharp contrast with the usual *metric* formulation of those same theories, where one finds fourth-order derivatives of the metric (see, for instance, [2] for a detailed analysis of the cosmology of the quadratic model (82) in metric formalism). The absence of higher-order derivatives in the Palatini formulation is a remarkable point that seems not to have been sufficiently appreciated in the literature. In fact, having second-order field equations is very important because it automatically implies the absence of ghosts and other dynamical instabilities. In this sense, it should be noted that Lovelock[3] theories [40], which are generally regarded as the natural extension of the Einstein-Hilbert Lagrangian to higher dimensions, have received a lot of attention in the literature because they are seen as the most general actions for gravity that give at most second-order field equations for the metric. As we have seen here, this property is shared (at least) by all Palatini theories of the $f(R,Q)$ type (with or without torsion). This puts forward that Palatini theories, are natural candidates to explore new dynamics beyond GR.

Before concluding, we would like to stress the fact that the quadratic Palatini model (82) is able to avoid the big bang singularity in very natural situations, such as in pure radiation, pure dust, or dust plus radiation universes with or without anisotropies (see Fig.5). This observation has been possible thanks to the formulas presented in section 5, where we have

[3] We would like to mention that when Lovelock theories are formulated à la Palatini, the resulting field equations are exactly the same as one finds in their usual *metric* formulation [7].

extended the analysis carried out in [5] for a single perfect fluid with constant equation of state to include several perfect fluids with arbitrary equation of state $w(\rho)$. This allows to explore the dynamics of realistic cosmological models with several fluids and is a necessary step prior to the consideration of the growth and evolution of inhomogeneities in these nonsingular backgrounds. Though the model (82) has been proposed on grounds of mathematical simplicity and motivated by the form of the effective action provided by perturbative quantization schemes in curved backgrounds, its ability to successfully deal with cosmological [5] and black hole singularities [27] as well as other aspects of quantum gravity phenomenology [26] demands further theoretical work to provide a more solid ground to it. In this sense, we note that the effective dynamics of loop quantum cosmology [4] in a Friedmann-Robertson-Walker background filled with a massless scalar can be exactly reproduced by a Palatini $f(R)$ theory [22]. The extension of that result to more general spacetimes and matter sources could shed new light on the potential relation of (82) with a more fundamental theory of quantum gravity. All these open questions will be considered in detail elsewhere.

This work has been supported by the Spanish grants FIS2008-06078-C03-02, FIS2011-29813-C02-02, the Consolider Program CPAN (CSD2007-00042), and the JAE-doc program of the Spanish Research Council (CSIC).

Author details

Gonzalo J. Olmo

Departamento de Física Teórica & IFIC, Centro Mixto Universidad de Valencia & CSIC, Facultad de Física, Universidad de Valencia, Burjassot, Valencia, Spain

References

[1] Agullo, I. and Parker, L. , Gen. Rel. Grav. 43, 2541 (2011) [Int. J. Mod. Phys. D 20, 2861 (2011)] [arXiv:1106.4240 [astro-ph.CO]].

[2] Anderson, P., Phys. Rev. D 28, 271 (1983).

[3] Ashtekar, A. and Lewandowski, J. , *Class. Quant. Grav.* 21 (2004) R53.

[4] Ashtekar, A. and Singh, P., Class. Quant. Grav. 28, 213001 (2011) [arXiv:1108.0893 [gr-qc]].

[5] Barragan, C. and Olmo, G. J. , Phys. Rev. D 82, 084015 (2010), [arXiv:1005.4136 [gr-qc]].

[6] Birrel N.D. and Davies P.C.W., *Quantum fields in curved space*, (Cambridge University Press, Cambridge, England, 1982).

[7] Borunda, M., Janssen, B. and Bastero-Gil, M., JCAP 0811, 008 (2008) [arXiv:0804.4440 [hep-th]].

[8] Clarke, C. J. S. and Krolak, A. , Journ. Geom. Phys. 2, 17 (1985).

[9] Dodelson S., *Modern Cosmology*, Academic Press, (2003).

[10] Feigl, H. and Brodbeck, M. (eds.), *Readings in the phylosophy of science*, Appleton-Century-Crofts, New York (1953). (See page 193).

[11] Green, M. , Schwarz, J. , and Witten, E. , "Superstring Theory", Cambridge University Press, Cambridge, England (1987).

[12] Hehl, F. W., McCrea, J. D., Mielke, E. W., and Ne'eman, Y., Phys. Rept. 258, 1 (1995) [gr-qc/9402012].

[13] Kajantie, K., Laine, M. , Rummukainen, K. and Schroder, Y. , Phys. Rev. D 67, 105008 (2003).

[14] Kibble, T. W. B., J. Math. Phys. 2, 212 (1961).

[15] Kolb, E.W. and Turner, M.S. *The early universe*, Westview Press (1990).

[16] Krolak, A. , Class. Quant. Grav. 3, 267 (1998).

[17] Lambiase, G. , Mohanty, S. , JCAP 0712 (2007) 008.

[18] Liddle A. R. and Lyth D.H., *Cosmological inflation and large-scale structure*, Cambridge University Press, (2000).

[19] Lyth D.H. and Liddle A.R. The primordial density perturbation, Cambridge University Press,(2009)

[20] Nakahara, M. , *Geometry, Topology and Physics*, Adam Hilger, Bristol (1990).

[21] Novello, M. and Perez Bergliaffa, S.E. , *Phys.Rep.* 463, 127-213 (2008).

[22] Olmo, G. J. , and Singh, P., JCAP 0901, 030 (2009) [arXiv:0806.2783 [gr-qc]].

[23] Olmo, G. J. , Sanchis-Alepuz, H., and Tripathi, S., Phys. Rev. D80, 024013 (2009). [arXiv:0907.2787 [gr-qc]].

[24] Olmo, G. J. , Int. J. Mod. Phys. D 20, 413 (2011).

[25] Olmo, G. J. , AIP Conf. Proc. 1458, (2011) 222-237.

[26] Olmo, G. J. , JCAP 1110, 018 (2011) [arXiv:1101.2841 [gr-qc]].

[27] Olmo, G. J. , and Rubiera-Garcia,D. , Phys.Rev. D86 (2012) 044014; Int.J.Mod.Phys. D21 (2012) 1250067; Eur.Phys.J. C72 (2012) 2098.

[28] Olmo, G. J. , and Rubiera-Garcia,D. , to appear (2012).

[29] Parker L. and Toms D.J., *Quantum field theory in curved spacetime: quantized fields and gravity*, (Cambridge University Press, Cambridge, England, 2009).

[30] Poplawski, N. J., Phys. Lett. B 694, 181 (2010) [Erratum-ibid. B 701, 672 (2011)] [arXiv:1007.0587 [astro-ph.CO]].

[31] Rovelli, C. , *Quantum Gravity*, (Cambridge U. Press, 2004).

[32] Schouten, J.A. , *Tensor analysis for physicists*, Oxford University Press, London (1951).

[33] Sciama, D. W. , in: Recent Developments in General Relativity, p. 415 (Pergamon, 1962).

[34] Sciama, D. W., Rev. Mod. Phys. 36, 463 (1964).

[35] Thiemann, T. ,*Modern canonical quantum general relativity*, (Cambridge U. Press, 2007).

[36] Tipler, F. J. , Phys. Lett. A 64, 8 (1977).

[37] Wald, R.M., *General Relativity*, (The University of Chicago Press, Chicago, 1984).

[38] Weinberg S., Cosmology, Oxford University Press, (2008).

[39] Will, C.M. (1993). *Theory and Experiment in Gravitational Physics*, Cambridge University Press, Cambridge; *Living Rev.Rel.* 9,3,(2005), gr-qc/0510072 .

[40] Zanelli, J. (2005). arXiv:hep-th/0502193v4 .

Chapter 9

Leptogenesis and Neutrino Masses in an Inflationary SUSY Pati-Salam Model

C. Pallis and N. Toumbas

Additional information is available at the end of the chapter

1. Introduction

One of the most promising and well-motivated mechanisms for the generation of the *Baryon Asymmetry of the Universe* (BAU) is via an initial generation of a lepton asymmetry, which can be subsequently converted to BAU through sphaleron effects – see e.g. Ref. [1, 2]. *Non-Thermal Leptogenesis* (nTL) [3, 4] is a variant of this proposal, in which the necessitated departure from equilibrium is achieved by construction. Namely, the *right-handed* (RH) neutrinos, ν_i^c, whose decay produces the lepton asymmetry, are out-of-equilibrium at the onset, since their masses are larger than the reheating temperature. Such a set-up can be achieved by the direct production of ν_i^c through the inflaton decay, which can also take place out-of-equilibrium. Therefore, such a leptogenesis paradigm largely depends on the inflationary stage, which it follows.

In a recent paper [5] – for similar attempts, see Ref. [6–8] –, we investigate an inflationary model where a *Standard Model* (SM) singlet component of the Higgs fields involved in the spontaneous breaking of a *supersymmetric* (SUSY) *Pati-Salam* (PS) *Grand Unified Theory* (GUT) can produce inflation of chaotic-type, named *non-minimal Higgs Inflation* (nMHI), since there is a relatively strong non-minimal coupling of the inflaton field to gravity [9–12]. This GUT provides a natural framework to implement our leptogenesis scenario, since the presence of the $SU(2)_{\rm R}$ gauge symmetry predicts the existence of three ν_i^c. In its simplest realization this GUT leads to third family *Yukawa unification* (YU), and does not suffer from the doublet-triplet splitting problem since both Higgs doublets are contained in a bidoublet other than the GUT scale Higgs fields. Although this GUT is not completely unified – as, e.g., a GUT based on the $SO(10)$ gauge symmetry group – it emerges in standard weakly coupled heterotic string models [13] and in recent D-brane constructions [14].

The inflationary model relies on renormalizable superpotential terms and does not lead to overproduction of magnetic monopoles. It is largely independent of the one-loop radiative corrections [15], and it can become consistent with the fitting [16] of the seven-year data of the *Wilkinson Microwave Anisotropy Probe Satellite* (WMAP7) combined with the *baryon-acoustic oscillation* (BAO) and the measurement of the *Hubble constant* (H_0). At the same time the GUT symmetry breaking scale attains its SUSY value and the μ problem of the *Minimal SUSY SM* (MSSM) is resolved via a Peccei-Quinn (PQ) symmetry, solving also the strong CP problem. Inflation can be followed by non-thermal leptogenesis, compatible with the gravitino ($\widetilde{G}$) limit [17–19] on the reheating temperature, leading to efficient baryogenesis. In Ref. [5] we connect non-thermal leptogenesis with neutrino data, implementing a two-generation formulation of the see-saw [20–22] mechanism and imposing extra restrictions from the data on the light neutrino masses and the GUT symmetry on the heaviest Dirac neutrino mass. There we [5] assume that the mixing angle between the first and third generation, θ_{13}, vanishes. However, the most updated [23, 24] analyses of the low energy neutrino data suggest that non-zero values for θ_{13} are now preferred, while the zero value can be excluded at 8 standard deviations. Therefore, a revision of our results, presented in Ref. [5], is worth pursuing.

The three-generation implementation of the see-saw mechanism is here adopted, following a bottom-up approach, along the lines of Ref. [25–28]. In particular, we use as input parameters the low energy neutrino observables considering several schemes of neutrino masses. Using also the third generation Dirac neutrino mass predicted by the PS GUT, assuming a mild hierarchy for the two residual generations and imposing the restriction from BAU, we constrain the masses of ν^c_i's and the residual neutrino Dirac mass spectrum. Renormalization group effects [28, 29] are also incorporated in our analysis.

We present the basic ingredients of our model in Sec. 2. In Sec. 3 we describe the inflationary potential and derive the inflationary observables. In Sec. 4 we outline the mechanism of non-thermal leptogenesis, while in Sec. 5 we exhibit the relevant imposed constraints and restrict the parameters of our model. Our conclusions are summarized in Sec. 6. Throughout the text, we use natural units for Planck's and Boltzmann's constants and the speed of light ($\hbar = c = k_{\rm B} = 1$); the subscript of type $,\chi$ denotes derivation *with respect to* (w.r.t) the field χ (e.g., $_{,\chi\chi} = \partial^2/\partial\chi^2$); charge conjugation is denoted by a star and log [ln] stands for logarithm with basis 10 [e].

2. The Pati-Salam SUSY GUT model

In this section, we present the particle content (Sec. 2.1), the structure of the superpotential and the Kähler potential(Sec. 2.2) and describe the SUSY limit (Sec. 2.3) of our model.

2.1. Particle content

We focus on a SUSY PS GUT model described in detail in Ref. [5, 30]. The representations and the transformation properties of the various superfields under $G_{\rm PS} = SU(4)_{\rm C} \times SU(2)_{\rm L} \times SU(2)_{\rm R}$, their decomposition under $G_{\rm SM} = SU(3)_{\rm C} \times SU(2)_{\rm L} \times U(1)_Y$, as well as their extra global charges are presented in Table 1.

The ith generation ($i = 1, 2, 3$) *left-handed* (LH) quark and lepton superfields, $u_{i\rm a}$, $d_{i\rm a}$ (a $= 1, 2, 3$ is a color index), e_i and ν_i are accommodated in a superfield F_i. The LH antiquark and antilepton superfields $u^c_{i\rm a}$, $d^c_{i\rm a}$, e^c_i and ν^c_i are arranged in another superfield F^c_i. The gauge symmetry $G_{\rm PS}$ can be spontaneously broken down to $G_{\rm SM}$ through v.e.vs which the superfields H^c and $\bar{H}^c$ acquire in the directions ν^c_H and $\bar{\nu}^c_H$. The model also contains a gauge singlet S, which triggers the breaking of $G_{\rm PS}$, as well as an $SU(4)_{\rm C}$ **6**-plet G, which splits into $g^c_{\rm a}$ and $\bar{g}^c_{\rm a}$ under G_{SM} and gives [13] superheavy masses to $d^c_{H\rm a}$ and $\bar{d}^c_{H\rm a}$. In the simplest

Super-fields	Representations under G_{PS}	Trasformations under G_{PS}	Decompositions under G_{SM}	Global Charges R	PQ	$\mathbb{Z}_2^{mp}$
			MATTER SUPERFIELDS			
F_i	$(\mathbf{4},\mathbf{2},\mathbf{1})$	$F_i U_L^\dagger U_C^T$	$Q_{ia}(\mathbf{3},\mathbf{2},1/6)$	1	-1	$-$
			$L_i(\mathbf{1},\mathbf{2},-1/2)$			
F_i^c	$(\bar{\mathbf{4}},\mathbf{1},\mathbf{2})$	$U_C^* U_R^* F_i^c$	$u_{ia}^c(\bar{\mathbf{3}},\mathbf{1},-2/3)$	1	0	$-$
			$d_{ia}^c(\bar{\mathbf{3}},\mathbf{1},1/3)$			
			$\nu_i^c(\mathbf{1},\mathbf{1},0)$			
			$e_i^c(\mathbf{1},\mathbf{1},1)$			
			HIGGS SUPERFIELDS			
H^c	$(\bar{\mathbf{4}},\mathbf{1},\mathbf{2})$	$U_C^* U_R^* H^c$	$u_{Ha}^c(\bar{\mathbf{3}},\mathbf{1},-2/3)$	0	0	$+$
			$d_{Ha}^c(\bar{\mathbf{3}},\mathbf{1},1/3)$			
			$\nu_H^c(\mathbf{1},\mathbf{1},0)$			
			$e_H^c(\mathbf{1},\mathbf{1},1)$			
$\bar{H}^c$	$(\mathbf{4},\mathbf{1},\mathbf{2})$	$\bar{H}^c U_R^T U_C^T$	$\bar{u}_{Ha}^c(\mathbf{3},\mathbf{1},2/3)$	0	0	$+$
			$\bar{d}_{Ha}^c(\mathbf{3},\mathbf{1},-1/3)$			
			$\bar{\nu}_H^c(\mathbf{1},\mathbf{1},0)$			
			$\bar{e}_H^c(\mathbf{1},\mathbf{1},-1)$			
S	$(\mathbf{1},\mathbf{1},\mathbf{1})$	S	$S(\mathbf{1},\mathbf{1},0)$	2	0	$+$
G	$(\mathbf{6},\mathbf{1},\mathbf{1})$	$U_C G U_C^T$	$\bar{g}_a^c(\mathbf{3},\mathbf{1},-1/3)$	2	0	$+$
			$g_a^c(\bar{\mathbf{3}},\mathbf{1},1/3)$			
$\mathbb{H}$	$(\mathbf{1},\mathbf{2},\mathbf{2})$	$U_L \mathbb{H} U_R^T$	$H_u(\mathbf{1},\mathbf{2},1/2)$	0	1	$+$
			$H_d(\mathbf{1},\mathbf{2},-1/2)$			
P	$(\mathbf{1},\mathbf{1},\mathbf{1})$	P	$P(\mathbf{1},\mathbf{1},0)$	1	-1	$+$
$\bar{P}$	$(\mathbf{1},\mathbf{1},\mathbf{1})$	$\bar{P}$	$\bar{P}(\mathbf{1},\mathbf{1},0)$	0	1	$+$

Table 1. The representations, the transformations under G_{PS}, the decompositions under G_{SM} as well as the extra global charges of the superfields of our model. Here $U_C \in SU(4)_C$, $U_L \in SU(2)_L$, $U_R \in SU(2)_R$ and T, $\dagger$ and $*$ stand for the transpose, the hermitian conjugate and the complex conjugate of a matrix respectively.

realization of this model [13, 30], the electroweak doublets H_u and H_d, which couple to the up and down quarks respectively, are exclusively contained in the bidoublet superfield $\mathbb{H}$.

In addition to G_{PS}, the model possesses two global $U(1)$ symmetries, namely a PQ and an R symmetry, as well as a discrete $\mathbb{Z}_2^{mp}$ symmetry ('matter parity') under which F, F^c change sign. The last symmetry forbids undesirable mixings of F and $\mathbb{H}$ and/or F^c and H^c and ensures the stability of the *lightest SUSY particle* (LSP). The imposed $U(1)$ R symmetry, $U(1)_R$, guarantees the linearity of the superpotential w.r.t the singlet S. Finally the $U(1)$ PQ symmetry, $U(1)_{PQ}$, assists us to generate the μ-term of the MSSM. The PQ breaking occurs at an intermediate scale through the v.e.vs of P, $\bar{P}$, and the μ-term is generated via a non-renormalizable coupling of P and $\mathbb{H}$. Following Ref. [30], we introduce into the scheme quartic (non-renormalizable) superpotential couplings of $\bar{H}^c$ to F_i^c, which generate intermediate-scale masses for the ν_i^c and, thus, masses for the light neutrinos, ν_i, via the seesaw mechanism [20–22]. Moreover, these couplings allow for the decay of the inflaton into ν_i^c, leading to a reheating temperature consistent with the $\widetilde{G}$ constraint with more or less natural values of the parameters. As shown finally in Ref. [30], the proton turns out to be practically stable in this model.

2.2 Superpotential and Kähler potential

The superpotential W of our model splits into three parts:

$$W = W_{\rm MSSM} + W_{\rm PQ} + W_{\rm HPS}, \tag{1}$$

which are analyzed in the following.

- $W_{\rm MSSM}$ is the part of W which contains the usual terms – except for the μ term – of the MSSM, supplemented by Yukawa interactions among the left-handed leptons and ν_i^c:

$$W_{\rm MSSM} = y_{ij} F_i \mathbb{H} F_j^c$$
$$= y_{ij} \left({H_d}^{\mathsf{T}} \varepsilon L_i e_j^c - {H_u}^{\mathsf{T}} \varepsilon L_i \nu_j^c + {H_d}^{\mathsf{T}} \varepsilon Q_{ia} d_{ja}^c - {H_u}^{\mathsf{T}} \varepsilon Q_{ia} u_{ja}^c \right), \text{ with } \varepsilon = \begin{pmatrix} 0 & 1 \\ -1 & 0 \end{pmatrix} \cdot \tag{2}$$

  Here $Q_{ia} = \begin{pmatrix} u_{ia} & d_{ia} \end{pmatrix}^{\mathsf{T}}$ and $L_i = \begin{pmatrix} \nu_i & e_i \end{pmatrix}^{\mathsf{T}}$ are the i-th generation $SU(2)_{\rm L}$ doublet LH quark and lepton superfields respectively. Summation over repeated color and generation indices is assumed. Obviously the model predicts YU at $M_{\rm GUT}$ since the fermion masses per family originate from a unique term of the PS GUT. It is shown [31, 32] that exact third family YU combined with non-universalities in the gaugino sector and/or the scalar sector can become consistent with a number of phenomenological and cosmological low-energy requirements. On the other hand, it is expected on generic grounds that the predictions of this simple model for the fermion masses of the two lighter generations are not valid. Usually this difficulty can be avoided by introducing [33] an abelian symmetry which establishes a hierarchy between the flavor dependent couplings. Alternatively, the present model can be augmented [34] with other Higgs fields so that H_u and H_d are not exclusively contained in $\mathbb{H}$, but receive subdominant contributions from other representations too. As a consequence, a moderate violation of exact YU can be achieved, allowing for an acceptable low-energy phenomenology, even with universal boundary conditions for the soft SUSY breaking terms. However, we prefer here to work with the simplest version of the PS model, using the prediction of the third family YU in order to determine the corresponding Dirac neutrino mass – see Sec. 5.1.

- $W_{\rm PQ}$, is the part of W which is relevant for the spontaneous breaking of $U(1)_{\rm PQ}$ and the generation of the μ term of the MSSM. It is given by

$$W_{\rm PQ} = \lambda_{\rm PQ} \frac{P^2 \bar{P}^2}{M_{\rm S}} - \lambda_\mu \frac{P^2}{2M_{\rm S}} \mathsf{Tr}\left(\mathbb{H} \varepsilon \mathbb{H}^{\mathsf{T}} \varepsilon \right), \tag{3}$$

where $M_S \simeq 5 \cdot 10^{17}$ GeV is the String scale. The scalar potential, which is generated by the first term in the RHS of Eq. (3), after gravity-mediated SUSY breaking, is studied in Ref. [30, 35]. For a suitable choice of parameters, the minimum lies at $|\langle P \rangle| = |\langle \bar{P} \rangle| \sim \sqrt{m_{3/2} M_S}$. Hence, the PQ symmetry breaking scale is of order $\sqrt{m_{3/2} M_S} \simeq (10^{10} - 10^{11})$ GeV. The μ-term of the MSSM is generated from the second term of the RHS of Eq. (3) as follows:

$$-\lambda_\mu \frac{\langle P \rangle^2}{2M_S} \mathsf{Tr}\left(\mathbb{H} \varepsilon \mathbb{H}^{\mathsf{T}} \varepsilon \right) = \mu H_d^{\mathsf{T}} \varepsilon H_u \;\Rightarrow\; \mu \simeq \lambda_\mu \frac{\langle P \rangle^2}{M_S}, \tag{4}$$

which is of the right magnitude if $\lambda_\mu \sim (0.001 - 0.01)$. Let us note that V_{PQ} has an additional local minimum at $P = \bar{P} = 0$, which is separated from the global PQ minimum by a sizable potential barrier, thus preventing transitions from the trivial to the PQ vacuum. Since this situation persists at all cosmic temperatures after reheating, we are obliged to assume that, after the termination of nMHI, the system emerges with the appropriate combination of initial conditions so that it is led [36] in the PQ vacuum.

- W_{HPS}, is the part of W which is relevant for nMHI, the spontaneous breaking of G_{PS} and the generation of intermediate Majorana [superheavy] masses for ν_i^c [d_H^c and $\bar{d}_H^c$]. It takes the form

$$W_{HPS} = \lambda S \left(\bar{H}^c H^c - M_{PS}^2 \right) + \lambda_H H^{c\mathsf{T}} G \varepsilon H^c + \lambda_{\bar{H}} \bar{H}^c \tilde{G} \varepsilon \bar{H}^{c\mathsf{T}} + \lambda_{i\nu^c} \frac{\left(\bar{H}^c F_i^c \right)^2}{M_S}, \tag{5}$$

where M_{PS} is a superheavy mass scale related to M_{GUT} – see Sec. 3.2 – and $\tilde{G}$ is the dual tensor of G. The parameters λ and M_{PS} can be made positive by field redefinitions.

According to the general recipe [11, 12], the implementation of nMHI within SUGRA requires the adoption of a Kähler potential, K, of the following type

$$K = -3m_P^2 \ln \left(1 - \frac{H^{c\dagger} H^c}{3m_P^2} - \frac{\bar{H}^c \bar{H}^{c\dagger}}{3m_P^2} - \frac{\mathsf{Tr}\left(G^\dagger G \right)}{6m_P^2} - \frac{|S|^2}{3m_P^2} + k_S \frac{|S|^4}{3m_P^4} + \frac{k_H}{2m_P^2} \left(\bar{H}^c H^c + \text{h.c.} \right) \right), \tag{6}$$

where $m_P = 2.44 \cdot 10^{18}$ GeV is the reduced Planck scale and the complex scalar components of the superfields $H^c, \bar{H}^c, G$ and S are denoted by the same symbol. The coefficients k_S and k_H are taken real. From Eq. (6) we can infer that we adopt the standard quadratic non-minimal coupling for Higgs-inflaton, which respects the gauge and global symmetries of the model. This non-minimal coupling of the Higgs fields to gravity is transparent in the Jordan frame. We also added the fifth term in the RHS of Eq. (6) in order to cure the tachyonic mass problem encountered in similar models [10–12] – see Sec. 3.1. In terms of the components of the various fields, K in Eq. (6) reads

$$K = -3m_P^2 \ln \left(1 - \frac{\phi^\alpha \phi^{*\bar{\alpha}}}{3m_P^2} + k_S \frac{|S|^4}{3m_P^4} + \frac{k_H}{2m_P^2} \left(\nu_H^c \bar{\nu}_H^c + e_H^c \bar{e}_H^c + u_H^c \bar{u}_H^c + d_H^c \bar{d}_H^c + \text{h.c.} \right) \right) \tag{7a}$$

with

$$\phi^{\alpha} = \nu^c_H, \bar{\nu}^c_H, e^c_H, \bar{e}^c_H, u^c_H, \bar{u}^c_H, d^c_H, \bar{d}^c_H, g^c, \bar{g}^c \text{ and } S \tag{7b}$$

and summation over the repeated Greek indices is implied.

2.3. The SUSY limit

In the limit where $m_{\rm P}$ tends to infinity, we can obtain the SUSY limit of the SUGRA potential. Assuming that the SM non-singlet components vanish, the F-term potential in this limit, $V_{\rm F}$, turns out to be

$$V_{\rm F} = \lambda^2 \left|\bar{\nu}^c_H \nu^c_H - M^2_{\rm PS}\right|^2 + \lambda^2 |S|^2 \left(|\nu^c_H|^2 + |\bar{\nu}^c_H|^2\right), \tag{8a}$$

while the D-term potential is

$$V_{\rm D} = \frac{5g^2}{16}\left(|\nu^c_H|^2 - |\bar{\nu}^c_H|^2\right)^2. \tag{8b}$$

Restricting ourselves to the D-flat direction $|\nu^c_H| = |\bar{\nu}^c_H|$, we find from $V_{\rm F}$ that the SUSY vacuum lies at

$$\langle S\rangle \simeq 0 \text{ and } |\langle \nu^c_H\rangle| = |\langle \bar{\nu}^c_H\rangle| = M_{\rm PS}. \tag{9}$$

Therefore, $W_{\rm HPS}$ leads to spontaneous breaking of $G_{\rm PS}$. As we shall see in Sec. 3, the same superpotential, $W_{\rm HPS}$, gives rise to a stage of nMHI . Indeed, along the D-flat direction $|\nu^c_H| = |\bar{\nu}^c_H| \gg M_{\rm PS}$ and $S = 0$, $V_{\rm SUSY}$ tends to a quartic potential, which can be employed in conjunction with K in Eq. (6) for the realization of nMHI along the lines of Ref. [12].

It should be mentioned that soft SUSY breaking and instanton effects explicitly break $U(1)_R \times U(1)_{\rm PQ}$ to $\mathbb{Z}_2 \times \mathbb{Z}_6$. The latter symmetry is spontaneously broken by $\langle P\rangle$ and $\langle \bar{P}\rangle$. This would lead to a domain wall problem if the PQ transition took place after nMHI. However, as we already mentioned above, $U(1)_{\rm PQ}$ is assumed already broken before or during nMHI. The final unbroken symmetry of the model is $G_{\rm SM} \times \mathbb{Z}_2^{\rm mp}$.

3. The inflationary scenario

Next we outline the salient features of our inflationary scenario (Sec. 3.1) and calculate a number of observable quantities in Sec. 3.2.

3.1. Structure of the inflationary potential

At tree-level the *Einstein Frame* (EF) SUGRA potential, $\widehat{V}_{\rm HI}$, is given by [11]

$$\widehat{V}_{\rm HI} = e^{K/m^2_{\rm P}}\left(K^{\alpha\bar{\beta}} {\rm F}_{\alpha} {\rm F}^*_{\bar{\beta}} - 3\frac{|W_{\rm HPS}|^2}{m^2_{\rm P}}\right) + \frac{1}{2} g^2 \sum_a D_a D_a, \tag{10a}$$

where g is the unified gauge coupling constant and the summation is applied over the 21 generators T_a of the PS gauge group – see Ref. [5]. Also, we have

$$K_{\alpha\bar{\beta}} = K_{,\phi^\alpha\phi^{*\beta}},\ K^{\bar{\beta}\alpha}K_{\alpha\bar{\gamma}} = \delta^{\bar{\beta}}_{\bar{\gamma}},\ F_\alpha = W_{\rm HPS,\phi^\alpha} + K_{,\phi^\alpha}W_{\rm HPS}/m_{\rm P}^2 \text{ and } D_a = \phi_\alpha\left(T_a\right)^\alpha_\beta K_{,\phi^\beta} \quad (10b)$$

The ϕ^α's are given in Eq. (7b). If we parameterize the SM singlet components of H^c and $\bar{H}^c$ by

$$\nu^c_H = he^{i\theta}\cos\theta_\nu/\sqrt{2} \text{ and } \bar{\nu}^c_H = he^{i\bar{\theta}}\sin\theta_\nu/\sqrt{2}, \quad (11)$$

we can easily deduce that a D-flat direction occurs at

$$\theta = \bar{\theta} = 0,\ \theta_\nu = \pi/4 \text{ and } e^c_H = \bar{e}^c_H = u^c_H = \bar{u}^c_H = d^c_H = \bar{d}^c_H = g^c = \bar{g}^c = 0. \quad (12)$$

Along this direction, the D-terms in Eq. (10a) – and, also, $V_{\rm D}$ in Eq. (8b) – vanish, and so $\widehat{V}_{\rm HI}$ takes the form

$$\widehat{V}_{\rm HI} = m_{\rm P}^4\frac{\lambda^2(x_h^2-4m_{\rm PS}^2)^2}{16f^2} \quad (13)$$

with
$$f = 1 + c_{\mathcal{R}}x_h^2,\ m_{\rm PS} = \frac{M_{\rm PS}}{m_{\rm P}},\ x_h = \frac{h}{m_{\rm P}} \text{ and } c_{\mathcal{R}} = -\frac{1}{6}+\frac{k_H}{4}. \quad (14)$$

From Eq. (13), we can verify that for $c_{\mathcal{R}} \gg 1$ and $m_{\rm PS} \ll 1$, $\widehat{V}_{\rm HI}$ takes a form suitable for the realization of nMHI, since it develops a plateau – see also Sec. 3.2. The (almost) constant potential energy density $\widehat{V}_{\rm HI0}$ and the corresponding Hubble parameter $\widehat{H}_{\rm HI}$ (along the trajectory in Eq. (12)) are given by

$$\widehat{V}_{\rm HI0} = \frac{\lambda^2h^4}{16f^2} \simeq \frac{\lambda^2m_{\rm P}^4}{16c_{\mathcal{R}}^2} \text{ and } \widehat{H}_{\rm HI} = \frac{\widehat{V}_{\rm HI0}^{1/2}}{\sqrt{3}m_{\rm P}} \simeq \frac{\lambda m_{\rm P}}{4\sqrt{3}c_{\mathcal{R}}}. \quad (15)$$

We next proceed to check the stability of the trajectory in Eq. (12) w.r.t the fluctuations of the various fields. To this end, we expand them in real and imaginary parts as follows

$$X = \frac{x_1+ix_2}{\sqrt{2}},\ \bar{X} = \frac{\bar{x}_1+i\bar{x}_2}{\sqrt{2}} \text{ where } X = e^c_H, u^c_H, d^c_H, g^c \text{ and } x = e, u, d, g. \quad (16)$$

Notice that the field S can be rotated to the real axis via a suitable R transformation. Along the trajectory in Eq. (12) we find

$$\left(K_{\alpha\bar{\beta}}\right) = \text{diag}\left(\frac{M_K}{f^2}, \underbrace{\frac{1}{f},\ldots,\frac{1}{f}}_{3+6\cdot3\text{ times}}\right) \text{ with } M_K = \begin{pmatrix}\kappa & \bar{\kappa}\\ \bar{\kappa} & \kappa\end{pmatrix},\ \bar{\kappa} = 3c_{\mathcal{R}}^2x_h^2 \text{ and } \kappa = f+\bar{\kappa}. \quad (17)$$

To canonically normalize the fields ν^c_H and $\bar{\nu}^c_H$, we first diagonalize the matrix M_K. This can be achieved via a similarity transformation involving an orthogonal matrix U_K as follows:

$$U_K M_K U_K^{\mathsf{T}} = \mathsf{diag}\left(\bar{f}, f\right), \text{ where } \bar{f} = f + 6c_{\mathcal{R}}^2 x_h^2 \text{ and } U_K = \frac{1}{\sqrt{2}}\begin{pmatrix} 1 & 1 \\ -1 & 1 \end{pmatrix}. \tag{18}$$

Utilizing U_K, the kinetic terms of the various fileds can be brought into the following form

$$K_{\alpha\bar{\beta}}\dot{\phi}^{\alpha}\dot{\phi}^{*\bar{\beta}} = \frac{\bar{f}}{2f^2}\left(\dot{h}^2 + \frac{1}{2}h^2\dot{\theta}_+^2\right) + \frac{h^2}{2f}\left(\frac{1}{2}\dot{\theta}_-^2 + \dot{\theta}_\nu^2\right) + \frac{1}{2f}\dot{\chi}_\alpha\dot{\chi}_\alpha = \frac{1}{2}\dot{\widehat{h}}^2 + \frac{1}{2}\dot{\widehat{\psi}}_\alpha\dot{\widehat{\psi}}_\alpha, \tag{19}$$

where $\theta_\pm = \left(\bar{\theta} \pm \theta\right)/\sqrt{2}$, $\chi_\alpha = x_1, x_2, \bar{x}_1, \bar{x}_2, S$ and $\psi_\alpha = \theta_+, \theta_-, \theta_\nu, \chi_\alpha$ and the dot denotes derivation w.r.t the cosmic time, t. In the last line, we introduce the EF canonically normalized fields, $\widehat{h}$ and $\widehat{\psi}$, which can be obtained as follows – cf. Ref. [5, 11, 12, 37]:

$$\frac{d\widehat{h}}{dh} = J = \frac{\sqrt{\bar{f}}}{f}, \ \widehat{\theta}_+ = \frac{Jh\theta_+}{\sqrt{2}}, \ \widehat{\theta}_- = \frac{h\theta_-}{\sqrt{2f}}, \ \widehat{\theta}_\nu = \frac{h}{\sqrt{f}}\left(\theta_\nu - \frac{\pi}{4}\right) \text{ and } \widehat{\chi}_\alpha = \frac{\chi_\alpha}{\sqrt{f}}. \tag{20}$$

Taking into account the approximate expressions for $\dot{h}$, J and the slow-roll parameters $\widehat{\epsilon}$, $\widehat{\eta}$, which are displayed in Sec. 3.2, we can verify that, during a stage of slow-roll inflation, $\dot{\widehat{\theta}}_+ \simeq Jh\dot{\theta}_+/\sqrt{2}$ since $Jh \simeq \sqrt{6}m_{\rm P}$, $\dot{\widehat{\theta}}_- \simeq h\dot{\theta}_-/\sqrt{2f}$ and $\dot{\widehat{\theta}}_\nu \simeq h\dot{\theta}_\nu/\sqrt{f}$ since $h/\sqrt{f} \simeq m_{\rm P}/\sqrt{c_{\mathcal{R}}}$. On the other hand, we can show that $\dot{\widehat{\chi}}_\alpha \simeq \dot{\chi}_\alpha/\sqrt{f}$, since the quantity $\dot{f}/2f^{3/2}\chi_\alpha$, involved in relating $\dot{\chi}_\alpha$ to $\dot{\widehat{\chi}}_\alpha$, turns out to be negligibly small compared with $\dot{\widehat{\chi}}_\alpha$. Indeed, the $\widehat{\chi}_\alpha$'s acquire effective masses $m_{\widehat{\chi}_\alpha} \gg \widehat{H}_{\rm HI}$ – see below – and therefore enter a phase of oscillations about $\widehat{\chi}_\alpha = 0$ with decreasing amplitude. Neglecting the oscillatory part of the relevant solutions, we find

$$\chi \simeq \widehat{\chi}_{\alpha 0}\sqrt{f}e^{-2\widehat{N}/3} \text{ and } \dot{\widehat{\chi}}_\alpha \simeq -2\chi_{\alpha 0}\sqrt{f}\widehat{H}_{\rm HI}\widehat{\eta}_{\chi_\alpha}e^{-2\widehat{N}/3}, \tag{21}$$

where $\widehat{\chi}_{\alpha 0}$ represents the initial amplitude of the oscillations, $\widehat{\eta}_{\chi_\alpha} = m^2_{\widehat{\chi}_\alpha}/3\widehat{H}_{\rm HI}$ and we assume $\dot{\widehat{\chi}}_\alpha(t=0) = 0$. Taking into account the approximate expressions for $\dot{h}$ and the slow-roll parameter $\widehat{\epsilon}$ in Sec. 3.2, we find

$$-\dot{f}/2f^{3/2}\chi_\alpha = \left(c_{\mathcal{R}}\widehat{\epsilon}\widehat{H}^2_{\rm HI}/m^2_{\widehat{\chi}_\alpha}\right)\dot{\widehat{\chi}}_\alpha \ll \dot{\widehat{\chi}}_\alpha. \tag{22}$$

Having defined the canonically normalized scalar fields, we can proceed in investigating the stability of the inflationary trajectory of Eq. (12). To this end, we expand $\widehat{V}_{\rm HI}$ in Eq. (10a) to quadratic order in the fluctuations around the direction of Eq. (12), as described in detail in Ref. [5]. In Table 2 we list the eigenvalues of the mass-squared matrices

$$M^2_{\alpha\beta} = \left.\frac{\partial^2\widehat{V}_{\rm HI}}{\partial\widehat{\psi}_\alpha\partial\widehat{\psi}_\beta}\right|_{\text{Eq. (12)}} \quad \text{with } \psi_\alpha = \theta_+, \theta_-, \theta_\nu, x_1, x_2, \bar{x}_1, \bar{x}_2 \text{ and } S \tag{23}$$

Fields	Masses Squared	Eigenstates
	THE $S - \nu^c_H - \bar{\nu}^c_H$ SECTOR	
2 real scalars	$m^2_{\widehat{\theta}_\nu} = m^2_{\rm P}x^2_h\left(2\lambda^2(x^2_h-6)+15g^2f\right)/24f^2$	$\widehat{\theta}_\nu$
	$m^2_{\widehat{\theta}+} = \lambda^2 m^4_{\rm P}x^2_h\left(1+6c_{\mathcal{R}}\right)/12J^2f^3 \simeq 4\widehat{H}^2_{\rm HI}$	$\widehat{\theta}_+$
1 complex scalar	$m^2_{\widehat{S}} = \lambda^2 m^2_{\rm P}x^2_h(12+x^2_h\bar{f})\left(6k_Sf-1\right)/6f^2\bar{f}$	$\widehat{S}$
	THE $u^c_{H{\rm a}} - \bar{u}^c_{H{\rm a}}$ (a $=1,2,3$) AND $e^c_H - \bar{e}^c_H$ SECTORS	
$2(3+1)$ real scalars	$m^2_{\widehat{u}-} = m^2_{\rm P}x^2_h\left(\lambda^2(x^2_h-3)+3g^2f\right)/12f^2$	$\widehat{u}^{\rm a}_{1-}$, $\widehat{u}^{\rm a}_{2+}$,
	$m^2_{\widehat{e}-} = m^2_{\widehat{u}-}$	$\widehat{e}_{1-}$, $\widehat{e}_{2+}$
	THE $d^c_{H{\rm a}} - \bar{d}^c_{H{\rm a}}$ AND $g^c_{\rm a} - \bar{g}^c_{\rm a}$ (a $=1,2,3$) SECTORS	
$3\cdot 8$ real scalars	$m^2_{\widehat{g}} = m^2_{\rm P}x^2_h\left(\lambda^2x^2_h+24\lambda^2_{\bar{H}}f\right)/24f^2$	$\widehat{g}^{\rm a}_1, \widehat{g}^{\rm a}_2$
	$m^2_{\widehat{\bar{g}}} = m^2_{\rm P}x^2_h\left(\lambda^2x^2_h+24\lambda^2_Hf\right)/24f^2$	$\widehat{\bar{g}}^{\rm a}_1, \widehat{\bar{g}}^{\rm a}_2$
	$m^2_{\widehat{d}+} = m^2_{\rm P}x^2_h\left(\lambda^2+4\lambda^2_Hf\right)/4f^2$	$\widehat{d}^{\rm a}_{1+}, \widehat{d}^{\rm a}_{2-}$
	$m^2_{\widehat{d}-} = m^2_{\rm P}x^2_h\left(\lambda^2\left(x^2_h-3\right)+12\lambda^2_Hf\right)/12f^2$	$\widehat{d}^{\rm a}_{1-}, \widehat{d}^{\rm a}_{2+}$

Table 2. The scalar mass spectrum of our model along the inflationary trajectory of Eq. (12). To avoid very lengthy formulas we neglect terms proportional to $m^2_{\rm PS}$ and we assume $\lambda_H \simeq \lambda_{\bar{H}}$ for the derivation of the masses of the scalars in the superfields d^c_H and $\bar{d}^c_H$.

involved in the expansion of $\widehat{V}_{\rm HI}$. We arrange our findings into three groups: the SM singlet sector, $S - \nu^c_H - \bar{\nu}^c_H$, the sector with the $u^c_H, \bar{u}^c_H$ and the $e^c_H, \bar{e}^c_H$ fields which are related with the broken generators of $G_{\rm PS}$ and the sector with the $d^c_H, \bar{d}^c_H$ and the $g^c, \bar{g}^c$ fields. Upon diagonalization of the relevant matrices we obtain the following mass eigenstates:

$$\widehat{x}_{1\pm} = \frac{1}{\sqrt{2}}\left(\widehat{\bar{x}}_1 \pm \widehat{x}_1\right) \text{ and } \widehat{x}_{2\mp} = \frac{1}{\sqrt{2}}\left(\widehat{\bar{x}}_2 \mp \widehat{x}_2\right) \text{ with } x = u, e, d \text{ and } g. \tag{24}$$

As we observe from the relevant eigenvalues, no instability – as the one found in Ref. [37] – arises in the spectrum. In particular, it is evident that $k_S \gtrsim 1$ assists us to achieve $m^2_{\widehat{S}} > 0$ – in accordance with the results of Ref. [12]. Moreover, the D-term contributions to $m^2_{\widehat{\theta}_\nu}$ and $m^2_{\widehat{u}-}$ – proportional to the gauge coupling constant $g \simeq 0.7$ – ensure the positivity of these masses squared. Finally the masses that the scalars $\widehat{d}_{1,2}$ acquire from the second and third term of the RHS of Eq. (5) lead to the positivity of $m^2_{\widehat{d}-}$ for λ_H of order unity. We have also numerically verified that the masses of the various scalars remain greater than the Hubble parameter during the last $50-60$ e-foldings of nMHI, and so any inflationary perturbations of the fields other than the inflaton are safely eliminated.

The 8 Goldstone bosons, associated with the modes $\widehat{x}_{1+}$ and $\widehat{x}_{2-}$ with $x = u^{\rm a}$ and e, are not exactly massless since $\widehat{V}_{{\rm HI},h} \neq 0$ – contrary to the situation of Ref. [30] where the direction with non vanishing $\langle \nu^c_H \rangle$ minimizes the potential. These masses turn out to be $m_{x0} = \lambda m_{\rm P}x_h/2f$. On the contrary, the angular parametrization in Eq. (11) assists us to isolate the massless mode $\widehat{\theta}_-$, in agreement with the analysis of Ref. [11]. Employing the well-known Coleman-Weinberg formula [15], we can compute the one-loop radiative corrections to the potential in our model. However, these have no significant effect on the inflationary dynamics and predictions, since the slope of the inflationary path is generated at the classical level – see the expressions for $\widehat{\epsilon}$ and $\widehat{\eta}$ below.

3.2. The inflationary observables

Based on the potential of Eq. (13) and keeping in mind that the EF canonically inflaton $\widehat{h}$ is related to h via Eq. (20), we can proceed to the analysis of nMHI in the EF, employing the standard slow-roll approximation. Namely, a stage of slow-roll nMHI is determined by the condition – see e.g. Ref. [38, 39]:

$$\max\{\widehat{\epsilon}(h), |\widehat{\eta}(h)|\} \leq 1,$$

where

$$\widehat{\epsilon} = \frac{m_{\rm P}^2}{2}\left(\frac{\widehat{V}_{{\rm HI},\widehat{h}}}{\widehat{V}_{\rm HI}}\right)^2 = \frac{m_{\rm P}^2}{2J^2}\left(\frac{\widehat{V}_{{\rm HI},h}}{\widehat{V}_{\rm HI}}\right)^2 \simeq \frac{4f_0^2 m_{\rm P}^4}{3c_{\mathcal{R}}^2 h^4} \tag{25a}$$

and

$$\widehat{\eta} = m_{\rm P}^2\,\frac{\widehat{V}_{{\rm HI},\widehat{h}\widehat{h}}}{\widehat{V}_{\rm HI}} = \frac{m_{\rm P}^2}{J^2}\left(\frac{\widehat{V}_{{\rm HI},hh}}{\widehat{V}_{\rm HI}} - \frac{\widehat{V}_{{\rm HI},h}}{\widehat{V}_{\rm HI}}\frac{J_{,h}}{J}\right) \simeq -\frac{4f_0 m_{\rm P}^2}{3c_{\mathcal{R}}h^2}\,, \tag{25b}$$

are the slow-roll parameters and $f_0 = f\left(\langle h\rangle = 2M_{\rm PS}\right) = 1 + 4c_{\mathcal{R}}m_{\rm PS}^2$ – see Sec. 4.1. Here we employ Eq. (15) and the following approximate relations:

$$J \simeq \sqrt{6}\frac{m_{\rm P}}{h},\;\; \widehat{V}_{{\rm HI},h} \simeq \frac{4\widehat{V}_{\rm HI}}{c_{\mathcal{R}}h^3}f_0 m_{\rm P}^2 \;\;\text{and}\;\; \widehat{V}_{{\rm HI},hh} \simeq -\frac{12\widehat{V}_{\rm HI}}{c_{\mathcal{R}}h^4}f_0 m_{\rm P}^2. \tag{26}$$

The numerical computation reveals that nMHI terminates due to the violation of the $\widehat{\epsilon}$ criterion at a value of h equal to $h_{\rm f}$, which is calculated to be

$$\widehat{\epsilon}\left(h_{\rm f}\right) = 1 \;\Rightarrow\; h_{\rm f} = (4/3)^{1/4} m_{\rm P}\sqrt{f_0/c_{\mathcal{R}}}. \tag{27}$$

The number of e-foldings, $\widehat{N}_*$, that the scale $k_* = 0.002/{\rm Mpc}$ suffers during nMHI can be calculated through the relation:

$$\widehat{N}_* = \frac{1}{m_{\rm P}^2}\int_{\widehat{h}_{\rm f}}^{\widehat{h}_*} d\widehat{h}\,\frac{\widehat{V}_{\rm HI}}{\widehat{V}_{{\rm HI},\widehat{h}}} = \frac{1}{m_{\rm P}^2}\int_{h_{\rm f}}^{h_*} dh\, J^2\frac{\widehat{V}_{\rm HI}}{\widehat{V}_{{\rm HI},h}}, \tag{28}$$

where h_* $[\widehat{h}_*]$ is the value of h $[\widehat{h}]$ when k_* crosses the inflationary horizon. Given that $h_{\rm f} \ll h_*$, we can write h_* as a function of $\widehat{N}_*$ as follows

$$\widehat{N}_* \simeq \frac{3c_{\mathcal{R}}}{4f_0}\frac{h_*^2 - h_{\rm f}^2}{m_{\rm P}^2} \Rightarrow h_* = 2m_{\rm P}\sqrt{\widehat{N}_* f_0/3c_{\mathcal{R}}}. \tag{29}$$

The power spectrum $\Delta_{\mathcal{R}}^2$ of the curvature perturbations generated by h at the pivot scale k_* is estimated as follows

$$\Delta_{\mathcal{R}} = \frac{1}{2\sqrt{3}\,\pi m_{\rm P}^3}\frac{\widehat{V}_{\rm HI}(\widehat{h}_*)^{3/2}}{|\widehat{V}_{{\rm HI},\widehat{h}}(\widehat{h}_*)|} \simeq \frac{\lambda h_*^2}{16\sqrt{2}\pi f_0 m_{\rm P}^2} \simeq \frac{\lambda\widehat{N}_*}{12\sqrt{2}\pi c_{\mathcal{R}}}. \tag{30}$$

Since the scalars listed in Table 2 are massive enough during nMHI, $\Delta_{\mathcal{R}}$ can be identified with its central observational value – see Sec. 5 – with almost constant $\widehat{N}_*$. The resulting relation reveals that λ is to be proportional to $c_{\mathcal{R}}$. Indeed we find

$$\lambda \simeq 8.4\cdot 10^{-4}\pi c_{\mathcal{R}}/\widehat{N}_* \Rightarrow c_{\mathcal{R}} \simeq 20925\lambda \;\text{ for }\; \widehat{N}_* \simeq 55. \tag{31}$$

The (scalar) spectral index $n_{\rm s}$, its running $a_{\rm s}$, and the scalar-to-tensor ratio r can be estimated through the relations:

$$n_{\rm s} = 1 - 6\widehat{\epsilon}_* + 2\widehat{\eta}_* \simeq 1 - 2/\widehat{N}_*, \tag{32a}$$

$$\alpha_{\rm s} = \frac{2}{3}\left(4\widehat{\eta}_*^2 - (n_{\rm s}-1)^2\right) - 2\widehat{\xi}_* \simeq -2\widehat{\xi}_* \simeq -2/\widehat{N}_*^2 \tag{32b}$$

and

$$r = 16\widehat{\epsilon}_* \simeq 12/\widehat{N}_*^2, \tag{32c}$$

where $\widehat{\xi} = m_{\rm P}^4\widehat{V}_{{\rm HI},h}\widehat{V}_{{\rm HI},\widehat{h}\widehat{h}\widehat{h}}/\widehat{V}_{\rm HI}^2 = m_{\rm P}\sqrt{2\widehat{\epsilon}}\,\widehat{\eta}_{,h}/J + 2\widehat{\eta}\widehat{\epsilon}$. The variables with subscript $*$ are evaluated at $h = h_*$ and Eqs. (25a) and (25b) have been employed.

4. Non-thermal leptogenesis

In this section, we specify how the SUSY inflationary scenario makes a transition to the radiation dominated era (Sec. 4.1) and give an explanation of the origin of the observed BAU (Sec. 4.2) consistently with the $\widetilde{G}$ constraint and the low energy neutrino data (Sec. 4.3).

4.1. The inflaton's decay

When nMHI is over, the inflaton continues to roll down towards the SUSY vacuum, Eq. (9). There is a brief stage of tachyonic preheating [40] which does not lead to significant particle production [41]. Soon after, the inflaton settles into a phase of damped oscillations initially around zero – where $\widehat{V}_{\rm HI0}$ has a maximum – and then around one of the minima of $\widehat{V}_{\rm HI0}$. Whenever the inflaton passes through zero, particle production may occur creating mostly superheavy bosons via the mechanism of instant preheating [42]. This process becomes more efficient as λ decreases, and further numerical investigation is required in order to check the viability of the non-thermal leptogenesis scenario for small values of λ. For this reason, we restrict to λ's larger than 0.001, which ensures a less frequent passage of the inflaton through zero, weakening thereby the effects from instant preheating and other parametric resonance effects – see Appendix B of Ref. [5]. Intuitively the reason is that larger λ's require larger $c_{\mathcal{R}}$'s, see Eq. (31), diminishing therefore $h_{\rm f}$ given by Eq. (29), which sets the amplitude of the very first oscillations.

Nonetheless the standard perturbative approach to the inflaton decay provides a very efficient decay rate. Namely, at the SUSY vacuum ν^c_H and $\bar\nu^c_H$ acquire the v.e.vs shown in Eq. (9) giving rise to the masses of the (canonically normalized) inflaton $\widehat{\delta h} = (h - 2M_{\rm PS})\,/\,J_0$ and RH neutrinos, $\widehat{\nu}^c_i$, which are given, respectively, by

$$\text{(a)}\;\; m_{\rm I} = \sqrt{2}\frac{\lambda M_{\rm PS}}{\langle J\rangle f_0} \;\;\text{and}\;\; \text{(b)}\;\; M_{i\widehat{\nu}^c} = 2\frac{\lambda_{i\nu^c} M^2_{\rm PS}}{M_{\rm S}\sqrt{\bar f_0}}, \tag{33}$$

where f_0 is defined below Eq. (25b) and $\bar f_0 = f_0 + 24c^2_{\mathcal{R}}m^2_{\rm PS} \simeq J_0^2$. Here, we assume the existence of a term similar to the second one inside ln of Eq. (7a) for ν^c_i too.

For larger λ's $\langle J\rangle = J(h = 2M_{\rm PS})$ ranges from 3 to 90 and so $m_{\rm I}$ is kept independent of λ and almost constant at the level of 10^{13} GeV. Indeed, if we express $\widehat{\delta h}$ as a function of δh through the relation

$$\frac{\widehat{\delta h}}{\delta h} \simeq J_0 \;\;\text{where}\;\; J_0 = \sqrt{1 + \frac{3}{2}m^2_{\rm P} f^2_{\mathcal{R},h}(\langle h\rangle)} = \sqrt{1 + 24c^2_{\mathcal{R}}m^2_{\rm PS}} \tag{34}$$

we find

$$m_{\rm I} \simeq \frac{\sqrt{2}\lambda M_{\rm PS}}{f_0 J_0} \simeq \frac{\lambda m_{\rm P}}{2\sqrt{3}c_{\mathcal{R}}} \simeq \frac{10^{-4}m_{\rm P}}{4.2\sqrt{3}} \simeq 3\cdot 10^{13}\;\text{GeV for}\;\; \lambda \gtrsim \frac{10^{-4}}{4.2\sqrt{6}m_{\rm PS}} \simeq 1.3\cdot 10^{-3} \tag{35}$$

where we make use of Eq. (31) – note that $f_0 \simeq 1$. The derivation of the (s)particle spectrum, listed in Table 2, at the SUSY vaccum of the model reveals [5] that perturbative decays of $\widehat{\delta h}$ into these massive particles are kinematically forbidden and therefore, narrow parametric resonance [40] effects are absent. Also $\widehat{\delta h}$ can not decay via renormalizable interaction terms to SM particles.

The inflaton can decay into a pair of $\widehat{\nu}^c_i$'s through the following lagrangian terms:

$$\mathcal{L}_{\mathrm{I}\widehat{\nu}_i^c} = -\lambda_{i\nu^c} \frac{M_{\mathrm{PS}}}{M_{\mathrm{S}}} \frac{f_0}{J_0} \left(1 - 12c_{\mathcal{R}} m_{\mathrm{PS}}^2\right) \widehat{\delta h} \widehat{\nu}_i^c \widehat{\nu}_i^c + \mathrm{h.c.}. \tag{36}$$

From Eq. (36) we deduce that the decay of $\widehat{\delta h}$ into $\widehat{\nu}_i^c$ is induced by two lagrangian terms. The first one originates exclusively from the non-renormalizable term of Eq. (5) – as in the case of a similar model in Ref. [30]. The second term is a higher order decay channel due to the SUGRA lagrangian – cf. Ref. [43]. The interaction in Eq. (36) gives rise to the following decay width

$$\Gamma_{\mathrm{I}j\widehat{\nu}^c} = \frac{c_{\mathrm{I}j\widehat{\nu}^c}^2}{64\pi} m_{\mathrm{I}} \sqrt{1 - \frac{4M_{j\widehat{\nu}^c}^2}{m_{\mathrm{I}}^2}} \quad \text{with} \quad c_{\mathrm{I}j\widehat{\nu}^c} = \frac{M_{j\widehat{\nu}^c}}{M_{\mathrm{PS}}} \frac{f_0^{3/2}}{J_0} \left(1 - 12c_{\mathcal{R}} m_{\mathrm{PS}}^2\right), \tag{37}$$

where $M_{j\widehat{\nu}^c}$ is the Majorana mass of the $\widehat{\nu}_j^c$'s into which the inflaton can decay. The implementation – see Sec. 4.3 – of the seesaw mechanism for the derivation of the light-neutrinos masses, in conjunction with the G_{PS} prediction $m_{3\mathrm{D}} \simeq m_t$ and our assumption that $m_{1\mathrm{D}} < m_{2\mathrm{D}} \ll m_{3\mathrm{D}}$ – see Sec. 5.1 – results to $2M_{3\widehat{\nu}^c} > m_{\mathrm{I}}$. Therefore, the kinematically allowed decay channels of $\widehat{\delta h}$ are those into $\widehat{\nu}_j^c$ with $j = 1$ and 2. Note that the decay of the inflaton to the heaviest of the $\widehat{\nu}_j^c$'s ($\widehat{\nu}_3^c$) is also disfavored by the $\widetilde{G}$ constraint – see below.

In addition, there are SUGRA-induced [43] – i.e., even without direct superpotential couplings – decay channels of the inflaton to the MSSM particles via non-renormalizable interaction terms. For a typical trilinear superpotential term of the form $W_y = yXYZ$, we obtain the effective interactions described by the langrangian part

$$\mathcal{L}_{\mathrm{I}y} = 6yc_{\mathcal{R}} \frac{M_{\mathrm{PS}}}{m_{\mathrm{P}}^2} \frac{f_0^{3/2}}{2J_0} \widehat{\delta h} \left(\widehat{X}\widehat{\psi}_Y\widehat{\psi}_Z + \widehat{Y}\widehat{\psi}_X\widehat{\psi}_Z + \widehat{Z}\widehat{\psi}_X\widehat{\psi}_Y\right) + \mathrm{h.c.}, \tag{38}$$

where y is a Yukawa coupling constant and ψ_X, ψ_Y and ψ_Z are the chiral fermions associated with the superfields X, Y and Z. Their scalar components are denoted with the superfield symbol. Taking into account the terms of Eq. (2) and the fact that the adopted SUSY GUT predicts YU for the 3rd generation at M_{PS}, we conclude that the interaction above gives rise to the following 3-body decay width

$$\Gamma_{\mathrm{I}y} = \frac{14c_{\mathrm{I}y}^2}{512\pi^3} m_{\mathrm{I}}^3 \simeq \frac{3y_{33}^2}{64\pi^3} f_0^3 \left(\frac{m_{\mathrm{I}}}{m_{\mathrm{P}}}\right)^2 m_{\mathrm{I}} \quad \text{where} \quad c_{\mathrm{I}y} = 6y_{33}c_{\mathcal{R}} \frac{M_{\mathrm{PS}}}{m_{\mathrm{P}}^2} \frac{f_0^{3/2}}{J_0}, \tag{39}$$

with $y_{33} \simeq (0.55 - 0.7)$ being the common Yukawa coupling constant of the third generation computed at the m_{I} scale, and summation is taken over color, weak and hypercharge degrees of freedom, in conjunction with the assumption that $m_{\mathrm{I}} < 2M_{3\widehat{\nu}^c}$.

Since the decay width of the produced $\widehat{\nu}^c_j$ is much larger than $\Gamma_{\rm I}$– see below – the reheating temperature, $T_{\rm rh}$, is exclusively determined by the inflaton decay and is given by [44]

$$T_{\rm rh} = \left(\frac{72}{5\pi^2 g_*}\right)^{1/4} \sqrt{\Gamma_{\rm I} m_{\rm P}} \;\; \text{with} \;\; \Gamma_{\rm I} = \Gamma_{{\rm I}1\widehat{\nu}^c} + \Gamma_{{\rm I}2\widehat{\nu}^c} + \Gamma_{{\rm I}y}, \tag{40}$$

where g_* counts the effective number of relativistic degrees of freedom at temperature $T_{\rm rh}$. For the MSSM spectrum plus the particle content of the superfields P and $\bar{P}$ we find $g_* \simeq 228.75 + 4(1+7/8) = 236.25$.

4.2. Lepton-number and gravitino abundances

If $T_{\rm rh} \ll M_{i\widehat{\nu}^c}$, the out-of-equilibrium condition [2] for the implementation of nTL is automatically satisfied. Subsequently $\widehat{\nu}^c_i$ decay into H_u and L^*_i via the tree-level couplings derived from the second term in the RHS of Eq. (2). Interference between tree-level and one-loop diagrams generates a lepton-number asymmetry (per $\widehat{\nu}^c_j$ decay) ε_j [2], when CP conservation is violated. The resulting lepton-number asymmetry after reheating can be partially converted through sphaleron effects into baryon-number asymmetry. In particular, the B yield can be computed as

$$({\sf a})\;\; Y_B = -0.35 Y_L \;\; \text{with} \;\; ({\sf b})\;\; Y_L = 2\frac{5}{4}\frac{T_{\rm rh}}{m_{\rm I}} \sum_{j=1}^{2} \frac{\Gamma_{{\rm I}j\widehat{\nu}^c}}{\Gamma_{\rm I}} \varepsilon_j. \tag{41}$$

The numerical factor in the RHS of Eq. (41a) comes from the sphaleron effects, whereas the one (5/4) in the RHS of Eq. (41b) is due to the slightly different calculation [44] of $T_{\rm rh}$ – cf. Ref. [1]. In the major part of our allowed parameter space – see Sec. 5.2 – $\Gamma_{\rm I} \simeq \Gamma_{{\rm I}y}$ and so the involved branching ratio of the produced $\widehat{\nu}^c_i$ is given by

$$\frac{\Gamma_{{\rm I}1\widehat{\nu}^c} + \Gamma_{{\rm I}2\widehat{\nu}^c}}{\Gamma_{\rm I}} \simeq \frac{\Gamma_{{\rm I}2\widehat{\nu}^c}}{\Gamma_{{\rm I}y}} = \frac{\pi^2 \left(1 - 12 c_{\mathcal{R}} m^2_{\rm PS}\right)^2}{72 c^2_{\mathcal{R}} y^2_{33} m^4_{\rm PS}} \frac{M^2_{2\widehat{\nu}^c}}{m^2_{\rm I}}. \tag{42}$$

For $M_{2\widehat{\nu}^c} \simeq \left(10^{11} - 10^{12}\right)$ GeV the ratio above takes adequately large values so that Y_L is sizable. Therefore, the presence of more than one inflaton decay channels does not invalidate the scenario of nTL.

It is worth emphasizing, however, that if $M_{1\nu^c} \lesssim 10 T_{\rm rh}$, part of the Y_L can be washed out due to $\widehat{\nu}^c_1$ mediated inverse decays and $\Delta L = 1$ scatterings – this possibility is analyzed in Ref. [27]. Trying to avoid the relevant computational complications we limit ourselves to cases with $M_{1\widehat{\nu}^c} \gtrsim 10 T_{\rm rh}$, so as any washout of the non-thermally produced Y_L is evaded. On the other hand, Y_L is not erased by the $\Delta L = 2$ scattering processes [45] at all temperatures T with $100\ {\rm GeV} \lesssim T \lesssim T_{\rm rh}$ since Y_L is automatically protected by SUSY [46] for $10^7\ {\rm GeV} \lesssim T \lesssim T_{\rm rh}$ and for $T \lesssim 10^7$ GeV these processes are well out of equilibrium provided that that mass of the heaviest light neutrino is 10 eV. This constraint, however, is overshadowed by a more stringent one induced by WMAP7 data [16] – see Sec. 5.1.

The required for successful nTL $T_{\rm rh}$ must be compatible with constraints on the $\widetilde{G}$ abundance, $Y_{\widetilde{G}}$, at the onset of *nucleosynthesis* (BBN). This is estimated to be [19]:

$$Y_{\widetilde{G}} \simeq c_{\widetilde{G}} T_{\rm rh} \text{ with } c_{\widetilde{G}} = 1.9 \cdot 10^{-22}/\text{GeV}, \tag{43}$$

where we assume that $\widetilde{G}$ is much heavier than the gauginos. Let us note that non-thermal $\widetilde{G}$ production within SUGRA is [43] also possible. However, we here prefer to adopt the conservative approach based on the estimation of $Y_{\widetilde{G}}$ via Eq. (43) since the latter $\widetilde{G}$ production depends on the mechanism of SUSY breaking.

Both Eqs. (41) and (43) yield the correct values of the B and $\widetilde{G}$ abundances provided that no entropy production occurs for $T < T_{\rm rh}$. This fact can be easily achieved within our setting. The mass spectrum of the P-$\bar{P}$ system is comprised by axion and saxion $P_- = (\bar{P} - P)/\sqrt{2}$, axino $\psi_- = (\psi_{\bar{P}} - \psi_P)/\sqrt{2}$, a higgs, $P_+ = (\bar{P} + P)/\sqrt{2}$, and a higgsino, $\psi_+ = (\psi_{\bar{P}} + \psi_P)/\sqrt{2}$, with mass of order 1 TeV and ψ denoting a Weyl spinor. The higgs and higgsinos can decay to lighter higgs and higgsinos before domination [36]. Regarding the saxion, P_-, we can assume that its decay mode to axions is suppressed (w.r.t the ones to gluons, higgses and higgsinos [47, 48]) and the initial amplitude of its oscillations is equal to $f_a \simeq 10^{12}$ GeV. Under these circumstances, it can [47] decay before domination too, and evades [48] the constraints from the effective number of neutrinos for the f_a's and $T_{\rm rh}$'s encountered in our model. As a consequence of its relatively large decay temperature, the LSPs produced by the saxion decay are likely to be thermalized and therefore, no upper bound on the saxion abundance is [48] to be imposed. Finally, axino can not play the role of LSP due to its large expected mass and the relatively high $T_{\rm rh}$'s encountered in our set-up which result to a large *Cold Dark Matter* (CDM) abundance. Nonetheless, it may enhance non-thermally the abundance of a higgsino-like neutralino-LSP, rendering it a successful CDM candidate.

4.3. Lepton-number asymmetry and neutrino masses

As mentioned above, the decay of $\widehat{\nu}^c_2$ and $\widehat{\nu}^c_1$, emerging from the $\widehat{\delta h}$ decay, can generate a lepton asymmetry, ε_i (with $i = 1, 2$) caused by the interference between the tree and one-loop decay diagrams, provided that a CP-violation occurs in h_{Nij}'s. The produced ε_i can be expressed in terms of the Dirac mass matrix of ν_i, $m_{\rm D}$, defined in a basis (called ν^c_i*-basis* henceforth) where ν^c_i are mass eigenstates, as follows:

$$\varepsilon_i = \sum_{i \neq j} \frac{\text{Im}\left[(m_{\rm D}^\dagger m_{\rm D})^2_{ij}\right]}{8\pi \langle H_u \rangle^2 (m_{\rm D}^\dagger m_{\rm D})_{ii}} \left(F_{\rm S}\left(x_{ij}, y_i, y_j\right) + F_{\rm V}(x_{ij}) \right), \tag{44a}$$

where we take $\langle H_u \rangle \simeq 174$ GeV, for large $\tan\beta$ and

$$x_{ij} := \frac{M_{j\widehat{\nu}^c}}{M_{i\widehat{\nu}^c}} \text{ and } y_i := \frac{\Gamma_{i\nu^c}}{M_{i\widehat{\nu}^c}} = \frac{(m_{\rm D}^\dagger m_{\rm D})_{ii}}{8\pi \langle H_u \rangle^2} \tag{44b}$$

(with $i, j = 1, 2, 3$). Also $F_{\rm V}$ and $F_{\rm S}$ represent, respectively, the contributions from vertex and self-energy diagrams which in SUSY theories read [49]

$$F_{\rm V}(x) = -x\ln\left(1+x^{-2}\right) \quad\text{and}\quad F_{\rm S}(x,y,z) = \frac{-2x(x^2-1)}{\left(x^2-1\right)^2+\left(x^2z-y\right)^2}. \tag{44c}$$

Note that for strongly hierarchical $M_{\widehat{\nu}^c}$'s with $x_{ij} \gg 1$ and $x_{ij} \gg y_i, y_j$, we obtain the well-known approximate result [26, 27]

$$F_{\rm V} + F_{\rm S} \simeq -3/x_{ij}^2. \tag{45}$$

The involved in Eq. (44a) $m_{\rm D}$ can be diagonalized if we define a basis – called *weak basis* henceforth – in which the lepton Yukawa couplings and the $SU(2)_{\rm L}$ interactions are diagonal in the space of generations. In particular we have

$$U^\dagger m_{\rm D} U^{c\dagger} = d_{\rm D} = \mathsf{diag}\left(m_{1\rm D}, m_{2\rm D}, m_{3\rm D}\right), \tag{46}$$

where U and U^c are 3×3 unitary matrices which relate L_i and ν_i^c (in the ν_i^c-basis) with the ones L_i' and $\nu_i^{c\prime}$ in the weak basis as follows:

$$L' = LU \quad\text{and}\quad \nu^{c\prime} = U^c\nu^c. \tag{47}$$

Here, we write LH lepton superfields, i.e. $SU(2)_{\rm L}$ doublet leptons, as row 3-vectors in family space and RH anti-lepton superfields, i.e. $SU(2)_{\rm L}$ singlet anti-leptons, as column 3-vectors. Consequently, the combination $m_{\rm D}^\dagger m_{\rm D}$ appeared in Eq. (44a) turns out to be a function just of $d_{\rm D}$ and U^c. Namely,

$$m_{\rm D}^\dagger m_{\rm D} = U^{c\dagger} d_{\rm D}^\dagger d_{\rm D} U^c. \tag{48}$$

The connection of the nTL scenario with the low energy neutrino data can be achieved through the seesaw formula, which gives the light-neutrino mass matrix m_ν in terms of $m_{i\rm D}$ and $M_{i\widehat{\nu}^c}$. Working in the ν_i^c-basis, we have

$$m_\nu = -m_{\rm D}\, d_{\nu^c}^{-1}\, m_{\rm D}^{\mathsf T}, \quad\text{where}\quad d_{\nu^c} = \mathsf{diag}\left(M_{1\widehat{\nu}^c}, M_{2\widehat{\nu}^c}, M_{3\widehat{\nu}^c}\right) \tag{49}$$

with $M_{1\widehat{\nu}^c} \leq M_{2\widehat{\nu}^c} \leq M_{3\widehat{\nu}^c}$ real and positive. Solving Eq. (46) w.r.t $m_{\rm D}$ and inserting the resulting expression in Eq. (49) we extract the mass matrix

$$\bar{m}_\nu = U^\dagger m_\nu U^* = -d_{\rm D} U^c d_{\nu^c}^{-1} U^{c\mathsf T} d_{\rm D}, \tag{50}$$

which can be diagonalized by the unitary PMNS matrix satisfying

$$\bar{m}_\nu = U_\nu^*\ \mathsf{diag}\left(m_{1\nu}, m_{2\nu}, m_{3\nu}\right)\ U_\nu^\dagger \tag{51}$$

and parameterized as follows:

$$U_\nu = \begin{pmatrix} c_{12}c_{13} & s_{12}c_{13} & s_{13}e^{-i\delta} \\ -c_{23}s_{12} - s_{23}c_{12}s_{13}e^{i\delta} & c_{23}c_{12} - s_{23}s_{12}s_{13}e^{i\delta} & s_{23}c_{13} \\ s_{23}s_{12} - c_{23}c_{12}s_{13}e^{i\delta} & -s_{23}c_{12} - c_{23}s_{12}s_{13}e^{i\delta} & c_{23}c_{13} \end{pmatrix} \cdot \begin{pmatrix} e^{-i\varphi_1/2} & & \\ & e^{-i\varphi_2/2} & \\ & & 1 \end{pmatrix}, \quad (52)$$

with $c_{ij} := \cos\theta_{ij}$, $s_{ij} := \sin\theta_{ij}$, δ the CP-violating Dirac phase and φ_1 and φ_2 the two CP-violating Majorana phases.

Following a bottom-up approach, along the lines of Ref. [26–28], we can find $\bar{m}_\nu$ via Eq. (51) using as input parameters the low energy neutrino observables, the CP violating phases and adopting the normal or inverted hierarchical scheme of neutrino masses. Taking also $m_{i\mathrm{D}}$ as input parameters we can construct the complex symmetric matrix

$$W = -d_{\mathrm{D}}^{-1}\bar{m}_\nu d_{\mathrm{D}}^{-1} = U^c d_{\nu^c} U^{c\mathsf{T}} \quad (53)$$

– see Eq. (50) – from which we can extract d_{ν^c} as follows:

$$d_{\nu^c}^{-2} = U^{c\dagger} W W^\dagger U^c. \quad (54)$$

Note that $WW^\dagger$ is a 3×3 complex, hermitian matrix and can be diagonalized following the algorithm described in Ref. [50]. Having determined the elements of U^c and the $M_{i\widehat{\nu}^c}$'s we can compute m_{D} through Eq. (48) and the ε_i's through Eq. (44a).

5. Constraining the model parameters

We exhibit the constraints that we impose on our cosmological set-up in Sec. 5.1, and delineate the allowed parameter space of our model in Sec. 5.2.

5.1. Imposed constraints

The parameters of our model can be restricted once we impose the following requirements:

1. According to the inflationary paradigm, the horizon and flatness problems of the standard Big Bang cosmology can be successfully resolved provided that the number of e-foldings, $\widehat{N}_*$, that the scale $k_* = 0.002/\mathrm{Mpc}$ suffers during nMHI takes a certain value, which depends on the details of the cosmological model. Employing standard methods [51], we can easily derive the required $\widehat{N}_*$ for our model, consistently with the fact that the $P - \bar{P}$ system remains subdominant during the post-inflationary era. Namely we obtain

$$\widehat{N}_* \simeq 22.5 + 2\ln\frac{V_{\mathrm{HI}}(h_*)^{1/4}}{1\ \mathrm{GeV}} - \frac{4}{3}\ln\frac{V_{\mathrm{HI}}(h_{\mathrm{f}})^{1/4}}{1\ \mathrm{GeV}} + \frac{1}{3}\ln\frac{T_{\mathrm{rh}}}{1\ \mathrm{GeV}} + \frac{1}{2}\ln\frac{f(h_{\mathrm{f}})}{f(h_*)}. \quad (55)$$

2. The inflationary observables derived in Sec. 3.2 are to be consistent with the fitting [16] of the WMAP7, BAO and H_0 data. As usual, we adopt the central value of $\Delta_{\mathcal{R}}$, whereas we allow the remaining quantities to vary within the 95% *confidence level* (c.l.) ranges. Namely,

$$(a)\ \Delta_{\mathcal{R}} \simeq 1.93\cdot 10^{-5},\ (b)\ n_s = 0.968\pm 0.024,\ (c)\ -0.062 \leq a_s \leq 0.018 \text{ and } (d)\ r < 0.24 \tag{56}$$

3. The scale $M_{\rm PS}$ can be determined by requiring that the v.e.vs of the Higgs fields take the values dictated by the unification of the gauge couplings within the MSSM. As we now recognize – cf. Ref. [5] – the unification scale $M_{\rm GUT} \simeq 2\cdot 10^{16}$ GeV is to be identified with the *lowest* mass scale of the model in the SUSY vacuum, Eq. (9), in order to avoid any extra contribution to the running of the MSSM gauge couplings, i.e.,

$$(a)\ \frac{gM_{\rm PS}}{\sqrt{f_0}} = M_{\rm GUT} \Rightarrow m_{\rm PS} = \frac{1}{2\sqrt{2c_{\mathcal{R}}^{\rm max}-c_{\mathcal{R}}}} \text{ with } (b)\ c_{\mathcal{R}}^{\rm max} = \frac{g^2m_{\rm P}^2}{8M_{\rm GUT}^2} \tag{57}$$

The requirement $2c_{\mathcal{R}}^{\rm max} - c_{\mathcal{R}} > 0$ sets an upper bound $c_{\mathcal{R}} < 2c_{\mathcal{R}}^{\rm max} \simeq 1.8\cdot 10^3$, which however can be significantly lowered if we combine Eqs. (55) and (28) – see Sec. 5.2.1.

4. For the realization of nMHI , we assume that $c_{\mathcal{R}}$ takes relatively large values – see e.g. Eq. (17). This assumption may [52, 53] jeopardize the validity of the classical approximation, on which the analysis of the inflationary behavior is based. To avoid this inconsistency – which is rather questionable [11, 54] though – we have to check the hierarchy between the ultraviolet cut-off, $\Lambda = m_{\rm P}/c_{\mathcal{R}}$, of the effective theory and the inflationary scale, which is represented by $\widehat{V}_{\rm HI}(h_*)^{1/4}$ or, less restrictively, by the corresponding Hubble parameter, $\widehat{H}_* = \widehat{V}_{\rm HI}(h_*)^{1/2}/\sqrt{3}m_{\rm P}$. In particular, the validity of the effective theory implies [52, 53]

$$(a)\ \widehat{V}_{\rm HI}(h_*)^{1/4} \leq \Lambda \text{ or } (b)\ \widehat{H}_* \leq \Lambda \text{ for } (c)\ c_{\mathcal{R}} \geq 1. \tag{58}$$

5. As discussed in Sec. 4.2, to avoid any erasure of the produced Y_L and to ensure that the inflaton decay to $\widehat{\nu}_2$ is kinematically allowed we have to bound $M_{1\widehat{\nu}^c}$ and $M_{2\widehat{\nu}^c}$ respectively as follows:

$$(a)\ M_{1\widehat{\nu}^c} \gtrsim 10T_{\rm rh} \text{ and } (b)\ m_{\rm I} \geq 2M_{2\widehat{\nu}^c} \Rightarrow M_{2\widehat{\nu}^c} \lesssim \frac{\lambda m_{\rm P}}{4\sqrt{3}c_{\mathcal{R}}} \simeq 1.5\cdot 10^{13} \text{ GeV}, \tag{59}$$

where we make use of Eq. (35). Recall that we impose also the restriction $\lambda \geq 0.001$ which allows us to ignore effects of instant preheating [5, 42].

6. As discussed below Eq. (2), the adopted GUT predicts YU at $M_{\rm GUT}$. Assuming negligible running of $m_{3\rm D}$ from $M_{\rm GUT}$ until the scale of nTL, Λ_L, which is taken to be $\Lambda_L = m_{\rm I}$, we end up with the requirement:

$$m_{3\rm D}(m_{\rm I}) = m_t(m_{\rm I}) \simeq (100-120) \text{ GeV}. \tag{60}$$

Parameter	Best Fit $\pm 1\sigma$	
	Normal	Inverted
	Hierarchy	
$\Delta m^2_{21}/10^{-3}\text{eV}^2$	7.62 ± 0.19	
$\Delta m^2_{31}/10^{-3}\text{eV}^2$	$2.53^{+0.08}_{-0.10}$	$-2.4^{+0.10}_{-0.07}$
$\sin^2\theta_{12}$	$0.320^{+0.015}_{-0.017}$	
$\sin^2\theta_{13}$	$0.026^{+0.003}_{-0.004}$	$0.027^{+0.003}_{-0.004}$
$\sin^2\theta_{23}$	$0.49^{+0.08}_{-0.05}$	$0.53^{+0.05}_{-0.07}$
δ/π	$0.83^{+0.54}_{-0.64}$	0.07

Table 3. Low energy experimental neutrino data for normal or inverted hierarchical neutrino masses. In the second case the full range $(0 - 2\pi)$ is allowed at 1σ for the phase δ.

where m_t is the top quark mass and the numerical values correspond to $y_{33}(m_{\rm I}) = (0.55 - 0.7)$ – cf. Ref. [55] – found [32, 56] working in the context of several MSSM versions with $\tan\beta \simeq 50$ and taking into account the SUSY threshold corrections. As regards the lighter generation, we limit ourselves in imposing just a mild hierarchy between $m_{1\rm D}$ and $m_{2\rm D}$, i.e., $m_{1\rm D} < m_{2\rm D} \ll m_{3\rm D}$ since it is not possible to achieve a simultaneous fulfilment of all the residual constraints if we impose relations similar to Eq. (60) – cf. Ref. [25–27].

7. From the solar, atmospheric, accelerator and reactor neutrino experiments we take into account the inputs listed in Table 3 on the neutrino mass-squared differences Δm^2_{21} and Δm^2_{31}, on the mixing angles θ_{ij} and on the CP-violating Dirac phase, δ for normal [inverted] neutrino mass hierarchy [23] – see also Ref. [24]. In particular, $m_{i\nu}$'s can be determined via the relations:

$$m_{2\nu} = \sqrt{m^2_{1\nu} + \Delta m^2_{21}} \text{ and } \begin{cases} m_{3\nu} = \sqrt{m^2_{1\nu} + \Delta m^2_{31}}, & \text{for } \textit{normally ordered} \text{ (NO) } m_\nu\text{'s} \\ \text{or} & \\ m_{1\nu} = \sqrt{m^2_{3\nu} + |\Delta m^2_{31}|}, & \text{for } \textit{invertedly ordered} \text{ (IO) } m_\nu\text{'s} \end{cases} \tag{61}$$

The sum of $m_{i\nu}$'s can be bounded from above by the WMAP7 data [16]

$$\Sigma_i m_{i\nu} \leq 0.58 \text{ eV} \tag{62}$$

at 95% c.l. This is more restrictive than the 95% c.l. upper bound arising from the effective electron neutrino mass in β-decay [57]:

$$m_\beta := \left|\Sigma_i U^2_{1i\nu} m_{i\nu}\right| \leq 2.3 \text{ eV}. \tag{63}$$

However, in the future, the KATRIN experiment [58] expects to reach the sensitivity of $m_\beta \simeq 0.2$ eV at 90% c.l.

8. The interpretation of BAU through nTL dictates [16] at 95% c.l.

$$Y_B = (8.74 \pm 0.42) \cdot 10^{-11} \;\Rightarrow\; 8.32 \leq 10^{11} Y_B \leq 9.16. \tag{64}$$

9. In order to avoid spoiling the success of the BBN, an upper bound on $Y_{\widetilde{G}}$ is to be imposed depending on the $\widetilde{G}$ mass, $m_{\widetilde{G}}$, and the dominant $\widetilde{G}$ decay mode. For the conservative case where $\widetilde{G}$ decays with a tiny hadronic branching ratio, we have [19]

$$Y_{\widetilde{G}} \lesssim \begin{cases} 10^{-14} \\ 10^{-13} \\ 10^{-12} \end{cases} \text{for } m_{\widetilde{G}} \simeq \begin{cases} 0.69 \text{ TeV} \\ 10.6 \text{ TeV} \\ 13.5 \text{ TeV}. \end{cases} \tag{65}$$

As we see below, this bound is achievable within our model only for $m_{\widetilde{G}} \gtrsim 10$ TeV. Taking into account that the soft masses of the scalars are not necessarily equal to $m_{\widetilde{G}}$, we do not consider such a restriction as a very severe tuning of the SUSY parameter space. Using Eq. (43) the bounds on $Y_{\widetilde{G}}$ can be translated into bounds on $T_{\rm rh}$. Specifically we take $T_{\rm rh} \simeq (0.53-5.3) \cdot 10^{8}$ GeV [$T_{\rm rh} \simeq (0.53-5.3) \cdot 10^{9}$ GeV] for $Y_{\widetilde{G}} \simeq (0.1-1) \cdot 10^{-13}$ [$Y_{\widetilde{G}} \simeq (0.1-1) \cdot 10^{-12}$].

Let us, finally, comment on the axion isocurvature perturbations generated in our model. Indeed, since the PQ symmetry is broken during nMHI, the axion acquires quantum fluctuations as all the almost massless degrees of freedom. At the QCD phase transition, these fluctuations turn into isocurvature perturbations in the axion energy density, which means that the partial curvature perturbation in axions is different than the one in photons. The results of WMAP put stringent bounds on the possible CDM isocurvature perturbation. Namely, taking into account the WMAP7, BAO and H_0 data on the parameter α_0 we find the following bound for the amplitude of the CDM isocurvature perturbation

$$|\mathcal{S}_{\rm c}| = \Delta_{\mathcal{R}} \sqrt{\frac{\alpha_0}{1-\alpha_0}} \lesssim 1.5 \cdot 10^{-5} \text{ at } 95\% \text{ c.l.} \tag{66}$$

On the other, $|\mathcal{S}_{\rm c}|$ due to axion, can be estimated by

$$|\mathcal{S}_{\rm c}| = \frac{\Omega_a}{\Omega_{\rm c}} \frac{\widehat{H}_{\rm HI}}{\pi|\theta_{\rm I}|\widehat{\phi}_{P*}} \;\; \text{with} \;\; \frac{\Omega_a}{\Omega_{\rm c}} \simeq \theta_{\rm I}^2 \left(\frac{f_a}{1.56 \cdot 10^{11}\ \text{GeV}}\right)^{1.175} \tag{67}$$

where Ω_a [$\Omega_{\rm c}$] is the axion [CDM] density parameter, $\widehat{\phi}_{P*} \sim 10^{16}$ GeV [36] denotes the field value of the PQ scalar when the cosmological scales exit the horizon and $\theta_{\rm I}$ is the initial misalignment angle which lies [36] in the interval $[-\pi/6, \pi/6]$. Satisfying Eq. (66) requires $|\theta_{\rm I}| \lesssim \pi/70$ which is a rather low but not unacceptable value. Therefore, a large axion contribution to CDM is disfavored within our model.

5.2. Numerical results

As can be seen from the relevant expressions in Secs. 2 and 4, our cosmological set-up depends on the parameters:

$$\lambda,\ \lambda_H,\ \lambda_{\bar{H}},\ k_S,\ g,\ y_{33},\ m_{\ell\nu},\ m_{i\mathrm{D}},\ \varphi_1 \text{ and } \varphi_2,$$

where $m_{\ell\nu}$ is the low scale mass of the lightest of ν_i's and can be identified with $m_{1\nu}$ $[m_{3\nu}]$ for NO [IO] neutrino mass spectrum. Recall that we determine $M_{\rm PS}$ via Eq. (57) with $g = 0.7$. We do not consider $c_{\mathcal{R}}$ and $\lambda_{i\nu^c}$ as independent parameters since $c_{\mathcal{R}}$ is related to m via Eq. (31) while $\lambda_{i\nu^c}$ can be derived from the last six parameters above which affect exclusively the Y_L calculation and can be constrained through the requirements 5 - 9 of Sec. 5.1. Note that the $\lambda_{i\nu^c}$'s can be replaced by $M_{i\widehat{\nu}^c}$'s given in Eq. (33b) keeping in mind that perturbativity requires $\lambda_{i\nu^c} \leq \sqrt{4\pi}$ or $M_{i\widehat{\nu}^c} \leq 10^{16}$ GeV. Note that if we replace $M_{\rm S}$ with $m_{\rm P}$ in Eq. (5), we obtain a tighter bound, i.e., $M_{i\widehat{\nu}^c} \leq 2.3 \cdot 10^{15}$ GeV. Our results are essentially independent of λ_H, $\lambda_{\bar{H}}$ and k_S, provided that we choose some relatively large values for these so as $m^2_{\widehat{u}-}$, $m^2_{\widehat{d}-}$ and $m^2_{\widehat{S}}$ in Table 2 are positive for $\lambda < 1$. We therefore set $\lambda_H = \lambda_{\bar{H}} = 0.5$ and $k_S = 1$ throughout our calculation. Finally $T_{\rm rh}$ can be calculated self-consistently in our model as a function of $m_{\rm I}$, $M_{2\widehat{\nu}^c} \gg M_{1\widehat{\nu}^c}$ and the unified Yukawa coupling constant y_{33} – see Sec. 4.1 – for which we take $y_{33} = 0.6$.

Summarizing, we set throughout our calculation:

$$k_S = 1,\ \lambda_H = \lambda_{\bar{H}} = 0.5, g = 0.7 \text{ and } y_{33} = 0.6. \tag{68}$$

The selected values for the above quantities give us a wide and natural allowed region for the remaining fundamental parameters of our model, as we show below concentrating separately in the inflationary period (Sec. 5.2.1) and in the stage of nTL (Sec. 5.2.2).

5.2.1. The Inflationary Stage

In this part of our numerical code, we use as input parameters $h_*, m_{2\rm D} \gg m_{1\rm D}$ and $c_{\mathcal{R}}$. For every chosen $c_{\mathcal{R}} \geq 1$ and $m_{2\rm D}$, we restrict λ and h_* so that the conditions Eq. (55) and (56a) are satisfied. In our numerical calculations, we use the complete formulas for the slow-roll parameters and $\Delta_{\mathcal{R}}$ in Eqs. (25a), (25b) and (30) and not the approximate relations listed in Sec. 3.2 for the sake of presentation. Our results are displayed in Fig. 1, where we draw the allowed values of $c_{\mathcal{R}}$ (solid line), $T_{\rm rh}$ (dashed line), the inflaton mass, $m_{\rm I}$ (dot-dashed line) and $M_{2\widehat{\nu}^c}$ (dotted line) – see Sec. 4.1 – [$h_{\rm f}$ (solid line) and h_* (dashed line)] versus λ (a) [(b)] for the $m_{2\rm D}$'s required from Eq. (64) and for the parameters adopted along the black dashed line of Fig. 2 – see Sec. 5.2.2. The required via Eq. (55) $\widehat{N}_*$ remains almost constant and close to 54.5.

The lower bound of the depicted lines comes from the saturation of the Eq. (58c). The constraint of Eq. (58b) is satisfied along the various curves whereas Eq. (58a) is valid only along the gray and light gray segments of these. Along the light gray segments, though, we obtain $h_* \geq m_{\rm P}$. The latter regions of parameter space are not necessarily excluded, since the energy density of the inflaton remains sub-Planckian and so, corrections from quantum

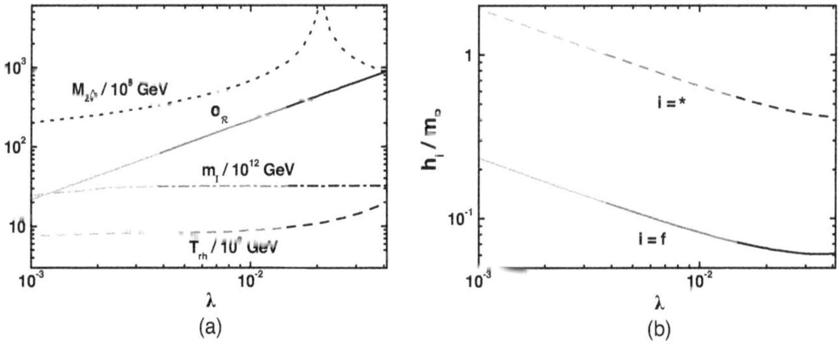

Figure 1. The allowed (by all the imposed constraints) values of $c_{\mathcal{R}}$ (solid line), $T_{\rm rh}$ – given by Eq. (40) – (dashed line), $m_{\rm I}$ (dot-dashed line) and $M_{2\rho^c}$ (dotted line) [$h_{\rm f}$ (solid line) and h_* (dashed line)] versus λ (a) [(b)] for $k_S = 1$, $\lambda_H = \lambda_{\bar H} = 0.5$ and $y_{33} = 0.6$. The light gray and gray segments denote values of the various quantities satisfying Eq. (58a) too, whereas along the light gray segments we obtain $h_* \geq m_{\rm P}$.

gravity can still be assumed to be small. As $c_{\mathcal{R}}$ increases beyond 906, f_0 becomes much larger than 1, $\hat{N}_*$ derived by Eq. (28) starts decreasing and therefore, nMHI fails to fulfil Eq. (55). This can be understood by the observation that $\hat{N}_*$, approximated fairly by Eq. (29), becomes monotonically decreasing function of $c_{\mathcal{R}}$ for $c_{\mathcal{R}} > c_{\mathcal{R}}^{\rm max}$ where $c_{\mathcal{R}}^{\rm max}$ can be found by the condition

$$\frac{d\hat{N}_*}{dc_{\mathcal{R}}} \simeq \frac{3h_*^2}{4m_{\rm P}^2}\frac{\left(c_{\mathcal{R}}^{\rm max} - c_{\mathcal{R}}\right)}{c_{\mathcal{R}}^{\rm max}} = 0 \;\Rightarrow\; c_{\mathcal{R}} \simeq c_{\mathcal{R}}^{\rm max}, \tag{69}$$

where $c_{\mathcal{R}}^{\rm max}$ is defined in Eq. (57b) and Eq. (57a) is also taken into account. As a consequence, the embedding of nMHI in a SUSY GUT provides us with a clear upper bound of $c_{\mathcal{R}}$. All in all, we obtain

$$0.001 \lesssim \lambda \lesssim 0.0412 \text{ and } 1 \lesssim c_{\mathcal{R}} \lesssim 907 \text{ for } 53.9 \lesssim \hat{N}_* \lesssim 54.7 \tag{70}$$

When $c_{\mathcal{R}}$ ranges within its allowed region, we take $M_{\rm PS} \simeq (2.87 - 4) \cdot 10^{16}$ GeV.

From Fig. 1-(a), we can verify our analytical estimation in Eq. (31) according to which λ is proportional to $c_{\mathcal{R}}$. On the other hand, the variation of $h_{\rm f}$ and h_* as a function of $c_{\mathcal{R}}$ – drawn in Fig. 1-(b) – is consistent with Eqs. (27) and (29). Letting λ or $c_{\mathcal{R}}$ vary within its allowed region in Eq. (70), we obtain

$$n_{\rm s} \simeq 0.964, \;\; -6.5 \lesssim \frac{\alpha_{\rm s}}{10^{-4}} \lesssim -6.2 \text{ and } 4.2 \gtrsim \frac{r}{10^{-3}} \gtrsim 3.5. \tag{71}$$

Clearly, the predicted $\alpha_{\rm s}$ and r lie within the allowed ranges given in Eq. (56b) and Eq. (56c) respectively, whereas $n_{\rm s}$ turns out to be impressively close to its central observationally favored value – see Eq. (56a) and cf. Ref. [12].

From Fig. 1-(a) we can conclude that $m_{\rm I}$ is kept independent of λ and almost constant at the level of 10^{13} GeV, as anticipated in Eq. (35). From the same plot we also remark that

Parameters	Cases						
	A	B	C	D	E	F	G
	Normal Hierarchy		Degenerate Masses			Inverted Hierarchy	
	Low Scale Parameters						
$m_{1\nu}/0.1$ eV	0.01	0.1	0.5	1.	0.7	0.5	0.49
$m_{2\nu}/0.1$ eV	0.088	0.13	0.5	1.	0.7	0.51	0.5
$m_{3\nu}/0.1$ eV	0.5	0.5	0.71	1.1	0.5	0.1	0.05
$\sum_i m_{i\nu}/0.1$ eV	0.6	0.74	1.7	3.1	1.9	1.1	1
$m_\beta/0.1$ eV	0.03	0.013	0.14	0.68	0.45	0.33	0.4
φ_1	$\pi/2$	$\pi/2$	$\pi/2$	$-\pi/2$	$-\pi/2$	$-\pi/2$	$-\pi/4$
φ_2	0	$-\pi/2$	$-\pi/2$	π	π	π	0
	Leptogenesis-Scale Parameters						
$m_{1\mathrm{D}}/$GeV	0.4	0.3	0.8	1	0.9	0.9	0.9
$m_{2\mathrm{D}}/$GeV	9.2	3	6.5	8.6	3.95	6.6	9.2
$m_{3\mathrm{D}}/$GeV	120	100	100	120	120	110	110
$M_{1\widehat{\nu}^c}/10^{10}$ GeV	4.4	4.7	3.6	1.2	1.5	1.6	1.7
$M_{2\widehat{\nu}^c}/10^{12}$ GeV	2.7	0.6	0.65	1.5	0.9	1.5	2.8
$M_{3\widehat{\nu}^c}/10^{14}$ GeV	27	0.28	0.46	0.4	0.4	3.8	10
	Resulting B-Yield						
$10^{11}Y_B^0$	8.75	8.9	8.6	8.63	8.9	9.	8.76
$10^{11}Y_B$	9.3	8.7	8.5	8.4	9.7	8.8	9.2
	Resulting T_{rh} and $Y_{\widetilde{G}}$						
$T_{\mathrm{rh}}/10^9$ GeV	1.2	0.89	0.89	0.99	0.91	0.99	1.2
$10^{13}Y_{\widetilde{G}}$	2.4	1.7	1.7	1.9	1.7	1.88	1.88

Table 4. Parameters yielding the correct BAU for various neutrino mass schemes for $\lambda = 0.01$ and $c_{\mathcal{R}} = 220$.

for $\lambda \lesssim 0.03$, T_{rh} remains almost constant since $\Gamma_{\mathrm{I}y}$ dominates over $\Gamma_{\mathrm{I}2\widehat{\nu}^c}$ and $f_0^3 \simeq 1$ – see Eq. (39). For $\lambda \gtrsim 0.03$, $f_0^3 \simeq 1 + 12c_{\mathcal{R}}m_{\mathrm{PS}}^2$ starts to deviate from unity and so, T_{rh} increases with $c_{\mathcal{R}}$ or λ as shown in Fig. 1. The required by Eq. (64) $M_{2\widehat{\nu}^c}$ follows the behavior of the required $m_{2\mathrm{D}}$ – see Fig. 2-(a) of Sec. 5.2.2.

5.2.2. *The Stage of non-Thermal Leptogenesis*

In this part of our numerical program, for a given neutrino mass scheme, we take as input parameters: $m_{\ell\nu}, m_{i\mathrm{D}}, \varphi_1, \varphi_2$ and the best-fit values of the neutrino parameters listed in

Table 3. We then find the *renormalization group* (RG) evolved values of these parameters at the scale of nTL, Λ_L, which is taken to be $\Lambda_L = m_{\rm I}$, integrating numerically the complete expressions of the RG equations – given in Ref. [29] – for $m_{i\nu}$, θ_{ij}, δ, φ_1 and φ_2. In doing this, we consider the MSSM with $\tan\beta \simeq 50$, favored by the preliminary LHC results – see, e.g., Ref. [32, 56] – as an effective theory between Λ_L and a SUSY-breaking scale, $M_{\rm SUSY} = 1.5$ TeV. Following the procedure described in Sec. 4.3, we evaluate $M_{i\widehat{\nu}^c}$ at Λ_L. We do not consider the running of $m_{i\rm D}$ and $M_{i\widehat{\nu}^c}$ and therefore, we give their values at Λ_L.

We start the exposition of our results arranging in Table 4 some representative values of the parameters leading to the correct BAU for $\lambda = 0.01$ and $c_{\mathcal{R}} = 220$ and normally hierarchical (cases A and B), degenerate (cases C, D and E) and invertedly hierarchical (cases F and G) m_ν's. For comparison we display the B-yield with (Y_B) or without (Y_B^0) taking into account the RG effects. We observe that the two results are more or less close with each other. In all cases the current limit of Eq. (62) is safely met – the case D approaches it –, while m_β turns out to be well below the projected sensitivity of KATRIN [58]. Shown are also the obtained $T_{\rm rh}$'s, which are close to 10^9 GeV in all cases, and the corresponding $Y_{\widetilde{G}}$'s, which are consistent with Eq. (65) for $m_{\widetilde{G}} \gtrsim 11$ TeV.

From Table 4 we also remark that the achievement of Y_B within the range of Eq. (64) dictates a clear hierarchy between the $M_{i\widehat{\nu}^c}$'s, which follows the imposed hierarchy in the sector of $m_{i\rm D}$'s – see paragraph 6 of Sec. 5.1. This is expected since, in the limit of hierarchical $m_{i\rm D}$'s, the $M_{i\widehat{\nu}^c}$'s can be approximated by the following expressions [25, 26]

$$(M_{1\widehat{\nu}^c}, M_{2\widehat{\nu}^c}, M_{3\widehat{\nu}^c}) \sim \begin{cases} \left(\frac{m_{1\rm D}^2}{m_{2\nu}s_{12}^2}, \frac{2m_{2\rm D}^2}{m_{3\nu}}, \frac{m_{3\rm D}^2 s_{12}^2}{2m_{1\nu}}\right) & \text{for NO } m_\nu\text{'s} \\ \left(\frac{m_{1\rm D}^2}{\sqrt{|\Delta m_{31}|}}, \frac{2m_{2\rm D}^2}{\sqrt{|\Delta m_{31}|}}, \frac{m_{3\rm D}^2}{2m_{3\nu}}\right) & \text{for IO } m_\nu\text{'s} \end{cases} \tag{72a}$$

Indeed, we see e.g. that for fixed j, the $M_{j\widehat{\nu}^c}$'s depends exclusively on the $m_{j\rm D}$'s and $M_{3\widehat{\nu}^c}$ increases when $m_{\ell\nu}$ decreases with fixed $m_{3\rm D}$. As a consequence, satisfying Eq. (59a) pushes the $m_{1\rm D}$'s well above the mass of the quark of the first generation. Similarly, the $m_{2\rm D}$'s required by Eq. (64) turns out to be heavier than the quark of the second generation. Also, the required by seesaw $M_{3\widehat{\nu}^c}$'s are lower in the case of degenerate ν_i spectra and can be as low as $3\cdot 10^{13}$ GeV in sharp contrast to our findings in Ref. [5], where much larger $M_{3\widehat{\nu}^c}$'s are necessitated. An order of magnitude estimation for the derived ε_L's can be achieved by [25, 26]

$$\varepsilon_2 \sim -\frac{3M_{2\widehat{\nu}^c}}{8\pi\langle H_u\rangle^2} \begin{cases} \left(\frac{m_{1\nu}}{s_{12}^2}\right) & \text{for NO } m_\nu\text{'s} \\ m_{3\nu} & \text{for IO } m_\nu\text{'s} \end{cases} \tag{72b}$$

which is rather accurate, especially in the case of IO m_ν's.

To highlight further our conclusions inferred from Table 4, we can fix $m_{\ell\nu}$ ($m_{1\nu}$ for NO $m_{i\nu}$'s or $m_{3\nu}$ for IO $m_{i\nu}$'s) $m_{1\rm D}$, $m_{3\rm D}$, φ_1 and φ_2 to their values shown in this table and vary $m_{2\rm D}$ so that the central value of Eq. (64) is achieved. This is doable since, according Eq. (72a), variation of $m_{2\rm D}$ induces an exclusive variation to $M_{2\widehat{\nu}^c}$ which, in turn, heavily influences ε_L – see Eqs. (44a) and (45) – and Y_L – see Eqs. (41) and (42). The resulting contours in the $\lambda - m_{2\rm D}$ plane are presented in Fig. 2 – since the range of Eq. (64) is very narrow the possible

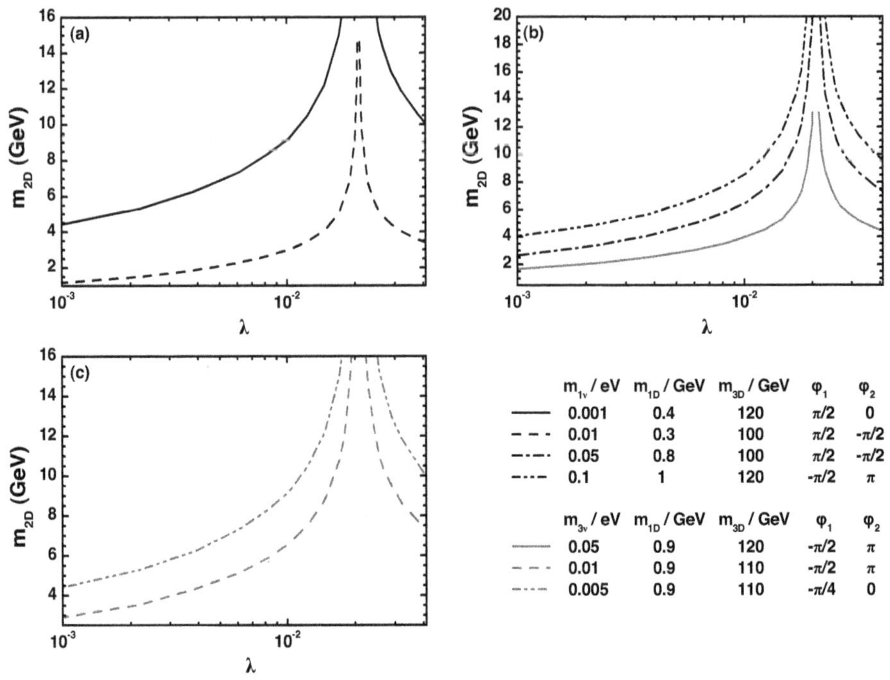

$m_{1\nu}$(eV)	λ (10^{-2})	m_{2D}(GeV)	$M_{1\widehat{\nu}^c}$ $(10^{10}$ GeV)	$M_{2\widehat{\nu}^c}$ $(10^{12}$ GeV)	$M_{3\widehat{\nu}^c}$ $(10^{14}$ GeV)
0.001	0.1 − 1.8	5.5 − 17	4.7	1 − 9.8	32
	2.4 − 4.1	17 − 10	4.7	17 − 10	32
0.01	0.1 − 2	1.5 − 13	3.5 − 6	0.26 − 8.75	0.33 − 0.4
	2 − 4.1	15 − 3.4	6 − 5.3	11 − 0.84	0.42 − 0.34
0.05	0.1 − 2.	3 − 28	3.5 − 4	2.1 − 11	0.5 − 0.6
	2.1 − 4.1	28 − 7.3	4 − 3.9	11 − 0.8	0.6 − 0.5
0.1	0.1 − 1.9	4 − 23	1.3	0.4 − 8	0.4 − 0.5
	2.2 − 4.1	23 − 9	1.3	10 − 2	0.5 − 0.4
$m_{3\nu}$(eV)	λ (10^{-2})	m_{2D}(GeV)	$M_{1\widehat{\nu}^c}$ $(10^{10}$ GeV)	$M_{2\widehat{\nu}^c}$ $(10^{12}$ GeV)	$M_{3\widehat{\nu}^c}$ $(10^{14}$ GeV)
0.05	0.1 − 2	2.1 − 12	1.5 − 1.6	0.28 − 7.8	0.48 − 0.56
	2.1 − 4.1	13 − 4.4	1.6	8.7 − 1.2	0.57 − 0.49
0.01	0.1 − 2	2.8 − 17	2.2	0.4 − 11	5.4 − 5.6
	2 − 4.1	17 − 7.3	2.2	11 − 2	5.6 − 5.4
0.005	0.1 − 1.8	5 − 17	1.8	0.7 − 10.1	12
	1.8 − 4.1	17.7 − 10	1.8	10.5 − 3.5	12

Figure 2. Contours on the $\lambda - m_{2D}$ plane, yielding the central Y_B in Eq. (64), consistently with the inflationary requirements, for $\lambda_H = \lambda_{\bar{H}} = 0.5, k_S = 1$ and $y_{33} = 0.6$ and various $(m_{\ell\nu}, m_{1D}, \varphi_1, \varphi_2)$'s indicated next to the graph (c) and NO [IO] $m_{i\nu}$'s (black [gray] lines). The corresponding ranges of $M_{i\widehat{\nu}^c}$'s are also shown in the table included.

variation of the drawn lines is negligible. The resulting $M_{j\widehat{\nu}^c}$'s are displayed in the table included. The conventions adopted for the types and the color of the various lines are also described next to the graph (c) of Fig. 2. In particular, we use black [gray] lines for NO [IO] $m_{i\nu}$'s. The black dashed and the solid gray line terminate at the values of $m_{2\mathrm{D}}$ beyond which Eq. (64) is non fulfilled due to the violation of Eq. (59b).

In all cases, two disconnected allowed domains arise according to which of the two contributions in Eq. (36) dominates. The critical point $(\lambda_{\mathrm{c}}, c_{\mathcal{R}\mathrm{c}})$ is extracted from:

$$1 - 12c_{\mathcal{R}\mathrm{c}}m_{\mathrm{PS}}^2 = 0 \Rightarrow c_{\mathcal{R}\mathrm{c}} = c_{\mathcal{R}}^{\max}/2 \simeq 453 \text{ or } \lambda_{\mathrm{c}} \simeq 10^{-4}c_{\mathcal{R}}^{\max}/4.2 \simeq 0.021 \tag{73}$$

where we make use of Eq. (57) and Eq. (31) in the intermediate and the last step respectively. From Eqs. (40), (41) and (42) one can deduce that for $\lambda < \lambda_c$, T_{rh} remains almost constant; $\Gamma_{\mathrm{I}2\widehat{\nu}^c}/\Gamma_{\mathrm{I}}$ decreases as $c_{\mathcal{R}}$ increases and so the $M_{2\widehat{\nu}^c}$'s, which satisfy Eq. (64), increase. On the contrary, for $\lambda > \lambda_c$, $\Gamma_{\mathrm{I}2\widehat{\nu}^c}/\Gamma_{\mathrm{I}}$ is independent of $c_{\mathcal{R}}$ but T_{rh} increases with $c_{\mathcal{R}}$ and so the fulfilling Eq. (64) $M_{2\widehat{\nu}^c}$'s decrease.

Summarizing, we conclude that our scenario prefers the following ranges for the $M_{i\widehat{\nu}^c}$'s:

$$1 \lesssim M_{1\widehat{\nu}^c}/10^{10}\ \mathrm{GeV} \lesssim 6,\ 0.6 \lesssim M_{2\widehat{\nu}^c}/10^{12}\ \mathrm{GeV} \lesssim 20,\ 0.3 \lesssim M_{3\widehat{\nu}^c}/10^{14}\mathrm{GeV} \lesssim 30, \tag{74a}$$

while the $m_{1\mathrm{D}}$ and $m_{2\mathrm{D}}$ are restricted in the ranges:

$$0.3 \lesssim m_{1\mathrm{D}}/\mathrm{GeV} \lesssim 1,\ 1.5 \lesssim m_{2\mathrm{D}}/\mathrm{GeV} \lesssim 20. \tag{74b}$$

6. Conclusions

We investigated the implementation of nTL within a realistic GUT, based on the PS gauge group. Leptogenesis follows a stage of nMHI driven by the radial component of the Higgs field, which leads to the spontaneous breaking of the PS gauge group to the SM one with the GUT breaking v.e.v identified with the SUSY GUT scale and without overproduction of monopoles. The model possesses also a resolution to the strong CP and the μ problems of the MSSM via a PQ symmetry which is broken during nMHI and afterwards. As a consequence the axion cannot be the dominant component of CDM, due to the present bounds on the axion isocurvature fluctuation. Moreover, we briefly discussed scenaria in which the potential axino and saxion overproduction problems can be avoided.

Inflation is followed by a reheating phase, during which the inflaton can decay into the lightest, $\widehat{\nu}_1^c$, and the next-to-lightest, $\widehat{\nu}_2^c$, RH neutrinos allowing, thereby for nTL to occur via the subsequent decay of $\widehat{\nu}_1^c$ and $\widehat{\nu}_2^c$. Although other decay channels to the SM particles via non-renormalizable interactions are also activated, we showed that the production of the required by the observations BAU can be reconciled with the observational constraints on the inflationary observables and the $\widetilde{G}$ abundance, provided that the (unstable) $\widetilde{G}$ masses are greater than 11 TeV. The required by the observations BAU can become consistent with the present low energy neutrino data, the restriction on $m_{3\mathrm{D}}$ due to the PS gauge group and the imposed mild hierarchy between $m_{1\mathrm{D}}$ and $m_{2\mathrm{D}}$. To this end, $m_{1\mathrm{D}}$ and $m_{2\mathrm{D}}$ turn out to be heavier than the ones of the corresponding quarks and lie in the ranges $(0.1-1)$ GeV and $(2-20)$ GeV while the obtained $M_{1\widehat{\nu}^c}$, $M_{2\widehat{\nu}^c}$ and $M_{3\widehat{\nu}^c}$ are restricted to the values 10^{10} GeV, $(10^{11}-10^{12})$ GeV and $(10^{13}-10^{15})$ GeV respectively.

Acknowledgements

We would like to thank A.B. Lahanas, G. Lazarides and V.C. Spanos for valuable discussions.

Author details

C. Pallis and N. Toumbas

Department of Physics, University of Cyprus, Nicosia, Cyprus

References

[1] Hamaguchi K. *Phd Thesis*, hep-ph/0212305.

[2] Buchmuller W. Peccei D.R and Yanagida T. *Ann. Rev. Nucl. Part. Sci.*, 55:311, 2005 [hep-ph/0502169].

[3] Lazarides G. and Shafi Q. *Phys. Lett. B* , 258:305, 1991.

[4] Kumekawa K. Moroi T. and Yanagida T. *Prog. Theor. Phys.*, 92:437, 1994 [hep-ph/9405337].

[5] Pallis C. and Toumbas N. *J. Cosmol. Astropart. Phys.*, 12:002, 2011 [arXiv:1108.1771].

[6] Cervantes-Cota J.L. and Dehnen H. *Nucl. Phys. B*, 442:391, 1995 [astro-ph/9505069].

[7] Arai M. Kawai S. and Okada N. *Phys. Rev. D*, 84:123515, 2011 [arXiv:1107.4767].

[8] Einhorn B.M. and Jones T.R.D. arXiv:1207.1710.

[9] Einhorn B.M. and Jones T.R.D. *J. High Energy Phys.*, 03:026, 2010 [arXiv:0912.2718].

[10] Lee M.H. *J. Cosmol. Astropart. Phys.*, 08:003, 2010 [arXiv:1005.2735].

[11] Ferrara S. Kallosh R. Linde D.A. Marrani A. and Proeyen Van A. *Phys. Rev. D* , 83:025008, 2011 [arXiv:1008.2942].

[12] Kallosh R. and Linde D.A. *J. Cosmol. Astropart. Phys.*, 11:011, 2010 [arXiv:1008.3375].

[13] Antoniadis I. and Leontaris K.G. *Phys. Lett. B* , 216:333, 1989.

[14] Shiu G. and Tye H.-H. S. *Phys. Rev. D* , 58:106007, 1998.

[15] Coleman R.S. and Weinberg J.E. *Phys. Rev. D* , 7:1888, 1973.

[16] Komatsu E. *et al.* [WMAP Collaboration] *Astrophys. J. Suppl.*, 192:18, 2011 [arXiv:1001. 4538].

[17] Khlopov Yu. M. and Linde D.A. *Phys. Lett. B* , 138:265, 1984.

[18] Bolz M. Brandenburg A. and Buchmüller W. *Nucl. Phys. B*, 606:518, 2001 [hep-ph/0012052].

[19] Kawasaki M. Kohri K. and Moroi T. *Phys. Lett. B* , 625:7, 2005 [astro-ph/0402490].

[20] Yanagida T. *Proceedings of the Workshop on the Unified Theory and the Baryon Number in the Universe (O. Sawada and A. Sugamoto, eds.), Tsukuba, Japan*, page 95, 1979.

[21] Gell-Mann M. Ramond P. and Slansky R. *Supergravity (P. van Nieuwenhuizen et al.. eds.), North Holland, Amsterdam*, page 315, 1979.

[22] Glashow L.S. *Proceedings of the 1979 Cargese Summer Institute on Quarks and Leptons (M. Levy et al.. eds.)*, page 687, 1980.

[23] Tortola M. Valle F.W.J. and Vanegas D. arXiv:1205.4018.

[24] Fogli L.G. Lisi E. Marrone A. Montanino D. Palazzo A. and Rotunno M.A. arXiv: 1205.5254.

[25] Branco C.G. *et al. Nucl. Phys. B*, 640:202, 2002.

[26] Akhmedov K.E. Frigerio M. and Smirnov Y.A. *J. High Energy Phys.*, 09:021, 2003 [hep-ph/0305322].

[27] Şenoğuz V.N. *Phys. Rev. D* , 76:013005, 2007 [arXiv:0704.3048].

[28] Pallis C. and Shafi Q. *Phys. Rev. D* , 86:023523, 2012 [arXiv:1204.0252].

[29] Antusch S. Kersten J. Lindner M. and Ratz M. *Nucl. Phys. B*, 674:401, 2003 [hep-ph/ 0305273].

[30] Jeannerot R. Khalil S. Lazarides G. and Shafi Q. *J. High Energy Phys.*, 10:012, 2000 [hep-ph/0002151].

[31] King F.S. and Oliveira M. *Phys. Rev. D* , 63:015010, 2001 [hep-ph/0008183].

[32] Gogoladze I. Khalid R. Raza S. and Shafi Q. *J. High Energy Phys.*, 12:055, 2010 [arXiv: 1008.2765].

[33] King F.S. and Oliveira M. *Phys. Rev. D* , 63:095004, 2001 [hep-ph/0009287].

[34] Gómez E.M. Lazarides G. and Pallis C. *Nucl. Phys. B*, 638:165, 2002 [hep-ph/0203131].

[35] Lazarides G. and Shafi Q. *Phys. Rev. D* , 58:071702.

[36] Dimopoulos K. *et al. J. High Energy Phys.*, 05:057, 2003.

[37] Pallis C. and Toumbas N. *J. Cosmol. Astropart. Phys.*, 02:019, 2011 [arXiv:1101.0325].

[38] Lyth H.D. and Riotto A. *Phys. Rept.*, 314:1, 1999 [hep-ph/9807278].

[39] Lazarides G. *J. Phys. Conf. Ser.*, 53:528, 2006 [hep-ph/0607032].

[40] Kofman L. Linde A.D. and Starobinsky A.D. *Phys. Rev. Lett.*, 73:3195, 1994 [hep-th/ 9405187].

[41] Garcia-Bellido J. Figueroa G.D. and Rubio J. *Phys. Rev. D*, 79:063531, 2009 [arXiv:0812. 4624].

[42] Felder N. Kofman L. and Linde A.D. *Phys. Rev. D*, 59:123523, 1999 [hep-ph/9812289].

[43] Endo M. Kawasaki M. Takahashi F. and Yanagida T.T. *Phys. Lett. B*, 642:518, 2006 [hep-ph/0607170].

[44] Pallis C. *Nucl. Phys. B*, 751:129, 2006 [hep-ph/0510234].

[45] Fukugita M. and Yanagida T. *Phys. Rev. D*, 42:1285, 1990.

[46] Ibáñez E.L. and Quevedo F. *Phys. Lett. B*, 283:261, 1992.

[47] Baer H. Kraml S. Lessa A. and Sekmen S. *J. Cosmol. Astropart. Phys.*, 04:039, 2011.

[48] Kawasaki M. Nakayama K. and Senami M. *J. Cosmol. Astropart. Phys.*, 03:009, 2008.

[49] Covi L. Roulet E. and Vissani F. *Phys. Lett. B*, 384:169, 1996 [hep-ph/9605319].

[50] Kopp J. *Int. J. Mod. Phys. C*, 19:523, 2008 [physics/0610206].

[51] Pallis C. *Phys. Lett. B*, 692:287, 2010.

[52] Burgess P.C. Lee M.H. and Trott M. *J. High Energy Phys.*, 09:103, 2009 [arXiv:0902. 4465].

[53] Barbon F.L.J. and Espinosa R.J. *Phys. Rev. D*, 79:081302, 2009 [arXiv:0903.0355].

[54] Bezrukov F. *et al.* *J. High Energy Phys.*, 016:01, 2011 [arXiv:1008.5157].

[55] Antusch S. and Spinrath M. *Phys. Rev. D*, 78:075020, 2008 [arXiv:0804.0717].

[56] Karagiannakis N. Lazarides G. and Pallis C. *Phys. Lett. B*, 638:165, 2011 [arXiv:1107. 0667].

[57] Klapdor-Kleingrothaus V. H. *et al.* *Eur. Phys. J. A*, 12:147, 2001.

[58] Osipowicz A. *et al.* [KATRIN Collaboration] hep-ex/0109033.

Chapter 10

Quintom Potential from Quantum Anisotropic Cosmological Models

J. Socorro, Paulo A. Rodríguez, O. Núñez-Soltero,
Rafael Hernández and Abraham Espinoza-García

Additional information is available at the end of the chapter

1. Introduction

At the present time, there are some paradigms to explain the observations for the accelerated expansion of the universe. Most of these paradigms are based on the dynamics of a scalar (quintessence) or multiscalar field (quintom) cosmological models of dark energy, (see the review [1–3]). The main discussion yields over the evolution of these models in the ω-ω' plane (ω is the equation of state parameter of the dark energy) [4–12, 14–19, 21–25]). In the present study we desire to perform our investigation in the case of quintom cosmology, constructed using both quintessence (σ) and phantom (ϕ) fields, mantaining a nonspecific potential form $V(\phi,\sigma)$. There are many works in the literature that deals with this type of problems, but in a general way, and not with a particular ansatz, one that considers dynamical systems [12, 13, 20]. One special class of potentials used to study this behaviour corresponds to the case of the exponential potentials [4, 6, 9, 26, 28] for each field, where the corresponding energy density of a scalar field has the range of scaling behaviors [29, 30], i.e, it scales exactly as a power of the scale factor like, $\rho_\phi \propto a^{-m}$, when the dominant component has an energy density which scales in a similar way. There are other works where other type of potentials are analyzed [1, 9, 15, 19, 20, 23, 24, 31].

How come that we claim that the analysis of general potentials using dynamical systems was made considering particular structures of them, in other words, how can we introduce this mathematical structure within a physical context?. We can partially answer this question, when the Bohmian formalism is introduced, i.e, many of them can be constructed using the Bohm formalism [32–34] of the quantum mechanics under the integral systems premise, which is known as the quantum potential approach. This approach makes possible to identity trajectories associated with the wave function of the universe [32] when we choose the superpotential function as the momenta associated to the coordinate field q^μ. This investigation was undertaken within the framework of the minisuperspace approximation of quantum theory when we investigate the models with a finite number of degrees of freedom. Considering the anisotropic Bianchi Class A cosmological models from canonical quantum cosmology under determined conditions in the evolution of our universe, and employing the Bohmian formalism, and in particular the Bianchi type I to obtain a family of potentials that correspond to the most probable to model the present day cosmic acceleration. In our analysis, we found this special class of potentials, however these appear mixed.

This work is arranged as follows. In section 2 we present the corresponding Einstein Klein Gordon equation for the quintom model. In section 3, we introduced the hamiltonian apparatus which is applied to Bianchi type I and the Bianchi Type IX in order to construct a master equation for all Bianchi Class A cosmological models with barotropic perfect fluid and cosmological constant. Furthermore, we present the classical equations for Bianchi type I, whose solutions are given in a quadrature form, which are presented in section 5 for particular scalar potentials. In section 4 we present the quantum scheme, where we use the Bohmian formalism and show its mathematical structure, also our approach is presented in a similar way. Our treatment is applied to build the mathematical structure of quintom scalar potentials using the integral systems formalism. For completeness we present the quantum solutions to the Wheeler-DeWitt equation. Moreover, this section represents our main objective for this work, and its where the utmost problem is treated. However it is important emphasize that the quantum potential from Bohm formalism will work as a constraint equation which restricts our family of potentials found. It is well known in the literature that in the Bohm formalism the imaginary part is never determined, however in this work such a problem is solved in order to find the quantum potentials, which is a more important matter for being able to find the classical trajectories, which is shown in section 5, that is devoted to obtain the classical solutions for particular scalar potentials, also we show through graphics how the classical trajectory is projected from its quantum counterpart. Finally we present the time dependence for the Ω, and quintom scalar fields (φ, ς).

2. The model

We begin with the construction of the quintom cosmological paradigm, which requires the simultaneous consideration of two fields, namely one canonical σ and one phantom ϕ and the implication that dark energy will be attributed to their combination. The action of a universe with the constitution of such a two fields, the cosmological term contribution and the matter as perfect fluid content, is

$$\mathcal{L} = \sqrt{-g}\left(R - 2\Lambda - \frac{1}{2}g^{\mu\nu}\nabla_\mu\phi\nabla_\nu\phi + \frac{1}{2}g^{\mu\nu}\nabla_\mu\sigma\nabla_\nu\sigma - V(\phi,\sigma)\right) + \mathcal{L}_{\text{matter}}, \tag{1}$$

and the corresponding field equations becomes

$$\begin{aligned} G_{\alpha\beta} + g_{\alpha\beta}\Lambda = & -\frac{1}{2}\left(\nabla_\alpha\phi\nabla_\beta\phi - \frac{1}{2}g_{\alpha\beta}g^{\mu\nu}\nabla_\mu\phi\nabla_\nu\phi\right) \\ & +\frac{1}{2}\left(\nabla_\alpha\sigma\nabla_\beta\sigma - \frac{1}{2}g_{\alpha\beta}g^{\mu\nu}\nabla_\mu\sigma\nabla_\nu\sigma\right) \end{aligned}$$

$$-\frac{1}{2}g_{\alpha\beta}V(\phi,\sigma) - 8\pi G T_{\alpha\beta}, \tag{2}$$

$$g^{\mu\nu}\phi_{,\mu\nu} - g^{\alpha\beta}\Gamma^{\nu}_{\alpha\beta}\nabla_{\nu}\phi + \frac{\partial V}{\partial\phi} = 0, \quad \Leftrightarrow \quad \Box\phi + \frac{\partial V}{\partial\phi} = 0$$

$$g^{\mu\nu}\sigma_{,\mu\nu} - g^{\alpha\beta}\Gamma^{\nu}_{\alpha\beta}\nabla_{\nu}\sigma - \frac{\partial V}{\partial\sigma} = 0, \quad \Leftrightarrow \quad \Box\sigma - \frac{\partial V}{\partial\sigma} = 0,$$

$$T^{\mu\nu}_{;\mu} = 0, \quad \text{with} \quad T_{\mu\nu} = P g_{\mu\nu} + (P+\rho)u_{\mu}u_{\nu}, \tag{3}$$

here ρ is the energy density, P the pressure, and u_μ the velocity, satisfying that $u_\mu u^\mu = -1$.

3. Hamiltonian approach

Let us recall here the canonical formulation in the ADM formalism of the diagonal Bianchi Class A cosmological models. The metric has the form

$$ds^2 = -N(t)dt^2 + e^{2\Omega(t)}\,(e^{2\beta(t)})_{ij}\,\omega^i\,\omega^j, \tag{4}$$

where $\beta_{ij}(t)$ is a 3x3 diagonal matrix, $\beta_{ij} = \mathrm{diag}(\beta_+ + \sqrt{3}\beta_-, \beta_+ - \sqrt{3}\beta_-, -2\beta_+)$, $\Omega(t)$ is a scalar and ω^i are one-forms that characterize each cosmological Bianchi type model, and obey the form $d\omega^i = \frac{1}{2}C^i_{jk}\omega^j \wedge \omega^k$, and C^i_{jk} are structure constants of the corresponding model.

The corresponding metric of the Bianchi type I in Misner's parametrization has the following form

$$ds^2_I = -N^2dt^2 + e^{2\Omega+2\beta_+ + 2\sqrt{3}\beta_-}dx^2 + e^{2\Omega+2\beta_+ - 2\sqrt{3}\beta_-}dy^2 + e^{2\Omega-4\beta_+}dz^2, \tag{5}$$

where the anisotropic radii are

$$R_1 = e^{\Omega+\beta_+ + \sqrt{3}\beta_-}, \qquad R_2 = e^{\Omega+\beta_+ - \sqrt{3}\beta_-}, \qquad R_3 = e^{\Omega-2\beta_+}.$$

We use the Bianchi type I and IX cosmological models as toy model to apply the formalism, and write a master equation for all Bianchi Class A models. The lagrangian density (1) for the Bianchi type I is written as (where the overdot denotes time derivative),

$$\mathcal{L}_I = e^{3\Omega}\left[6\frac{\dot{\Omega}^2}{N} - 6\frac{\dot{\beta}_+^2}{N} - 6\frac{\dot{\beta}_-^2}{N} + 6\frac{\dot{\varphi}^2}{N} - 6\frac{\dot{\varsigma}^2}{N} + N\left(-V(\varphi,\varsigma) + 2\Lambda + 16\pi G\rho\right)\right], \tag{6}$$

the fields were re-scaled as $\phi = \sqrt{12}\varphi, \sigma = \sqrt{12}\varsigma$ for simplicity in the calculations.

The momenta are defined as $\Pi_{q^i} = \frac{\partial \mathcal{L}}{\partial \dot{q}^i}$, where $q^i = (\beta_\pm, \Omega, \varphi, \varsigma)$ are the coordinates fields.

$$
\begin{aligned}
\Pi_\Omega &= \frac{\partial \mathcal{L}}{\partial \dot\Omega} = \frac{12e^{3\Omega}\dot\Omega}{N}, && \rightarrow \dot\Omega = \frac{N\Pi_\Omega}{12}e^{-3\Omega} \\
\Pi_\pm &= \frac{\partial \mathcal{L}}{\partial \dot\beta_\pm} = -12\frac{e^{3\Omega}\dot\beta_\pm}{N}, && \rightarrow \dot\beta_\pm = -\frac{N\Pi_\pm}{12}e^{-3\Omega} \\
\Pi_\varphi &= \frac{\partial \mathcal{L}}{\partial \dot\varphi} = 12\frac{e^{3\Omega}\dot\varphi}{N}, && \rightarrow \dot\varphi = \frac{N\Pi_\varphi}{12}e^{-3\Omega} \\
\Pi_\varsigma &= \frac{\partial \mathcal{L}}{\partial \dot\varsigma} = -12\frac{e^{3\Omega}\dot\varsigma}{N}, && \rightarrow \dot\varsigma = -\frac{N\Pi_\varsigma}{12}e^{-3\Omega}.
\end{aligned}
\tag{7}
$$

Writing (6) in canonical form, $\mathcal{L}_{\text{canonical}} = \Pi_q\dot{q} - N\mathcal{H}$ and substituting the energy density for the barotropic fluid, we can find the Hamiltonian density $\mathcal{H}$ in the usual way

$$
\mathcal{H}_I = \frac{e^{-3\Omega}}{24}\left[\Pi_\Omega^2 - \Pi_\varsigma^2 - \Pi_+^2 - \Pi_-^2 + \Pi_\varphi^2 + e^{6\Omega}\left\{24V(\varphi,\varsigma) - 48\left(\Lambda + 8\pi G M_\gamma e^{-3(\gamma+1)\Omega}\right)\right\}\right]. \tag{8}
$$

For the Bianchi type IX we have the Lagrangian and Hamiltonian density, respectively

$$
\begin{aligned}
\mathcal{L}_{IX} = e^{3\Omega}&\left[6\frac{\dot\Omega^2}{N} - 6\frac{\dot\beta_+^2}{N} - 6\frac{\dot\beta_-^2}{N} + 6\frac{\dot\varphi^2}{N} - 6\frac{\dot\varsigma^2}{N} + N\left(-V(\varphi,\varsigma) + 2\Lambda + 16\pi G\rho\right)\right] \\
&+ Ne^{-2\Omega}\left\{\frac{1}{2}\left(e^{4\beta_+ + 4\sqrt{3}\beta_-} + e^{4\beta_+ - 4\sqrt{3}\beta_-} + e^{-8\beta_+}\right)\right. \\
&\left.\qquad - \left(e^{-2\beta_+ + 2\sqrt{3}\beta_-} + e^{-2\beta_+ - 2\sqrt{3}\beta_-} + e^{4\beta_+}\right)\right\}\Big],
\end{aligned}
\tag{9}
$$

$$
\begin{aligned}
\mathcal{H}_{IX} = \frac{e^{-3\Omega}}{24}&\left[\Pi_\Omega^2 - \Pi_\varsigma^2 - \Pi_+^2 - \Pi_-^2 + \Pi_\varphi^2 + e^{6\Omega}\left\{24V(\varphi,\varsigma) - 48\left(\Lambda + 8\pi G M_\gamma e^{-3(\gamma+1)\Omega}\right)\right\}\right] \\
&-24e^{4\Omega}\left\{\frac{1}{2}\left(e^{4\beta_+ + 4\sqrt{3}\beta_-} + e^{4\beta_+ - 4\sqrt{3}\beta_-} + e^{-8\beta_+}\right)\right. \\
&\left.\qquad - \left(e^{-2\beta_+ + 2\sqrt{3}\beta_-} + e^{-2\beta_+ - 2\sqrt{3}\beta_-} + e^{4\beta_+}\right)\right\}\Big],
\end{aligned}
\tag{10}
$$

where we have used the covariant derivative of (3), obtaining the relation

$$
3\dot\Omega\rho + 3\dot\Omega p + \dot\rho = 0, \tag{11}
$$

whose solution becomes

$$
\rho = M_\gamma e^{-3(1+\gamma)\Omega}. \tag{12}
$$

where M_γ is an integration constant, in this sense we have all Bianchi Class A cosmological models, and their corresponding Hamiltonian density becomes

$$\mathcal{H}_A = \mathcal{H}_\mathrm{I} - \frac{1}{24}\mathrm{e}^{-3\Omega}\mathrm{U}_\mathrm{A}(\Omega, \beta_\pm), \tag{13}$$

where the gravitational potential can be seen in table I, in particular, the Bianchi Type IX is

$$\mathrm{U}_\mathrm{IX}(\Omega, \beta_\pm) = 12\mathrm{e}^{4\Omega}\left(\mathrm{e}^{4\beta_+ + 4\sqrt{3}\beta_-} + \mathrm{e}^{4\beta_+ - 4\sqrt{3}\beta_-} + \mathrm{e}^{4\beta_+}\right.$$
$$\left. -2\left\{\mathrm{e}^{4\beta_+} + \mathrm{e}^{2\beta_+ - 2\sqrt{3}\beta_-} + \mathrm{e}^{-2\beta_+ + 2\sqrt{3}\beta_-}\right\}\right).$$

Considering the inflationary phenomenon $\gamma = -1$, the Hamiltonian density is

$$\begin{aligned}\mathcal{H}_\mathrm{IX} = \frac{\mathrm{e}^{-3\Omega}}{24}&\left[\Pi_\Omega^2 - \Pi_\varsigma^2 - \Pi_+^2 - \Pi_-^2 + \Pi_\varphi^2 + \mathrm{e}^{6\Omega}\left\{24\mathrm{V}(\varphi, \varsigma) - \lambda_\mathrm{eff}\right\}\right. \\ &-24e^{4\Omega}\left\{\frac{1}{2}\left(e^{4\beta_+ + 4\sqrt{3}\beta_-} + e^{4\beta_+ - 4\sqrt{3}\beta_-} + e^{-8\beta_+}\right)\right. \\ &\left.\left. - \left(\mathrm{e}^{-2\beta_+ + 2\sqrt{3}\beta_-} + \mathrm{e}^{-2\beta_+ - 2\sqrt{3}\beta_-} + \mathrm{e}^{4\beta_+}\right)\right\}\right],\end{aligned} \tag{14}$$

where $\lambda_\mathrm{eff} = 48(\Lambda + 8\pi \mathrm{GM}_{-1})$.

The equation (13) can be considered as a master equation for all Bianchi Class A cosmological models in this formalism, where $\mathrm{U}(\Omega, \beta_\pm)$ is the potential term of the cosmological model under consideration, it can be read in table 1.

Bianchi type	Hamiltonian density $\mathcal{H}$
I	$\frac{\mathrm{e}^{-3\Omega}}{24}\left[\Pi_\Omega^2 - \Pi_\varsigma^2 - \Pi_+^2 - \Pi_-^2 + \Pi_\varphi^2 + \mathrm{e}^{6\Omega}\left\{24\mathrm{V}(\varphi, \varsigma) - \lambda_\mathrm{eff}\right\}\right]$
II	$\frac{\mathrm{e}^{-3\Omega}}{24}\left[\Pi_\Omega^2 - \Pi_\varsigma^2 - \Pi_+^2 - \Pi_-^2 + \Pi_\varphi^2 + \mathrm{e}^{6\Omega}\left\{24\mathrm{V}(\varphi, \varsigma) - \lambda_\mathrm{eff}\right\} -12\mathrm{e}^{4\Omega}\mathrm{e}^{4\beta_+ + 4\sqrt{3}\beta_-}\right]$
VI_{-1}	$\frac{\mathrm{e}^{-3\Omega}}{24}\left[\Pi_\Omega^2 - \Pi_\varsigma^2 - \Pi_+^2 - \Pi_-^2 + \Pi_\varphi^2 + \mathrm{e}^{6\Omega}\left\{24\mathrm{V}(\varphi, \varsigma) - \lambda_\mathrm{eff}\right\} -48\mathrm{e}^{4\Omega}\mathrm{e}^{4\beta_+}\right]$
VII_0	$\frac{e^{-3\Omega}}{24}\left[\Pi_\Omega^2 - \Pi_\varsigma^2 - \Pi_+^2 - \Pi_-^2 + \Pi_\varphi^2 + e^{6\Omega}\left\{24V(\varphi, \varsigma) - \lambda_{eff}\right\} -12e^{4\Omega}\left(e^{4\beta_+ + 4\sqrt{3}\beta_-} - e^{4\beta_+} + e^{4\beta_+ - 4\sqrt{3}\beta_-}\right)\right]$
VIII	$\frac{\mathrm{e}^{-3\Omega}}{24}\left[\Pi_\Omega^2 - \Pi_\varsigma^2 - \Pi_+^2 - \Pi_-^2 + \Pi_\varphi^2 + \mathrm{e}^{6\Omega}\left\{24\mathrm{V}(\varphi, \varsigma) - \lambda_\mathrm{eff}\right\} -12e^{4\Omega}\left(e^{4\beta_+ + 4\sqrt{3}\beta_-} + e^{4\beta_+ - 4\sqrt{3}\beta_-} + e^{-8\beta_+} +2\left\{\mathrm{e}^{4\beta_+} - \mathrm{e}^{-2\beta_+ - 2\sqrt{3}\beta_-} - \mathrm{e}^{-2\beta_+ + 2\sqrt{3}\beta_-}\right\}\right)\right]$
IX	$\frac{\mathrm{e}^{-3\Omega}}{24}\left[\Pi_\Omega^2 - \Pi_\varsigma^2 - \Pi_+^2 - \Pi_-^2 + \Pi_\varphi^2 + \mathrm{e}^{6\Omega}\left\{24\mathrm{V}(\varphi, \varsigma) - \lambda_\mathrm{eff}\right\} -12e^{4\Omega}\left(e^{4\beta_+ + 4\sqrt{3}\beta_-} + e^{4\beta_+ - 4\sqrt{3}\beta_-} + e^{-8\beta_+} +2\left\{\mathrm{e}^{4\beta_+} + \mathrm{e}^{2\beta_+ - 2\sqrt{3}\beta_-} + \mathrm{e}^{-2\beta_+ + 2\sqrt{3}\beta_-}\right\}\right)\right]$

Table 1. Hamiltonian density for the Bianchi Class A models in the quintom approach for the inflationary phenomenon.

3.1. Classical field equation for Bianchi type I

On the other hand, the Einstein field equations (2,3) for the Bianchi type I, are

$$3\frac{\dot{\Omega}^2}{N^2} - 3\frac{\dot{\beta}_+^2}{N^2} - 3\frac{\dot{\beta}_-^2}{N^2} = 8\pi G\rho - 3\frac{\dot{\varphi}^2}{N^2} + 3\frac{\dot{\varsigma}^2}{N^2} + \Lambda + \frac{V(\varphi,\varsigma)}{2}, \tag{15}$$

$$2\frac{\ddot{\Omega}}{N^2} + 3\frac{\dot{\Omega}^2}{N^2} - 3\frac{\dot{\Omega}\dot{\beta}_+}{N^2} - 3\sqrt{3}\frac{\dot{\Omega}\dot{\beta}_-}{N^2} - 2\frac{\dot{\Omega}\dot{N}}{N^3} - \frac{\ddot{\beta}_+}{N^2} + 3\frac{\dot{\beta}_+^2}{N^2} + \frac{\dot{\beta}_+\dot{N}}{N^3} - \sqrt{3}\frac{\ddot{\beta}_-}{N^2} + 3\frac{\dot{\beta}_-^2}{N^2}$$
$$+\sqrt{3}\frac{\dot{\beta}_-\dot{N}}{N^3} = -8\pi GP + 3\frac{\dot{\varphi}^2}{N^2} - 3\frac{\dot{\varsigma}^2}{N^2} + \Lambda + \frac{V(\varphi,\varsigma)}{2}, \tag{16}$$

$$2\frac{\ddot{\Omega}}{N^2} + 3\frac{\dot{\Omega}^2}{N^2} - 3\frac{\dot{\Omega}\dot{\beta}_+}{N^2} + 3\sqrt{3}\frac{\dot{\Omega}\dot{\beta}_-}{N^2} - 2\frac{\dot{\Omega}\dot{N}}{N^3} - \frac{\ddot{\beta}_+}{N^2} + 3\frac{\dot{\beta}_+^2}{N^2} + \frac{\dot{\beta}_+\dot{N}}{N^3} + \sqrt{3}\frac{\ddot{\beta}_-}{N^2} + 3\frac{\dot{\beta}_-^2}{N^2}$$
$$-\sqrt{3}\frac{\dot{\beta}_-\dot{N}}{N^3} = -8\pi GP + 3\frac{\dot{\varphi}^2}{N^2} - 3\frac{\dot{\varsigma}^2}{N^2} + \Lambda + \frac{V(\varphi,\varsigma)}{2}, \tag{17}$$

$$2\frac{\ddot{\Omega}}{N^2} + 3\frac{\dot{\Omega}^2}{N^2} + 6\frac{\dot{\Omega}\dot{\beta}_+}{N^2} - 2\frac{\dot{\Omega}\dot{N}}{N^3} + 2\frac{\ddot{\beta}_+}{N^2} + 3\frac{\dot{\beta}_+^2}{N^2} - 2\frac{\dot{\beta}_+\dot{N}}{N^3}$$
$$+3\frac{\dot{\beta}_-^2}{N^2} = -8\pi GP + 3\frac{\dot{\varphi}^2}{N^2} - 3\frac{\dot{\varsigma}^2}{N^2} + \Lambda + \frac{V(\varphi,\varsigma)}{2}, \tag{18}$$

$$-3\frac{\dot{\Omega}\dot{\varsigma}}{N^2} + \frac{\dot{N}\dot{\varsigma}}{N^3} - \frac{\ddot{\varsigma}}{N^2} + \frac{\partial V(\varsigma,\varphi)}{\partial\varsigma} = 0, \tag{19}$$

$$-3\frac{\dot{\Omega}\dot{\varphi}}{N^2} + \frac{\dot{N}\dot{\varphi}}{N^3} - \frac{\ddot{\varphi}}{N^2} - \frac{\partial V(\varsigma,\varphi)}{\partial\varphi} = 0, \tag{20}$$

which can be written as

$$8\pi GP - \Lambda + \frac{1}{2}\left(-6\varphi'^2 + 6\varsigma'^2 - V(\varphi,\varsigma)\right) = -\frac{2}{3}\frac{a''}{a} - 3H^2, \tag{21}$$

$$8\pi G\rho + \Lambda + \frac{1}{2}\left(-6\varphi'^2 + 6\varsigma'^2 + V(\varphi,\varsigma)\right) = 3H^2, \tag{22}$$

$$-3\Omega'\varsigma' - \varsigma'' - \frac{\partial V(\varsigma,\varphi)}{\partial\varsigma} = 0, \tag{23}$$

$$-3\Omega'\varphi' - \varphi'' + \frac{\partial V(\varsigma,\varphi)}{\partial\varphi} = 0, \tag{24}$$

where H^2 is defined as $H^2 = H_1H_2 + H_1H_3 + H_2H_3$, $a = R_1R_2R_3$, and $H_i = \frac{\dot{R}_i}{R_i}$. We have done the time transformation $\frac{d}{d\tau} = \frac{d}{Ndt} = \prime$. Adding (21) and (22) we arrive

$$-\frac{a''}{a} = 12\pi G\left[\rho + \rho_\varphi + \rho_\varsigma + P + P_\varphi + P_\varsigma\right], \tag{25}$$

where

$$P_\varphi = \frac{1}{16\pi G}\left(-6\varphi'^2 - V(\varphi,\varsigma)|_\varsigma\right), \qquad P_\varsigma = \frac{1}{16\pi G}\left(6\varsigma'^2 - V(\varphi,\varsigma)|_\varphi\right),$$

$$\rho_\varphi = \frac{1}{16\pi G}\left(-6\varphi'^2 + V(\varphi,\varsigma)|_\varsigma\right), \qquad \rho_\varsigma = \frac{1}{16\pi G}\left(6\varsigma'^2 + V(\varphi,\varsigma)|_\varphi\right),$$

which are useful when we study the behavior of dynamical systems. Additionally we can introduce the total quintom energy density and pressure as:

$$\rho_{DE} = \rho_\varsigma + \rho_\varphi, \qquad P_{DE} = P_\varsigma + P_\varphi, \qquad P_{DE} = \omega_{DE}\rho_{DE} \tag{26}$$

where

$$\omega_{DE} = \frac{6\dot{\varsigma}^2 - 6\dot{\varphi}^2 - V(\varsigma,\varphi)}{6\dot{\varsigma}^2 - 6\dot{\varphi}^2 + V(\varsigma,\varphi)} \tag{27}$$

To solve the set of differential equation $(\beta_\pm, \Omega, \varphi, \varsigma)$ we begin with the equations (16, 17) where we obtain the relation between the functions β_- and Ω as

$$\beta_- = \beta_0 \int e^{-3\Delta\Omega} d\tau, \tag{28}$$

similar to equations (17,18) we find

$$\beta_+ = \beta_1 \int e^{-3\Delta\Omega} d\tau, \tag{29}$$

then there is the relation between the anisotropic functions $\beta_- = \beta_2\beta_+$ with $\beta_2 = \frac{\beta_0}{\beta_1}$.

For separable potentials, equations (24,23) can be solved in some cases in terms of the Ω function, then, using equation (15) we can obtain in a quadrature form, the structure of Ω as

$$\int \frac{d\Omega}{\sqrt{h(\Omega)}} = \Delta\tau, \tag{30}$$

where the function $h(\Omega)$ has the corresponding information of all functions presented in this equation (15). For instance, when the potential $V(\varphi,\varsigma)$ becomes null or constant, the formalism is like the one formulated by Sáez and Ballester in 1986 [35] because both field are equivalent, see equations (24,23), where

$$\varsigma' = \varsigma_0 e^{-3\Delta\Omega}, \qquad \varsigma(\tau) = \varsigma_0 \int e^{-3\Delta\Omega(\tau)} d\tau + \varsigma_1, \tag{31}$$

$$\varphi' = \varphi_0 e^{-3\Delta\Omega}, \qquad \varphi(\tau) = \varphi_0 \int e^{-3\Delta\Omega(\tau)} d\tau + \varphi_1,, \tag{32}$$

and the function $h(\Omega)$ is

$$h(\Omega) = \frac{\frac{8\pi G M_\gamma}{3} e^{-3(1+\gamma)\Omega} + \left(-\varphi_0^2 + \varsigma_0^2\right) e^{-6\Delta\Omega} + \frac{\Lambda_{eff}}{3}}{1 - 9\left(1+\beta_2^2\right)\beta_1^2 e^{-6\Delta\Omega}}, \tag{33}$$

where $\Lambda_{eff} = \Lambda + V_0/2$. For particular values in the γ parameter, the equation (30) has a solution. This formalism was studied by one of the author and collaborators, in the FRW and Bianchi type Class A cosmological models, [36–38].

4. Quantum approach

On the Wheeler-DeWitt (WDW) equation there are a lot of papers dealing with different problems, for example in [39], they asked the question of what a typical wave function for the universe is. In Ref. [40] there appears an excellent summary of a paper on quantum cosmology where the problem of how the universe emerged from big bang singularity can no longer be neglected in the GUT epoch. On the other hand, the best candidates for quantum solutions become those that have a damping behavior with respect to the scale factor, in the sense that we obtain a good classical solution using the WKB approximation in any scenario in the evolution of our universe [41, 42]. Our goal in this paper deals with the problem to build the appropriate scalar potential in the inflationary scenario.

The Wheeler-DeWitt equation for this model is achieved by replacing $\Pi_{q^\mu} = -i\partial_{q^\mu}$ in (8). The factor $e^{-3\Omega}$ may be a factor ordered with $\hat{\Pi}_\Omega$ in many ways. Hartle and Hawking [41] have suggested what might be called a semi-general factor ordering which in this case would order $e^{-3\Omega}\hat{\Pi}^2_\Omega$ as

$$-e^{-(3-Q)\Omega}\partial_\Omega e^{-Q\Omega}\partial_\Omega = -e^{-3\Omega}\partial^2_\Omega + Q\,e^{-3\Omega}\partial_\Omega, \tag{34}$$

where Q is any real constant that measure the ambiguity in the factor ordering in the variable Ω. In the following we will assume this factor ordering for the Wheeler-DeWitt equation, which becomes

$$\Box\Psi + Q\frac{\partial\Psi}{\partial\Omega} + e^{6\Omega}U(\Omega, \beta_\pm, \varphi, \varsigma, \lambda_{\rm eff})\Psi = 0, \tag{35}$$

where $\Box = -\frac{\partial^2}{\partial\Omega^2} + \frac{\partial^2}{\partial\varsigma^2} - \frac{\partial^2}{\partial\varphi^2} + \frac{\partial^2}{\partial\beta_-^2} + \frac{\partial^2}{\partial\beta_+^2}$ is the d'Alambertian in the coordinates $q^\mu = (\Omega, \varsigma, \beta_\pm, \varphi)$. In the following, we introduce the main idea of the Bohm formalism, and why we choose the phase in the wave function to be real and not imaginary.

4.1. Mathematical structure in the Bohm formalism

In this section we will explain how the quantum potential approach or as is also known, the Bohm formalism [34], works in the context of quantum cosmology. For the cases that will be object of our investigation in the sections to come, it is sufficient to consider the simplest model, for which the whole quantum dynamics resides in the single equation,

$$\mathcal{H}\psi = \left(g^{\mu\nu}\nabla_{\mu}\nabla_{\nu} - V(q^{\mu})\right)\psi = 0, \tag{36}$$

where the metric may be q^{μ} dependent. The ψ is called the wave function of the universe, and we consider that ψ has the following traditional decomposition

$$\psi = R(q^{\mu})\, e^{\frac{i}{\hbar}S(q^{\mu})}, \tag{37}$$

with R and S as real functions. Inserting (37) into (36), we obtain two equations corresponding to the real and imaginary parts, respectively, which are

$$\Box R - R\left[\frac{1}{\hbar^2}(\nabla S)^2 + V\right] = \Box R - R\left[H(S)\right] = 0, \tag{38}$$

$$2\nabla R \cdot \nabla S + R\Box S = 0, \tag{39}$$

when we consider the problem of factor ordering, usually in cosmological problems, as we indicated in the beginning of this section, equation (34), must be included as linear term of $Q\frac{\partial\psi}{\partial q}$, where Q is a real parameter that measures the ambiguity in this factor ordering. So, the equations (38,39) are written as

$$\Box R + Q\frac{\partial R}{\partial q} - R\left[\frac{1}{\hbar^2}(\nabla S)^2 + V\right] = 0, \tag{40}$$

$$2\nabla R \cdot \nabla S + R\Box S + R\frac{\partial S}{\partial q} = 0, \tag{41}$$

where q is a single field coordinate.

We assume that the wave function ψ is a solution of equation (36), and thus this equation is equally satisfied. Considering the Hamiton-Jacobi analysis, we can identify the equation (40) as the most important equation of this treatment, because with this equation we can derive the time dependence, and thus, it serves as the evolutionary equation in this formalism. Following the Hamilton-Jacobi procedure, the Π_q momenta is related to the superpotential function S, as $\Pi_{q^{\mu}} = \frac{\partial S}{\partial q^{\mu}}$, which are related with the classical momenta (8) written in the previous section, thus,

$$\frac{dq^{\mu}}{dt} = g^{\mu\nu}\frac{\delta H(S)}{\delta\frac{\partial S}{\partial q^{\nu}}}, \tag{42}$$

which defines the trajectory q^{μ} in terms of the phase of the wave function S. We substitute this equation into (40), and we find (using $\dot{q}^{\mu} = \frac{dq^{\mu}}{dt}$ and $\hbar = 1$),

$$\left[\Box R + Q\frac{\partial R}{\partial q}\right] = R\left[g_{\mu\nu}\dot{q}^{\mu}\dot{q}^{\nu} + V\right]. \tag{43}$$

Therefore we see that the quantum evolution differs from the classical one only by the presence of the quantum potential term

$$\left[\Box R + Q\frac{\partial R}{\partial q}\right]$$

on the left-hand side of the equation of motion. Since we assume that the wave function is known, the quantum potential term is also known.

In the next subsection we will choose the $\psi = We^{-S}$ ansatz for the wave function, it was first remarked by Kodama [43, 44] that the solutions to the Wheeler-DeWitt (WDW) equation in the formulation of Arnowitt-Deser and Misner (ADM) and the Ashtekar formulation (in the connection representation) are related by $\psi_{ADM} = \psi_A e^{\pm i\Phi_A}$, where Ψ_A is the homogeneous specialization for the generating functional of the canonical transformation between ADM variables to Ashtekar's, [45]. This function was calculated explicitly for the diagonal Bianchi type IX model by Kodama, who also found $\Psi_A = const$ as a solution, and Ψ_A is pure imaginary, for a certain factor ordering, one expects a solution of the form $\psi = We^{\pm\Phi}$, where W is a constant, and $\Phi = i\Phi_A$. In fact this type of solution has been found for the diagonal Bianchi Class A cosmological models [46, 47], but W in some cases is a function, as we will see in our present study.

4.2. Our treatment

Using the ansatz

$$\Psi = \mathrm{e}^{\pm a_1\beta_+}\mathrm{e}^{\pm a_2\beta_-}\Xi(\Omega, \varsigma, \varphi), \tag{44}$$

the WDW equation is read as

$$\left[\Box + \mathrm{Q}\frac{\partial}{\partial\Omega} + \mathrm{e}^{6\Omega}\mathrm{U}(\varphi, \varsigma, \lambda_{\mathrm{eff}}) + \mathrm{c}^2\right]\Xi = 0, \tag{45}$$

where $c^2 = a_1^2 + a_2^2$ and now $\Box$ is written in the reduced coordinates $\ell^\mu = (\Omega, \varsigma, \varphi)$

We find that the WDW equation is solved when we choose an ansatz similar to the one employed in the Bohmian formalism of quantum mechanics [34], so we make the following Ansatz for the wave function

$$\Xi(\ell^\mu) = \mathrm{W}(\ell^\mu)\mathrm{e}^{-\mathrm{S}(\ell^\mu)}, \tag{46}$$

where $S(\ell^\mu)$ is known as the superpotential function, and W is the amplitude of probability that is employed in Bohmian formalism [34]. Then (45) transforms into

$$\Box W - W\Box S - 2\nabla W \cdot \nabla S - Q\frac{\partial W}{\partial\Omega} + QW\frac{\partial S}{\partial\Omega} + W\left[(\nabla S)^2 - \mathcal{U}\right] = 0, \tag{47}$$

where now, $\Box = \mathrm{G}^{\mu\nu}\frac{\partial^2}{\partial\ell^\mu\partial\ell^\nu}$, $\nabla \mathrm{W} \cdot \nabla \Phi = \mathrm{G}^{\mu\nu}\frac{\partial \mathrm{W}}{\partial\ell^\mu}\frac{\partial\Phi}{\partial\ell^\nu}$, $(\nabla)^2 = \mathrm{G}^{\mu\nu}\frac{\partial}{\partial\ell^\mu}\frac{\partial}{\partial\ell^\nu} = -(\frac{\partial}{\partial\varsigma})^2 + (\frac{\partial}{\partial\Omega})^2 + (\frac{\partial}{\partial\varphi})^2$, with $\mathrm{G}^{\mu\nu} = \mathrm{diag}(-1,1,1)$, $\mathcal{U} = \mathrm{e}^{6\Omega}\mathrm{U}(\varsigma, \varphi, \lambda_{\mathrm{eff}}) + \mathrm{c}^2$ is the potential term of the cosmological model under consideration.

Eq (47) can be written as the following set of partial differential equations

$$(\nabla S)^2 - \mathcal{U} = 0, \tag{48a}$$

$$\Box W \quad Q\frac{\partial W}{\partial \Omega} = 0 \tag{48b}$$

$$W\left(\Box S - Q\frac{\partial S}{\partial \Omega}\right) + 2\nabla W \cdot \nabla S = 0,. \tag{48c}$$

The first two equations correspond to the real part in a separated way, also, the first equation is called the Einstein-Hamilton-Jacobi equation (EHJ), and the third equation is the imaginary part, such as the equations presented in previous section (40, 41).

Following the references [32, 33], first, we shall choose to solve Eqs. (48a) and (48c), whose solutions at the end will have to fulfill Eq. (48b), which will play the role of a constraint equation.

Taking the ansatz

$$S(\Omega, \varsigma, \varphi) = \frac{e^{3\Omega}}{\mu} g(\varphi)h(\varsigma) + c\,(b_1\Omega + b_2\Delta\varphi + b_3\Delta\varsigma)\,, \tag{49}$$

where $\Delta\varphi = \varphi - \varphi_0$, $\Delta\varsigma = \varsigma - \varsigma_0$ with φ_0 and ς_0 as constant scalar fields, and b_i as arbitrary constants. Then, Eq (48a) is transformed as

$$e^{6\Omega}\left[\frac{h^2}{\mu^2}\left(\frac{dg}{d\varphi}\right)^2 - \frac{g^2}{\mu^2}\left(\frac{dh}{d\varsigma}\right)^2 + \frac{9}{\mu^2}g^2h^2 - U(\varphi, \varsigma, \lambda_{\mathrm{eff}})\right]$$
$$+\frac{6ce^{3\Omega}}{\mu}\left[b_1gh + \frac{b_2}{3}h\frac{dg}{d\varphi} - \frac{b_3}{3}g\frac{dh}{d\varsigma}\right] + c^2\left(b_1^2 + b_2^2 - b_3^2 - 1\right) = 0. \tag{50}$$

At this point we question ourselves how to solve this equation in relation to the constant c, implying the behavior of the universe with the anisotropic parameter $\beta_{\pm}$.

1. When we consider this equation as an expansion in powers of e^{Ω}, then each term is null in a separated way, but maintaining that the constant $c \neq 0$,

$$b_1^2 - b_3^2 + b_2^2 - 1 = 0, \tag{51}$$

$$b_1gh + \frac{b_2}{3}h\frac{dg}{d\varphi} - \frac{b_3}{3}g\frac{dh}{d\varsigma} = 0, \tag{52}$$

$$h^2\left(\frac{dg}{d\varphi}\right)^2 - g^2\left(\frac{dh}{d\varsigma}\right)^2 + 9g^2h^2 - U(\varphi, \varsigma, \lambda_{\mathrm{eff}}) = 0, \tag{53}$$

these set of equations do not have solutions in closed form, because the first equation is not satisfied. So, it is necessary to take c=0, implying that the wave function in the anisotropic coordinates have an oscilatory and hyperbolic behavior

2. For the case c=0, we have the following.
   The constants a_i are related as $a_2 = \pm i a_1$, hence the wave function corresponding to the anisotropic behavior becomes $e^{\pm a_1\beta_+ \pm i a_1 \beta_-}$, i.e, one part goes as oscillatory in the anisotropic parameter.

4.3. Mathematical structure of potential fields

To solve the Hamilton-Jacobi equation (48a)

$$-\left(\frac{\partial S}{\partial \varsigma}\right)^2 + \left(\frac{\partial S}{\partial \Omega}\right)^2 + \left(\frac{\partial S}{\partial \varphi}\right)^2 = e^{6\alpha} U(\varphi, \varsigma, \lambda_{\rm eff})$$

we propose that the superpotential function has the form

$$S = \frac{e^{3\Omega}}{\mu} g(\varphi) h(\varsigma), \tag{54}$$

and the potential

$$U = g^2 h^2 \left[a_0 G(g) + b_0 H(h)\right], \tag{55}$$

where $g(\varphi)$, $h(\varsigma)$, $G(g)$ and $H(h)$ are generic functions of the arguments, which will be determined under this process. When we introduce the ansatz in (48a) we find the following master equations for the fields (φ, ς), (here $c_1 = \mu a_0$ and $c_0 = \mu_0 b_0$)

$$d\varphi = \pm \frac{dg}{g\sqrt{p^2 + c_1 G}}, \qquad \text{with} \quad p^2 = \nu^2 - \frac{9}{2}, \tag{56a}$$

$$d\varsigma = \pm \frac{dh}{h\sqrt{\ell^2 - c_0 H}}, \qquad \text{with} \quad \ell^2 = \nu^2 + \frac{9}{2}, \tag{56b}$$

where ν is the separation constant.

For a particular structure of functions G and H we can solve the $g(\varphi)$ and $h(\varsigma)$ functions, and then use the expression for the potential term (55) over again to find the corresponding scalar potential that leads to an exact solution to the Hamilton-Jacobi equation (48a). Some examples are shown in Tables 2 and 3.

Thereby, the superpotential $S(\Omega, \varphi)$ is known, and the possible quintom potentials are shown in table 3

To solve (48c) we assume that

$$W = e^{[\eta(\Omega) + \xi(\varphi) + \lambda(\varsigma)]}, \tag{57}$$

and introducing the corresponding superpotential function S (54) into the equation (48c), it follow the equation

$H(h)$	$h(\varsigma)$	$G(g)$	$g(\varphi)$
0	$h_0 e^{\pm \ell \Delta\varsigma}$	0	$g_0 e^{\pm p \Delta\varphi}$
H_0	$h_0 e^{\pm\sqrt{\ell^2 - c_0 H_0}\Delta\varsigma}$	G_0	$g_0 e^{\pm\sqrt{p^2 + c_1 G_0}\Delta\varphi}$
$H_0 h^{-2}$	$\frac{\sqrt{c_0 H_0}}{\ell}\cosh[\ell\Delta\varsigma]$	$G_0 g^{-2}$	$\frac{\sqrt{c_1 G_0}}{p}\sinh[p\Delta\varphi]$
$H_0 h^{-n}\ (n \neq 2)$	$\left[\frac{c_0 H_0}{\ell^2}\cosh^2\left(\frac{n\ell\Delta\varsigma}{2}\right)\right]^{1/n}$	$G_0 g^{-n}\ (n \neq 2)$	$\left[\frac{c_1 G_0}{p^2}\sinh^2\left(\frac{np\Delta\varphi}{2}\right)\right]^{1/n}$
$H_0 \ln h$	$e^{u(\varsigma)}$, $u(\varsigma) = \frac{\ell^2 - \left(\frac{c_0 H_0}{2}\Delta\varsigma\right)^2}{c_0 H_0}$	$G_0 \ln g$	$e^{v(\varphi)}$, $v(\varphi) = \frac{-p^2 + \left(\frac{c_1 G_0}{2}\Delta\varphi\right)^2}{c_1 G_0}$
$H_0(\ln h)^2$	$e^{r(\varsigma)}$, $r(\varsigma) = \frac{\ell}{\sqrt{c_0 H_0}}\sin\left(\sqrt{c_0 H_0}\,\Delta\varsigma\right)$	$G_0(\ln g)^2$	$e^{\omega(\varphi)}$, $\omega(\varphi) = \frac{p}{\sqrt{c_1 G_0}}\sinh\left(\sqrt{c_1 G_0}\,\Delta\varphi\right)$

Table 2. Some exact solutions to eqs. (56a,56b), where n is any real number, G_0 and H_0 are arbitrary constants.

$U(\varphi,\varsigma)$	**Relation between all constants**
0	$\ell^2(s - p^2 - 3k - 9)^2 - p^2(s - \ell^2)^2 + \ell^2 p^2 (k^2 - Q^2) = 0$
$U_0 e^{\pm 2\left[\sqrt{\ell^2 - c_0 H_0}\Delta\varsigma + \sqrt{p^2 + c_1 G_0}\Delta\varphi\right]}$	$(\ell^2 - c_0 H_0)(s - p^2 - 3k - 9 - c_1 G_0)^2 - (p^2 + c_1 G_0)(s - \ell^2 + c_0 H_0)^2 + (\ell^2 - c_0 H_0)(p^2 + c_1 G_0)(k^2 - Q^2) = 0$
$U_0 \sinh^2(p\Delta\varphi) + U_1 \cosh^2(\ell\Delta\varsigma)$	$k(k-6) - Q^2,\ \varepsilon = \ell^2,$ $6k(9 + p^2) + 9Q^2 - p^4 + (\ell^2 - 9)^2 = 0$
$b_0 H_0 \left[\frac{c_1 G_0}{p^2}\sinh^2\left(\frac{np\Delta\varphi}{2}\right)\right]^{\frac{2}{n}} \left[\frac{c_0 H_0}{\ell^2}\cosh^2\left(\frac{n}{2}\ell\Delta\varsigma\right)\right]^{\frac{2-n}{n}} + a_0 G_0 \left[\frac{c_1 G_0}{p^2}\sinh^2\left(\frac{np\Delta\varphi}{2}\right)\right]^{\frac{2-n}{n}} \left[\frac{c_0 H_0}{\ell^2}\cosh^2\left(\frac{n}{2}\ell\Delta\varsigma\right)\right]^{\frac{2}{n}}$	quantum constraint is not satisfied
$e^{2u(\varsigma) + 2v(\varphi)}\left[b_0 H_0 u(\varsigma) + a_0 G_0 v(\varphi)\right]$	quantum constraint is not satisfied
$e^{2r(\varsigma) + 2\omega(\varphi)}\left[b_0 H_0 r^2 + a_0 G_0 \omega^2\right]$	quantum constraint is not satisfied

Table 3. The corresponding quintom potentials that emerge from quantum cosmology in direct relation with the table (2). Also we present the relation between all constant that satisfy the eqn. (48b).

$$\frac{1}{g}\frac{d^2 g}{d\varphi^2} + \frac{2}{g}\frac{dg}{d\varphi}\frac{d\xi}{d\varphi} + 9 - \frac{1}{h}\frac{d^2 h}{d\varsigma^2} - \frac{2}{h}\frac{d\lambda}{d\varsigma}\frac{dh}{d\varsigma} + 6\frac{d\eta}{d\Omega} - 3Q = 0, \tag{58}$$

and using the method of separation of variables, we arrive to a set of ordinary differential equations for the functions $\eta(\Omega)$, $\xi(\varphi)$ and $\lambda(\varsigma)$ (however, this decomposition is not unique, because it depend as we put the constants in the equations).

$$2\frac{d\eta}{d\Omega} - Q = k, \tag{59}$$

$$\frac{d^2 g}{d\varphi^2} + 2\frac{dg}{d\varphi}\frac{d\xi}{d\varphi} = [s - 3(k+3)]g, \tag{60}$$

$$\frac{d^2 h}{d\varsigma^2} + 2\frac{dh}{d\varsigma}\frac{d\lambda}{d\varsigma} = sh, \tag{61}$$

whose solutions in the generic fields g and h are

$$\eta(\Omega) = \frac{Q+k}{2}\Omega,$$

$$\lambda(\varsigma) = \frac{s}{2}\int \frac{d\varsigma}{\partial_\varsigma(\ln h)} - \frac{1}{2}\int \frac{\frac{d^2h}{d\varsigma^2}}{\partial_\varsigma h}d\varsigma,$$

$$\xi(\varphi) = \left(\frac{s}{2} - \frac{3k}{2} - \frac{9}{2}\right)\int \frac{d\varphi}{\partial_\varphi(\ln g)} - \frac{1}{2}\int \frac{\frac{d^2g}{d\varphi^2}}{\partial_\varphi g}d\varphi,$$

then

$$W = e^{\frac{s}{2}\int\left(\frac{d\varsigma}{\partial_\varsigma(\ln h)} + \frac{d\varphi}{\partial_\varphi(\ln g)}\right)} e^{-\frac{1}{2}\int\left(\frac{\frac{d^2h}{d\varsigma^2}}{\partial_\varsigma h}d\varsigma + \frac{\frac{d^2g}{d\varphi^2}}{\partial_\varphi g}d\varphi\right)} e^{\frac{k}{2}\left(\Omega - 3\int\frac{d\varphi}{\partial_\varphi(\ln g)}\right)} e^{\frac{1}{2}\left(Q\Omega - 9\int\frac{d\varphi}{\partial_\varphi(\ln g)}\right)}. \tag{62}$$

In a similar way, the constraint (48b) can be written as

$$\partial^2_\varphi\xi + (\partial_\varphi\xi)^2 - \partial^2_\varsigma\lambda - (\partial_\varsigma\lambda)^2 + \frac{k^2 - Q^2}{4} = 0, \tag{63}$$

or in other words (here $\mu_0 = s - 3(3+\kappa)$)

$$2\frac{\partial^3_\varsigma h}{\partial_\varsigma h} - 2\frac{\partial^3_\varphi g}{\partial_\varphi g} + 4sh\frac{\partial^2_\varsigma h}{(\partial_\varsigma h)^2} - 4\mu_0 g\frac{\partial^2_\varphi g}{(\partial_\varphi g)^2} - 3\frac{(\partial^2_\varsigma h)^2}{(\partial_\varsigma h)^2} + 3\frac{(\partial^2_\varphi g)^2}{(\partial_\varphi g)^2} - \frac{s^2h^2}{(\partial_\varsigma h)^2}$$
$$+ \frac{\mu_0^2 g^2}{(\partial_\varphi g)^2} - 2s + 2\mu_0 + k^2 - Q^2 = 0.$$

when we use the different cases presented in table (2), the following relations between all constants were found, which we present in the same table II with the quintom potentials. So, the quantum solutions for each potential scalar fields are presented in quadrature form, using the equations (46, 54) and (57).

Thereby, under canonical quantization we were able to determine a family of potentials that are the most probable to characterize the inflation phenomenon in the evolution of our universe. The exact quantum solutions to the Wheeler-DeWitt equation were found using the Bohmian scheme [34] of quantum mechanics where the ansatz to the wave function $\Psi(\ell^\mu) = e^{a_1\beta_+ + ia_i\beta_-}W(\ell^\mu)e^{-S(\ell^\mu)}$ includes the superpotential function which plays an important role in solving the Hamilton-Jacobi equation. It was necessary to study the classical behavior in order to know when the Universe evolves from a quintessence dominated phase to a phantom dominated phase crossing the $w_{eff} = -1$ dividing line, as a transient stage. Also, this family of potentials can be studied within the dynamical systems framework to obtain useful information about the asymptotic properties of the model and give a classification of which ones are in agreement with the observational data [48]

5. Classical solutions a la WKB

For our study, we shall make use of a semi-classical approximation to extract the dynamics of the WDW equation. The semi-classical limit of the WDW equation is achieved by taking $\Psi = e^{-S}$, and imposing the usual WKB conditions on the superpotential function S, namely

$$\left(\frac{\partial S}{\partial q}\right)^2 >> \frac{\partial^2 S}{\partial q^2}$$

Hence, the WDW equation, under the particular factor ordering $Q = 0$, becomes exactly the afore-mentioned EHJ equation (48a) (this approximation is equivalent to a zero quantum potential in the Bohmian interpretation of quantum cosmology [53]). The EHJ equation is also obtained if we introduce the following transformation on the canonical momenta $\Pi_q \rightarrow \partial_q S$ in Eq. (8) and then Eq. (8) provides the classical solutions of the Einstein.Klein.Gordon (EKG) equations. Moreover, for the particular cases shown in Table 1, the classical solutions of the EKG, in terms of $q(\tau)$, arising from Eqs. (8) and (50) are given by

$$gh = 4\mu \frac{d\Omega}{d\tau}, \qquad \frac{d\varphi}{\partial_\varphi Lng} + \frac{d\varsigma}{\partial_\varsigma Lnh} = 0, \tag{64}$$

the second equation appears in the W function (57), then the W is simplified by, we also have the corresponding relation with the time τ

$$d\tau = 12\mu \frac{1}{h}\frac{d\varphi}{\partial_\varphi g}, \qquad d\tau = -12\mu \frac{1}{g}\frac{d\varsigma}{\partial_\varsigma h}. \tag{65}$$

In the following subsection, we will give details about the solutions corresponding to some of the scalar potential shown in Table (3).

To recover the solutions for the anisotropic function $\beta_\pm$, (28,29) we need to extend the superpotential function $\mathcal{S} = S - a_1\beta_+ - ia_2\beta_-$, remember that these functions were used in the ansatz for the wave function (44) in order to simplify the WDW equation (45). With this extension, we has

$$\frac{\partial \mathcal{S}}{\partial \beta_\pm} = -b_i = \text{constants}$$

and using the corresponding momenta (8), we obtain the corresponding solutions written in quadrature form in equations (28,29). In this subsection we calculate the solution for the Ω function, thence the classical solution will be complete.

5.0.1. *Free wave function*

This particular case corresponds to an null potential function $U(\varphi, \varsigma)$, (see first line in Table (3)). The particular exact solution for the wave function Ξ becomes

$$\Xi(\Omega,\varphi,\varsigma) = e^{\pm\frac{s}{2}\left(\frac{\Delta\varsigma}{\ell}+\frac{\Delta\varphi}{p}\right)} e^{\pm\frac{1}{2}(\ell\Delta\varsigma+p\Delta\varphi)} e^{\frac{\Omega}{2}(k+Q)\pm(3k-9)\frac{\Delta\varphi}{2p}} \,\mathrm{Exp}\left[-g_0 h_0 e^{3\Omega\pm\ell\Delta\varsigma\pm p\Delta\varphi}\right], \quad (66)$$

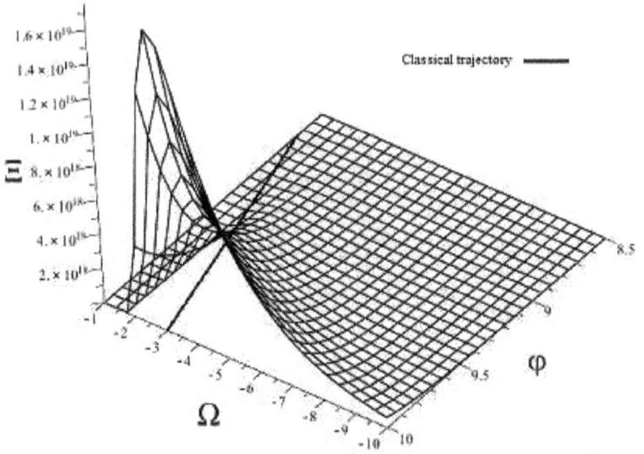

Figure 1. Exact wave function for the free case, i.e. for $U(\varphi,\varsigma) = 0$. The wave function (66) is peaked around the classical trajectory $\Delta\Omega \pm \frac{3}{p}\Delta\varphi = \nu_0 = \text{const}$, which is the solid line shown on the $\{\Omega,\varphi\}$ plane. For this case $\nu_0 = -1$ on equation (68a)

the classical trajectory implies that $\frac{\Delta\varsigma}{\ell} + \frac{\Delta\varphi}{p} = 0$, then there is the relation between the fields φ and ς, as

$$\Delta\varsigma = -\frac{\ell}{p}\Delta\varphi.$$

So, this wavefunction can be written in terms of φ and Ω solely,

$$\Xi(\Omega,\varphi) = e^{\frac{k}{2}\left(\Delta\Omega\pm\frac{3}{p}\Delta\varphi\right)+\frac{Q}{2}\Delta\Omega} \,\mathrm{Exp}\left[-g_0 h_0 e^{3\left(\Omega\pm\frac{3}{p}\Delta\varphi\right)}\right], \quad (67)$$

Using the equantion (65), we find the *classical trajectory* on the $\{\Omega,\varphi\}$ plane as

$$\Delta\Omega \pm \frac{3}{p}\Delta\varphi = \nu_0 = \text{const}, \quad (68a)$$

$$\Delta\varphi = \mathrm{Ln}\left[\frac{3g_0h_0}{4}\Delta\tau\right]^{\pm\frac{p}{9}}, \quad (68b)$$

$$\Delta\varsigma = \mathrm{Ln}\left[\frac{3g_0h_0}{4}\Delta\tau\right]^{\mp\frac{\ell}{9}}, \tag{68c}$$

$$\Delta\Omega = \mathrm{Ln}\left[\frac{3g_0h_0}{4}\Delta\tau\right]^{\frac{1}{3}}, \tag{68d}$$

which corresponds to constant phase of second exponential in the W function. The behavior of the scale factor correspond at stiff matter epoch in the evolution on the universe.

5.0.2. *Exponential scalar potential*

For an exponential scalar potential, see second line in Table (3), the exact solution of the WDW equation is similar to the last one, only we redefine the constants,

$$\ell \to \sqrt{\ell^2 - c_1 G_0}, \qquad p \to \sqrt{p^2 + c_0 H_0}$$

5.0.3. *Hyperbolic scalar potential*

This case corresponds to third line in Table ($U(\varphi,\varsigma) = U_0 \sinh^2(p\Delta\varphi) + U_1 \cosh^2(\ell\Delta\varsigma)$, the wave function for this is

$$\Upsilon = e^{-\frac{1}{2}\mathrm{Ln}|\cosh(\ell\Delta\varsigma)\sinh(p\Delta\varphi)|} e^{\frac{k}{2}\left[\Delta\Omega - \frac{3}{p^2}\mathrm{Ln}|\cosh(p\Delta\varphi)|\right]} e^{\frac{1}{2}\left(Q\Delta\Omega - \frac{9}{p^2}\mathrm{Ln}|\cosh(p\Delta\varphi)|\right)} e^{\left[-g_0h_0e^{3\Omega}\sinh(p\Delta\varphi)\cosh(\ell\Delta\varsigma)\right]} \tag{69}$$

and the classical trajectory in the plane $\{\Omega, \varphi\}$ reads as

$$\Delta\Omega - \frac{3}{p^2}\mathrm{Ln}|\cosh(p\Delta\varphi)| = \epsilon = \mathrm{const.} \tag{70}$$

Using the second equation in (64), we find the relation between the quintom fields

$$\Delta\varsigma = \frac{1}{\ell}\mathrm{arcsinh}\left[\frac{F_0}{\cosh^{\frac{1}{p^2}}(p\Delta\varphi)}\right]^{\ell^2},$$

then, the time dependence of the quintom fields only were possible to write in quadrature form, having the following structure

$$\sqrt{\frac{c_0H_0}{c_1G_0}}\frac{\Delta\tau}{12\ell} = \int \frac{d\varphi}{\cosh^{\frac{9}{p^2}}(p\Delta\varphi)\sqrt{F_0^{2\ell^2} + \cosh^{\frac{2\ell^2}{p^2}}(p\Delta\varphi)}}. \tag{71}$$

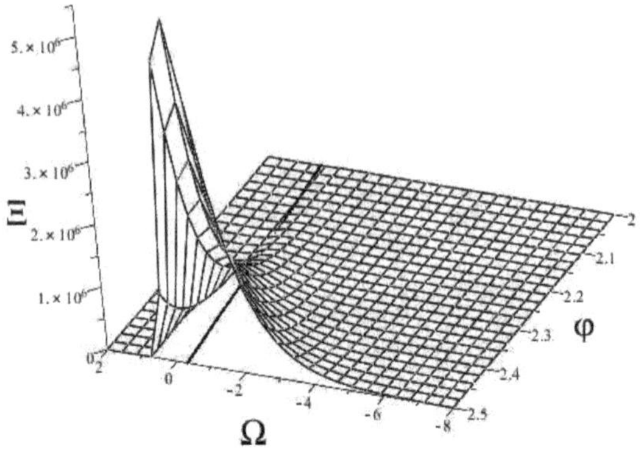

Figure 2. Wave Function for the potential $U(\varphi,\varsigma) = U_0 \sinh^2(p\Delta\varphi) + U_1 \cosh^2(\ell\Delta\varsigma)$. The wave function (69) is peaked around the classical trajectory $\Delta\Omega - \frac{3}{p^2}\mathrm{Ln}|\cosh(p\Delta\varphi)| = \epsilon$, which is the solid line shown on the $\{\Omega, \varphi\}$ plane. For this case $\epsilon = 2.4$ on equation (70)

6. Conclusions

In summary, we presented the corresponding Einstein Klein Gordon equation for the quintom model, which is applied to the Bianchi Type I cosmological model including as a matter content a barotropic perfect fluid and cosmological constant, and the classical solutions are given in a quadrature form for null and constant scalar potentials, these solutions are related to the Sáez-Ballester formalism, [35–38].

The quantum scheme in the Bohmian formalism and its mathematical structure, and our approach were applied to the Bianchi type I cosmological model in order to build the mathematical structure of quintom scalar potentials using the integral systems formalism.

Also, we presented the quantum solutions to the Wheeler-DeWitt equation, which is the main equation to be solved and such a subject is our principal objective in this work to obtain the family of scalar potential in the inflation phenomenon.

We emphasize that the quantum potential from the Bohm formalism will work as a constraint equation which restricts our family of potentials found, see Table (3), [33].

It is well known in the literature that in the Bohm formalism the imaginary part is never determined, however in this work such a problem has been solved in order to find the quantum potentials, which was a more important matter for being able to find the classical trajectories, which were showed through graphics how the classical trajectory is projected from its quantum counterpart. We include some steps how we solve the imaginary like equation (48c) when we found the superpotential function S (54) and particular ansatz for the function W, being the equation (58), and using the separation variables method we find the set of equations that is necessary to solve. Also we give some explanation why this decomposition in not unique.

Finally, we do the comment that solution to the equation (50), when we write this equation as a quadratic equation, $C_2x^2 + C_1x + C_0 = 0$ with $x = e^{\Omega}$, is not possible because the set of equations that appear , only a subset have solution in closed form. The algebraic one is not fulfill, making that the c parameter become null. In forthcoming work we will analyze under the scheme of dynamical systems, the relation between the corresponding critical points and its stability properties with these classical solutions for separable potentials obtained in this work [54].

7. Acknowledgments

This work was partially supported by CONACYT 179881 grant. DAIP (2011-2012) and PROMEP grants UGTO-CA-3, UAM-I-43. PRB and MA were partially supported by UAEMex grant FEO1/2012 103.5/12/2126. This work is part of the collaboration within the Instituto Avanzado de Cosmología, and Red PROMEP: Gravitation and Mathematical Physics under project *Quantum aspects of gravity in cosmological models, phenomenology and geometry of space-time*. Many calculations were done by Symbolic Program REDUCE 3.8.

Author details

J. Socorro[1,2], Paulo A. Rodríguez [1], O. Núñez-Soltero[1],
Rafael Hernández[1] and Abraham Espinoza-García[1]

1 Departamento de Física de la DCeI de la Universidad de Guanajuato-Campus León, Guanajuato, México
2 Departamento de Física, Universidad Autónoma Metropolitana, Apartado Postal 55-534, C.P. 09340 México, DF, México

References

[1] E.J. Copeland, M. Sami and S. Tsujikawa *Dynamics of dark energy Int. J. Mod. Phys. D* 15 1753, (2006) [arXiv:hep-th 0603057].

[2] B. Feng *The Quintom Model of Dark Energy* [ArXiv:astro-ph/0602156].

[3] Y.F. Cai, E.N. Saridakis, M.R. Setare and J.Q. Xia *Quintom cosmology: Theoretical implications and observations Phys. Rep.* 493, 1 (2010).

[4] Z.K. Guo, Y.S. Piao, X. Zhang and Y.Z. Zhang *Cosmological evolution of a quintom model of dark energy Phys. Lett. B* 608, 177 (2005).

[5] B. Feng, X. Wang and X. Zhang *Dark energy constraints from the cosmic age and supernova Phys. Lett. B* 607, 35 (2005).

[6] H.M. Sadjadi and M. Alimohammadi *Transition from quintessence to the phantom phase in the quintom model Phys. Rev. D* 74 (4), 043506 (2006).

[7] Z.K. Guo, Y.S. Piao, X. Zhang and Y.Z. Zhang *Two-field quintom models in the $\omega - \omega'$ plane Phys. Rev. D* 74 (12), 127304 (2006).

[8] B. Feng, M. Li, Y.S. Piao and X. Zhang *Oscillating quintom and the recurrent universe Phys. Lett. B* 634, 101 (2006).

[9] W. Zhao and Y. Zhang *Quintom models with an equation of state crossing -1 Phys. Rev. D* 73 (12) 123509 (2006).

[10] Y.F. Cai, M. Li,J.X. Lu, Y.S. Piao, T. Qiu and X. Zhang *A string-inspired quintom model of dark energy Phys. Lett. B* 651,1 (2007)

[11] Y.F. Cai, T. Qiu, X. Zhang, Y.S. Piao and M. Li *Bouncing universe with Quintom matter J. of High Energy Phys.* 10 71 (2007).

[12] M. Alimohammadi and H.M. Sadjadi, *Phys. Lett. B* 648, 113 (2007).

[13] R. Lazkoz, G. León and I. Quiros *Quintom cosmologies with arbitrary potentials Phys. Lett. B* 649, 103 (2007).

[14] Y.F. Cai, T. Qiu, R. Brandenberger, Y.S. Piao, and X. Zhang *On perturbations of a quintom bounce J of Cosmology and Astroparticle Phys.* 3, 13 (2008).

[15] Y.F. Cai and J. Wang *Dark energy model with spinor matter and its quintom scenario Classical and Quantum Gravity* 25 (16), 165014 (2008).

[16] S. Zhang and B. Chen *Reconstructing a string-inspired quintom model of dark energy Phys. Lett. B* 669, 4 (2008).

[17] J. Sadeghi, M.R. Setare, A. Banijamali and F. Milani *Non-minimally coupled quintom model inspired by string theory Phys. Lett. B* 662, 92, (2008).

[18] K. Nozari. M.R. Setare, T. Azizi and N. Behrouz *A non-minimally coupled quintom dark energy model on the warped DGP brane Physica Scripta,* 80 (2), 025901 (2009).

[19] M.R. Setare and E.N. Saridakis *Quintom dark energy models with nearly flat potentials Phys. Rev. D* 79 (4), 043005 (2009).

[20] M. R. Setare and E. N. Saridakis *Quintom Cosmology with general potentials. Int. Jour. of Mod. Phys. D* 18 (4), 549 (2009).

[21] J. Sadeghi, M.R. Setare and A. Banijamali *String inspired quintom model with non-minimally coupled modified gravity Phys. Lett. B* 678,164 (2009)

[22] T. Qiu *Theoretical Aspects of Quintom Models Mod. Phys. Lett.A* 25, 909 (2010).

[23] E.N. Saridakis *Quintom evolution in power-law potentials Nuclear Phys. B* 830, 374 (2010).

[24] A.R. Amani *Stability of Quintom Model of Dark Energy in (ω, ω') Phase Plane Int. J. of Theor. Phys.* 50, 3078 (2011).

[25] H. Farajollahi, A. Shahabi, A. Salehi *Dynamical stability in scalar-tensor cosmology Astronomy & Astrophysics, Supplement* 338, 205 (2012).

[26] Song-Kuan Guo, Yun-Song Piao, Xinmin Zhang and Yuan-Zhong Zhang *Cosmological Evolution of a Quintom Model of Dark Energy Phys. Lett. B* 608, 177 (2005). [arXiv:astro ph/0410654]

[27] R. Lazkoz and G. León *Quintom cosmologies admitting either tracking or phantom attractors Phys. Lett. B* 638, 303 (2006).

[28] L.P. Chimento, M. Forte, R. Lazkoz and M.G. Richarte *Internal space structure generalization of the quintom cosmological scenario Phys. Rev. D* 79 (4) 043502(2009).

[29] A.R. Liddle, and R.J. Scherrer *Classification of scalar field potential with cosmological scaling solutions Phys. Rev. D* 59, 023509 (1998).

[30] P.G. Ferreira & M. Joyce *Cosmology with a primordial scaling field, Phys. Rev. D,* 58, 023503 (1998).

[31] D. Adak, A. Bandyopadhyay and D. Majumdar *Quintom scalar field dark energy model with a Gaussian potential* (2011) [arXiv:1103.1533]

[32] W. Guzmán, M. Sabido, J. Socorro and L. Arturo Urena-Lopez *Scalar potentials out of canonical quantum cosmology Int. J. Mod. Phys. D* 16 (4), 641-653 (2007).

[33] J. Socorro and Marco D'oleire *Inflation from supersymmetric quantum cosmology Phys. Rev. D* 82(4), 044008 (2010).

[34] D. Bohm *Suggested interpretation of the quantum theory in terms of "Hidden" variables I Phys. Rev.* 85 (2), 166 (1952).

[35] D. Sáez and V.J. Ballester *Physics Letters A* 113, 467 (1986).

[36] J. Socorro, M. Sabido, M.A. Sánchez G. and M.G. Frías Palos *Anisotropic cosmology in Sáez-Ballester theory: classical and quantum solutions Rev. Mex. Fís.* 56(2), 166-171 (2010), [arxiv:1007.3306].

[37] M. Sabido, J. Socorro and L. Arturo Ureña-López *Classical and quantum Cosmology of the Sáez-Ballester theory Fizika B* 19 (4), 177-186 (2010), [arXiv:0904.0422].

[38] J. Socorro, Paulo A. Rodríguez, Abraham Espinoza-García, Luis O. Pimentel and P. Romero *Cosmological Bianchi Class A Models in Sáez-Ballester Theory: in Aspects of Today's Cosmology* InTech, Antonio Alfonso-Faus (Ed.) *pages 185-204*. Available from: http://www.intechopen.com/articles/show/title/

[39] G.W. Gibbons and L. P. Grishchuk *Nucl. Phys. B* 313, 736 (1989).

[40] Li Zhi Fang and Remo Ruffini, Editors, *Quantum Cosmology, Advances Series in Astrophysics and Cosmology Vol. 3* (World Scientific, Singapore, 1987).

[41] J. Hartle, & S.W. Hawking *Phys. Rev. D,* 28, 2960 (1983).

[42] S.W. Hawking *Nucl. Phys. B* 239, 257 (1984).

[43] H. Kodama *Progress of Theor. Phys.* 80, 1024 (1988).

[44] H. Kodama *Phys. Rev D* 42, 2548 (1990).

[45] A. Ashtekar *Phys. Rev. D* 36,1587 (1989).

[46] V. Moncrief and M.P. Ryan *Phys. Rev. D* 44, 2375 (1991).

[47] O. Obregón and J. Socorro $\Psi = We^{\pm\Phi}$ *quantum cosmological solutions for Class A Bianchi models Int. J. of Theor. Phys.* 35 (7), 1381 (1995).

[48] Private communication with Luis Ureña, DCI-Universidad de Guanajuato.

[49] M.P, Ryan *Hamiltonian cosmology*, (Springer, Berlin, (1992)).

[50] Andrei C. Polyanin & Valentin F. Zaitsev *Handbook of Exact solutions for ordinary differential equations, Second edition*, Chapman & Hall/CRC (2003).

[51] M.P. M.P. & L.C. Shepley *Homogeneous Relativistic Cosmologies* (Princenton (1985))

[52] P.V. Moniz *Quantum cosmology -the supersymmetric perspective- Vol. 1 & 2*, Lecture Notes in Physics 803 & 804, (Springer, Berlin) (2010).

[53] G. D. Barbosa and N. Pinto-Neto, *Noncommutative geometry and cosmology, Phys. Rev. D* 70 103512 (2004).

[54] León, G., Leyva, Y., and Socorro, J.: [gr-qc.1208-0061]

Chapter 11

Light Cold Dark Matter in a Two-Singlet Model

Abdessamad Abada and Salah Nasri

Additional information is available at the end of the chapter

1. Introduction

Understanding the nature of dark matter (DM) as well as the origin of baryon asymmetry are two of the most important questions in both Cosmology and Particle Physics. The failure of the Standard Model (SM) of the electroweak and strong interactions in accommodating such questions motivates the search for an explanation in the realm of new physics beyond it. The most attractive strategy so far has been to look into extensions of the SM that incorporate electrically neutral and colorless weakly interacting massive particles (WIMPs), with masses from one to a few hundred GeV, coupling constants in the milli-weak scale and lifetimes longer than the age of the Universe.

The most popular extension of the SM is the minimal supersymmetric standard model (MSSM) in which the neutral lightest supersymmetric particle (LSP) is seen as a candidate for dark matter. Indeed, neutralinos are odd under R -parity and are only produced or destroyed in pairs, thus making the LSP stable [1]. They can annihilate through a t-channel sfermion exchange into Standard Model fermions, or via a t-channel chargino-mediated process into W^+W^-, or through an s-channel pseudoscalar Higgs exchange into fermion pairs. They can also undergo elastic scattering with nuclei through mainly a scalar Higgs exchange [2].

However, most particularly in light of the recent signal reported by CoGeNT [3], which favors a light dark matter (LDM) with a mass in the range $7-9$GeV and nucleon scattering cross section $\sigma_{\rm det} \sim 10^{-4}$ pb, having a neutralino as a LDM candidate can be challenging. Indeed, systematic studies show that an LSP with a mass around 10GeV and an elastic scattering cross-section off a nucleus larger than $\sim 10^{-5}$ pb requires a very large $\tan\beta$ and a relatively light CP-odd Higgs [4]. This choice of parameters leads to a sizable contribution to the branching ratios of some rare decays, which then disfavors the scenario of light neutralinos [5]. Also, in the next-to-minimal supersymmetric standard model (NMSSM) with 12 input parameters [6], realizing a LDM with an elastic scattering cross section capable of generating the CoGeNT signal is possible only in a finely-tuned region of the parameters where the neutralino is mostly singlino and the light CP-even Higgs is singlet-like with a mass below a few GeV. In such a situation, it is very difficult to detect such a light Higgs at the collider. It is clear then that other alternative scenarios for LDM are needed [7].

The simplest of scenarios is then to extend the Standard Model by a real $\mathbb{Z}_2$ symmetric scalar field, the dark matter, which has to be a SM gauge singlet that interacts with visible particles via the Higgs field only. Such an extension was first proposed in [8] and further studied in [9] where the $\mathbb{Z}_2$ symmetry is extended to a global U(1) symmetry, more extensively in [10]. Specific implications on Higgs detection and LHC physics were discussed in [11] and one-loop vacuum stability and perturbativity bounds discussed in [12]. However, the work of [13] uses constraints from the experiments XENON10 [14] and CDMSII [15] to exclude DM masses smaller than 50, 70 and 75GeV for Higgs masses equal to 120, 200 and 350 GeV respectively. Also, the Fermi-LAT data on the isotropic diffuse gamma-ray emission can potentially exclude this one-singlet dark-matter model for masses as low as 6GeV, assuming a NFW profile for the dark-matter distribution [16].

Therefore, a two-singlet extension of the Standard Model was proposed in [17] as a simple model for light cold dark matter. Both scalar fields are $\mathbb{Z}_2$ symmetric with one undergoing spontaneous symmetry breaking. The behavior of the model under the DM relic-density constraint and the restrictions from experimental direct detection was studied. It was concluded that the model was capable of bearing a light dark-matter WIMP.

The present chapter describes how we can further the study of the two-singlet scalar model (2SSM) by discussing some of its phenomenological implications. We limit ourselves to small DM masses, from 0.1GeV to 10GeV. We discuss the implications of the model on the meson factories and the Higgs search at the LHC. In fact, it is pertinent at this stage to mention that there are more than one motivation for scalar-singlet extensions of the SM. Indeed, besides providing a possible account for the dark matter in the Universe consistent with the CoGeNT signal, they also provide a solution to the mu problem in the supersymmetric standard model. They can explain the matter-anti matter asymmetry in the Universe [18], and account for the possible occurrence of a light Higgs with a mass less or equal to 100GeV while still in agreement with the electroweak precision tests [19] and potential signatures at B-factories [20].

This chapter is organized as follows. The next section introduces briefly the model and summarizes the results of [17] regarding relic-density constraint and direct detection. The two following sections investigate the rare decays of Υ (section three) and B (section four) mesons, most particularly their invisible channels. Section five looks into the decay channels of the Higgs particle. In each of these situations, we try, when possible, to deduce preferred regions of the parameter space and excluded ones. The last section is devoted to concluding remarks. Results presented here are found in [20].

2. The 2SSM Model

The Standard Model is extended with two real, spinless and $\mathbb{Z}_2$-symmetric fields. One is the dark-matter field S_0 with unbroken symmetry to ensure the stability of the dark matter, the other is an auxiliary field χ_1 for which the symmetry is spontaneously broken. Both fields are Standard-Model gauge singlets and can interact with SM particles only via the Higgs doublet H. This latter is taken in the unitary gauge such that $H^\dagger = 1/\sqrt{2}\,(0 \;\; h')$, where h' is a real scalar. We assume all processes calculable in perturbation theory. The potential function that incorporates S_0, h' and χ_1 is:

$$U=\frac{\tilde{m}_0^2}{2}S_0^2-\frac{\mu^2}{2}h'^2-\frac{\mu_1^2}{2}\chi_1^2+\frac{\eta_0}{24}S_0^4+\frac{\lambda}{24}h'^4+\frac{\eta_1}{24}\chi_1^4+\frac{\lambda_0}{4}S_0^2h'^2+\frac{\eta_{01}}{4}S_0^2\chi_1^2+\frac{\lambda_1}{4}h'^2\chi_1^2. \quad (1)$$

The mass parameters squared $\tilde{m}_0^2$, $\mu^2$and μ_1^2 and all the coupling constants are real positive numbers. Electroweak spontaneous symmetry breaking occurs for the Higgs field with the vacuum expectation value $v = 246\text{GeV}$. The field χ_1 will oscillate around the vacuum expectation value v_1 which we take at the electroweak scale 100GeV. Writing $h' = v + \tilde{h}$ and $\chi_1 = v_1 + \tilde{S}_1$, the potential function becomes, up to an irrelevant zero-field energy:

$$U = U_{\text{quad}} + U_{\text{cub}} + U_{\text{quar}}, \quad (2)$$

where the quadratic terms are given by:

$$U_{\text{quad}} = \frac{1}{2}m_0^2S_0^2 + \frac{1}{2}M_h^2\tilde{h}^2 + \frac{1}{2}M_1^2\tilde{S}_1^2 + M_{1h}^2\tilde{h}\tilde{S}_1, \quad (3)$$

with the mass-squared coefficients related to the original parameters of the theory by the following relations:

$$\begin{aligned} m_0^2 &= \tilde{m}_0^2 + \frac{\lambda_0}{2}v^2 + \frac{\eta_{01}}{2}v_1^2; \quad M_h^2 = -\mu^2 + \frac{\lambda}{2}v^2 + \frac{\lambda_1}{2}v_1^2; \\ M_1^2 &= -\mu_1^2 + \frac{\lambda_1}{2}v^2 + \frac{\eta_1}{2}v_1^2; \quad M_{1h}^2 = \lambda_1 v\, v_1. \end{aligned} \quad (4)$$

Clearly, we need to diagonalize the mass-squared matrix. Denoting the physical mass-squared field eigenmodes by h and S_1, we rewrite:

$$U_{\text{quad}} = \frac{1}{2}m_0^2S_0^2 + \frac{1}{2}m_h^2h^2 + \frac{1}{2}m_1^2S_1^2, \quad (5)$$

where the physical fields are related to the mixed ones by a 2×2 rotation:

$$\begin{pmatrix} h \\ S_1 \end{pmatrix} = \begin{pmatrix} \cos\theta & \sin\theta \\ -\sin\theta & \cos\theta \end{pmatrix} \begin{pmatrix} \tilde{h} \\ \tilde{S}_1 \end{pmatrix}. \quad (6)$$

Here θ is the mixing angle, given by the relation $\tan 2\theta = 2M_{1h}^2/\left(M_1^2 - M_h^2\right)$,and the physical masses in (5) by the two relations:

$$\begin{aligned} m_h^2 &= \frac{1}{2}\left[M_h^2 + M_1^2 + \varepsilon\left(M_h^2 - M_1^2\right)\sqrt{\left(M_h^2 - M_1^2\right)^2 + 4M_{1h}^4}\right]; \\ m_1^2 &= \frac{1}{2}\left[M_h^2 + M_1^2 - \varepsilon\left(M_h^2 - M_1^2\right)\sqrt{\left(M_h^2 - M_1^2\right)^2 + 4M_{1h}^4}\right], \end{aligned} \quad (7)$$

where ε is the sign function.

Written now directly in terms of the physical fields, the cubic interaction terms are expressed as follows:

$$U_{\text{cub}} = \frac{\lambda_0^{(3)}}{2} S_0^2 h + \frac{\eta_{01}^{(3)}}{2} S_0^2 S_1 + \frac{\lambda^{(3)}}{6} h^3 + \frac{\eta_1^{(3)}}{6} S_1^3 + \frac{\lambda_1^{(3)}}{2} h^2 S_1 + \frac{\lambda_2^{(3)}}{2} h S_1^2, \tag{8}$$

where the cubic physical coupling constants are related to the original parameters via the following relations:

$$\begin{aligned}
\lambda_0^{(3)} &= \lambda_0 v \cos\theta + \eta_{01} v_1 \sin\theta, \\
\eta_{01}^{(3)} &= \eta_{01} v_1 \cos\theta - \lambda_0 v \sin\theta; \\
\lambda^{(3)} &= \lambda v \cos^3\theta + \frac{3}{2}\lambda_1 \sin 2\theta\, (v_1 \cos\theta + v \sin\theta) + \eta_1 v_1 \sin^3\theta; \\
\eta_1^{(3)} &= \eta_1 v_1 \cos^3\theta - \frac{3}{2}\lambda_1 \sin 2\theta\, (v \cos\theta - v_1 \sin\theta) - \lambda v \sin^3\theta; \\
\lambda_1^{(3)} &= \lambda_1 v_1 \cos^3\theta + \frac{1}{2} \sin 2\theta\, [(2\lambda_1 - \lambda)\, v \cos\theta - (2\lambda_1 - \eta_1)\, v_1 \sin\theta] - \lambda_1 v \sin^3\theta; \\
\lambda_2^{(3)} &= \lambda_1 v \cos^3\theta - \frac{1}{2} \sin 2\theta\, [(2\lambda_1 - \eta_1)\, v_1 \cos\theta + (2\lambda_1 - \lambda)\, v \sin\theta] + \lambda_1 v_1 \sin^3\theta.
\end{aligned} \tag{9}$$

In the same way, in terms of the physical fields too, the quartic interactions are given by:

$$\begin{aligned}
U_{\text{quar}} = {} & \frac{\eta_0}{24} S_0^4 + \frac{\lambda^{(4)}}{24} h^4 + \frac{\eta_1^{(4)}}{24} S_1^4 + \frac{\lambda_0^{(4)}}{4} S_0^2 h^2 + \frac{\eta_{01}^{(4)}}{4} S_0^2 S_1^2 + \frac{\lambda_{01}^{(4)}}{2} S_0^2 h S_1 \\
& + \frac{\lambda_1^{(4)}}{6} h^3 S_1 + \frac{\lambda_2^{(4)}}{4} h^2 S_1^2 + \frac{\lambda_3^{(4)}}{6} h S_1^3,
\end{aligned} \tag{10}$$

with the physical quartic coupling constants written as:

$$\begin{aligned}
\lambda^{(4)} &= \lambda \cos^4\theta + \frac{3}{2}\lambda_1 \sin^2 2\theta + \eta_1 \sin^4\theta, \\
\eta_1^{(4)} &= \eta_1 \cos^4\theta + \frac{3}{2}\lambda_1 \sin^2 2\theta + \lambda \sin^4\theta; \\
\lambda_0^{(4)} &= \lambda_0 \cos^2\theta + \eta_{01} \sin^2\theta, \\
\eta_{01}^{(4)} &= \eta_{01} \cos^2\theta + \lambda_0 \sin^2\theta; \\
\lambda_{01}^{(4)} &= \frac{1}{2} (\eta_{01} - \lambda_0) \sin 2\theta, \\
\lambda_1^{(4)} &= \frac{1}{2} \left[(3\lambda_1 - \lambda) \cos^2\theta - (3\lambda_1 - \eta_1) \sin^2\theta\right] \sin 2\theta; \\
\lambda_2^{(4)} &= \lambda_1 \cos^2 2\theta - \frac{1}{4} (2\lambda_1 - \eta_1 - \lambda) \sin^2 2\theta; \\
\lambda_3^{(4)} &= \frac{1}{2} \left[(\eta_1 - 3\lambda_1) \cos^2\theta - (\lambda - 3\lambda_1) \sin^2\theta\right] \sin 2\theta.
\end{aligned} \tag{11}$$

Finally, after spontaneous breaking of the electroweak and $\mathbb{Z}_2$ symmetries, the part of the Standard Model Lagrangian that is relevant to dark matter annihilation writes, in terms of the physical fields h and S_1, as follows:

$$\begin{aligned} U_{SM} = & \sum_f \left(\lambda_{hf} h\bar{f}f + \lambda_{1f} S_1 \bar{f}f\right) + \lambda_{hw}^{(3)} hW_\mu^- W^{+\mu} + \lambda_{1w}^{(3)} S_1 W_\mu^- W^{+\mu} \\ & + \lambda_{hz}^{(3)} h \left(Z_\mu\right)^2 + \lambda_{1z}^{(3)} S_1 \left(Z_\mu\right)^2 + \lambda_{hw}^{(4)} h^2 W_\mu^- W^{+\mu} + \lambda_{1w}^{(4)} S_1^2 W_\mu^- W^{+\mu} \\ & + \lambda_{h1w} h S_1 W_\mu^- W^{+\mu} + \lambda_{hz}^{(4)} h^2 \left(Z_\mu\right)^2 + \lambda_{1z}^{(4)} S_1^2 \left(Z_\mu\right)^2 + \lambda_{h1z} h S_1 \left(Z_\mu\right)^2 . \end{aligned} \quad (12)$$

The quantities m_f, m_w and m_z are the masses of the fermion f , the W and the Z gauge bosons respectively, and the above coupling constants are given by the following relations:

$$\begin{aligned} \lambda_{hf} &= -\frac{m_f}{v} \cos\theta; & \lambda_{1f} &= \frac{m_f}{v} \sin\theta; & & \\ \lambda_{hw}^{(3)} &= 2\frac{m_w^2}{v} \cos\theta; & \lambda_{1w}^{(3)} &= -2\frac{m_w^2}{v} \sin\theta; & & \\ \lambda_{hz}^{(3)} &= \frac{m_z^2}{v} \cos\theta; & \lambda_{1z}^{(3)} &= -\frac{m_z^2}{v} \sin\theta; & & \\ \lambda_{hw}^{(4)} &= \frac{m_w^2}{v^2} \cos^2\theta; & \lambda_{1w}^{(4)} &= \frac{m_w^2}{v^2} \sin^2\theta; & \lambda_{h1w} &= -\frac{m_w^2}{v^2} \sin 2\theta; \\ \lambda_{hz}^{(4)} &= \frac{m_z^2}{2v^2} \cos^2\theta; & \lambda_{1z}^{(4)} &= \frac{m_z^2}{2v^2} \sin^2\theta; & \lambda_{h1z} &= -\frac{m_z^2}{2v^2} \sin 2\theta. \end{aligned} \quad (13)$$

2.1. Dark Matter relic density constraint

The field S_0 is odd under the unbroken $\mathbb{Z}_2$ symmetry, and so is a stable relic and can therefore constitute the dark matter of the Universe. Its relic density can be obtained using the standard approximate solutions to the Boltzmann equations [21]:

$$\Omega_D \bar{h}^2 = \frac{1.07 \times 10^9 x_f}{\sqrt{g_*} M_{Pl} \left\langle v_{12}\sigma_{ann}\right\rangle \mathrm{GeV}}, \quad (14)$$

where $\bar{h}$ is the normalized Hubble constant,$M_{Pl} = 1.22 \times 10^{19}$GeV is the Planck mass, g_* the number of relativistic degrees of freedom at the freeze-out temperature T_f, and $x_f = m_0/T_f$ which, for m_0 in the range $1 - 20$GeV, is in the range $18.2 - 19.4$. The quantity $\langle v_{12}\sigma_{ann}\rangle$ is the thermally averaged annihilation cross section of S_0 into fermion pairs $f\bar{f}$ for $m_f < m_0/2$, and into S_1S_1 for $m_1 < m_0/2$. The annihilation cross-section into fermions proceeds via s-channel exchange of h and S_1 and is given by:

$$v_{12}\sigma_{S_0S_0\to f\bar{f}} = \frac{\sqrt{\left(m_0^2-m_f^2\right)^3}}{4\pi m_0^3}\Theta\left(m_0-m_f\right)\left[\frac{\left(\lambda_0^{(3)}\lambda_{hf}\right)^2}{\left(4m_0^2-m_h^2\right)^2+\epsilon_h^2}+\frac{\left(\eta_{01}^{(3)}\lambda_{1f}\right)^2}{\left(4m_0^2-m_1^2\right)^2+\epsilon_1^2}\right.$$
$$\left.+\frac{2\lambda_0^{(3)}\eta_{01}^{(3)}\lambda_{hf}\lambda_{1f}\left(4m_0^2-m_h^2\right)\left(4m_0^2-m_1^2\right)}{\left[\left(4m_0^2-m_h^2\right)^2+\epsilon_h^2\right]\left[\left(4m_0^2-m_1^2\right)^2+\epsilon_1^2\right]}\right], \tag{15}$$

and the annihilation into S_1 pairs given by:

$$v_{12}\sigma_{S_0S_0\to S_1S_1} = \frac{\sqrt{m_0^2-m_1^2}}{64\pi m_0^3}\Theta(m_0-m_1)\left[\left(\eta_{01}^{(4)}\right)^2+\frac{4\eta_{01}^{(4)}\left(\eta_{01}^{(3)}\right)^2}{m_1^2-2m_0^2}+\frac{2\eta_{01}^{(4)}\eta_{01}^{(3)}\eta_1^{(3)}}{4m_0^2-m_1^2}\right.$$
$$+\frac{2\eta_{01}^{(4)}\lambda_0^{(3)}\lambda_2^{(3)}(4m_0^2-m_h^2)}{(4m_0^2-m_h^2)^2+\epsilon_h^2}+\frac{4\left(\eta_{01}^{(3)}\right)^4}{(m_1^2-2m_0^2)^2}+\frac{4\left(\eta_{01}^{(3)}\right)^3\eta_1^{(3)}}{(4m_0^2-m_1^2)(m_1^2-2m_0^2)}$$
$$+\frac{4\left(\eta_{01}^{(3)}\right)^2\lambda_0^{(3)}\lambda_2^{(3)}(4m_0^2-m_h^2)}{[(4m_0^2-m_h^2)^2+\epsilon_h^2]\,(m_1^2-2m_0^2)}+\frac{\left(\eta_{01}^{(3)}\right)^2\left(\eta_1^{(3)}\right)^2}{(4m_0^2-m_1^2)^2}$$
$$\left.+\frac{\left(\lambda_0^{(3)}\right)^2\left(\lambda_2^{(3)}\right)^2}{(4m_0^2-m_h^2)^2+\epsilon_h^2}+\frac{2\eta_{01}^{(3)}\eta_1^{(3)}\lambda_0^{(3)}\lambda_2^{(3)}(4m_0^2-m_h^2)}{[(4m_0^2-m_h^2)^2+\epsilon_h^2]\,(4m_0^2-m_1^2)}\right]. \tag{16}$$

Solving (14) with the current value for the dark matter relic density $\Omega_D\bar{h}^2 = 0.1123 \pm 0.0035$ [22] translates into a relation between the parameters of a given theory entering the calculated expression of $\langle v_{12}\sigma_{\text{ann}}\rangle$, hence imposing a constraint on these parameters which will limit the intervals of possible dark matter masses. This constraint is exploited to examine aspects of the theory like perturbativity, while at the same time reducing the number of parameters by one.

Indeed, the model starts with eight parameters. The spontaneous breaking of the electroweak and $\mathbb{Z}_2$ symmetries introduces the two vacuum expectation values v and v_1 respectively, which means we are left with six. Four of the parameters are the three physical masses m_0 (dark-matter singlet S_0), m_1 (the second singlet S_1) and m_h (Higgs h), plus the mixing angle θ between h and S_1. We will fix the Higgs mass to $m_h = 125$GeV [23, 24], except in the section discussing Higgs decays where we let m_h vary in the interval 100GeV $-$ 200GeV. We will take both m_0 and m_1 in the interval 0.1GeV $-$ 10GeV. For the purpose of our discussions, it is sufficient to let θ vary in the interval $1^\circ - 40^\circ$. The last parameters are the two physical mutual coupling constants $\lambda_0^{(4)}$ (dark matter – Higgs) and $\eta_{01}^{(4)}$ (dark matter – S_1 particle). But $\eta_{01}^{(4)}$ is not free as it is the smallest real and positive solution to the dark-matter relic density constraint (14), which will be implemented systematically throughout [17]. Thus we are left with four parameters, namely, m_0, m_1, θ and $\lambda_0^{(4)}$. To ensure applicability of perturbation theory, the requirement $\eta_{01}^{(4)} < 1$ will also be imposed throughout, as well as a choice of rather small values for $\lambda_0^{(4)}$.

Studying the effects of the relic-density constraint for large ranges of the parameters through the behavior of the physical mutual coupling constant $\eta_{01}^{(4)}$ between S_0 and S_1 as a function of the DM mass m_0 shows that, apart from forbidden regions and others where perturbativity is lost, viable solutions in the small-moderate mass ranges of the DM sector exist for most values of the parameters [17]. Forbidden regions are found for most of the ranges of the parameters whereas perturbativity is lost mainly for larger values of m_1.

2.2. Direct detection

On the other hand, experiments like CDMS II [15], XENON 10/100 [14, 25], DAMA/LIBRA [26] and CoGeNT [3] search directly for a dark matter signal, which would typically come from the elastic scattering of a dark matter WIMP off a non-relativistic nucleon target. However, throughout the years, such experiments have not yet detected an unambiguous signal, but rather yielded increasingly stringent exclusion bounds on the dark matter – nucleon elastic scattering total cross-section σ_{det} in terms of the DM mass m_0. Any viable theoretical dark-matter model has to satisfy these bounds. In the 2SSM, σ_{det} is found to be given by the relation [17]:

$$\sigma_{\text{det}} \equiv \sigma_{S_0 N \to S_0 N} = \frac{m_N^2 \left(m_N - \frac{7}{9} m_B\right)^2}{4\pi \left(m_N + m_0\right)^2 v^2} \left[\frac{\lambda_0^{(3)} \cos\theta}{m_h^2} - \frac{\eta_{01}^{(3)} \sin\theta}{m_1^2}\right]^2, \quad (17)$$

in which m_N is the nucleon mass and m_B the baryon mass in the chiral limit [13, 27, 28]. This relation was compared against the experimental bounds from CDMSII and XENON100. We found that strong constraints were imposed on m_0 in the range between 10 to 20GeV. We found also that for small values of m_1, very light dark matter is viable, with m_0 as small as 1GeV.

3. Upsilon decays

We now further the analysis of the two-singlet model and start by looking at the constraints on the parameter space of the model coming from the decay of the meson Υ in the state nS $(n = 1, 3)$ into one photon γ and one particle S_1. For $m_1 \lesssim 8$GeV, the branching ratio for this process is given by the relation:

$$\text{Br}\left(\Upsilon_{nS} \to \gamma + S_1\right) = \frac{G_F m_b^2 \sin^2\theta}{\sqrt{2}\pi\alpha}\, x_n \left(1 - \frac{4\alpha_s}{3\pi} f(x_n)\right) \text{Br}^{(\mu)}\, \Theta\left(m_{\Upsilon_{nS}} - m_1\right). \quad (18)$$

In this expression, $x_n \equiv \left(1 - m_1^2/m_{\Upsilon_{nS}}^2\right)$ with $m_{\Upsilon_{1(3)S}} = 9.46(10.355)$GeV the mass of $\Upsilon_{1(3)S}$, the branching ratio $\text{Br}^{(\mu)} \equiv \text{Br}\left(\Upsilon_{1(3)S} \to \mu^+\mu^-\right) = 2.48(2.18) \times 10^{-2}$ [29], α is the QED coupling constant, $\alpha_s = 0.184$ the QCD coupling constant at the scale $m_{\Upsilon_{nS}}$, the quantity G_F

is the Fermi coupling constant and m_b the b quark mass [22]. The function $f(x)$ incorporates the effect of QCD radiative corrections given in [30].

But the above expression is not sufficient because a rough estimate of the lifetime of S_1 indicates that this latter is likely to decay inside a typical particle detector, which means we ought to take into account its most dominant decay products. We first have a process by which S_1 decays into a pair of pions, with a decay rate given by:

$$\Gamma(S_1 \to \pi\pi) \simeq \frac{G_F m_1}{4\sqrt{2}\pi} \sin^2\theta \left[\frac{m_1^2}{27}\left(1 + \frac{11 m_\pi^2}{2 m_1^2}\right)^2 \times \left(1 - \frac{4m_\pi^2}{m_1^2}\right)^{\frac{1}{2}} \Theta[(m_1 - 2m_\pi)(2m_K - m_1)] + 3\left(M_u^2 + M_d^2\right)\left(1 - \frac{4m_\pi^2}{m_1^2}\right)^{\frac{3}{2}} \Theta(m_1 - 2m_K) \right]. \tag{19}$$

In the above decay rate, $m_{\pi(K)}$ is the pion (kaon) mass. Also, chiral perturbation theory is used below the kaon pair production threshold [31, 32], and the spectator quark model above up to roughly 3GeV, with the dressed u and d quark masses $M_u = M_d \simeq 0.05$GeV. Note that this rate includes all pions, charged and neutral. Above the $2m_K$ threshold, both pairs of kaons and η particles are produced. The decay rate for K production is:

$$\Gamma(S_1 \to KK) \simeq \frac{9}{13}\frac{3G_F M_s^2 m_1}{4\sqrt{2}\pi} \sin^2\theta \left(1 - \frac{4m_K^2}{m_1^2}\right)^{\frac{3}{2}} \Theta(m_1 - 2m_K). \tag{20}$$

In the above rate, $M_s \simeq 0.45$GeV is the s quark-mass in the spectator quark model [33, 34]. For η production, replace m_K by m_η and $\frac{9}{13}$ by $\frac{4}{13}$.

The particle S_1 can also decay into c and b quarks (mainly c). Including the radiative QCD corrections, the corresponding decay rates are given by:

$$\Gamma(S_1 \to q\bar{q}) \simeq \frac{3G_F \bar{m}_q^2 m_1}{4\sqrt{2}\pi} \sin^2\theta \left(1 - \frac{4\bar{m}_q^2}{m_h^2}\right)^{\frac{3}{2}} \left(1 + 5.67\frac{\bar{\alpha}_s}{\pi}\right) \Theta(m_1 - 2\bar{m}_q). \tag{21}$$

The dressed quark mass $\bar{m}_q \equiv m_q(m_1)$ and the running strong coupling constant $\bar{\alpha}_s \equiv \alpha_s(m_1)$ are defined at the energy scale m_1 [35]. Gluons can also be produced, with a corresponding decay rate given by the relation:

$$\Gamma(S_1 \to gg) \simeq \frac{G_F m_1^3 \sin^2\theta}{12\sqrt{2}\pi}\left(\frac{\alpha_s'}{\pi}\right)^2 \left[6 - 2\left(1 - \frac{4m_\pi^2}{m_1^2}\right)^{\frac{3}{2}} - \left(1 - \frac{4m_K^2}{m_1^2}\right)^{\frac{3}{2}}\right] \Theta(m_1 - 2m_K). \tag{22}$$

Here, $\alpha_s' = 0.47$ is the QCD coupling constant at the spectator-quark-model scale, between roughly 1GeV and 3GeV.

We then have the decay of S_1 into leptons, the corresponding rate given by:

$$\Gamma\left(S_1 \to \ell^+\ell^-\right) = \frac{G_F m_\ell^2 m_1}{4\sqrt{2}\pi}\sin^2\theta\left(1-\frac{4m_\ell^2}{m_1^2}\right)^{\frac{3}{2}}\Theta\left(m_1-2m_\ell\right), \tag{23}$$

where m_ℓ is the lepton mass. Finally, S_1 can decay into a pair of dark matter particles, with a decay rate:

$$\Gamma\left(S_1 \to S_0S_0\right) = \frac{\left(\eta_{01}^{(3)}\right)^2}{32\pi m_1}\sqrt{1-\frac{4m_0^2}{m_1^2}}\Theta\left(m_1-2m_0\right). \tag{24}$$

The branching ratio for Υ_{nS} decaying via S_1 into a photon plus X, where X represents any kinematically allowed final state, will be:

$$\mathrm{Br}\left(\Upsilon_{nS} \to \gamma + X\right) = \mathrm{Br}\left(\Upsilon_{nS} \to \gamma + S_1\right) \times \mathrm{Br}\left(S_1 \to X\right). \tag{25}$$

In particular, $X \equiv S_0S_0$ corresponds to a decay into invisible particles.

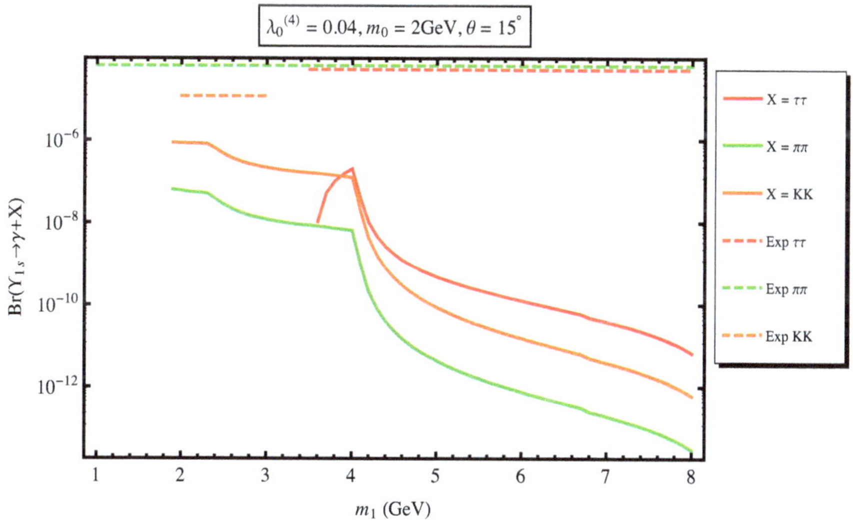

Figure 1. Typical branching ratios of Υ_{1S} decaying into τ's, charged pions and charged kaons as functions of m_1. The corresponding experimental upper bounds are shown.

The best available experimental upper bounds on 1S–state branching ratios are: (i) $\mathrm{Br}\left(\Upsilon_{1S} \to \gamma + \tau\tau\right) < 5\times10^{-5}$ for 3.5GeV $< m_1 <$ 9.2GeV [36]; (ii) $\mathrm{Br}\left(\Upsilon_{1S} \to \gamma + \pi^+\pi^-\right) <$

6.3×10^{-5} for $1\text{GeV} < m_1$ [37]; (iii) $\mathrm{Br}\left(\Upsilon_{1S} \to \gamma + K^+K^-\right) < 1.14 \times 10^{-5}$ for $2\text{GeV} < m_1 < 3\text{GeV}$ [38]. Figure 1 displays the corresponding branching ratios of Υ_{1S} decays via S_1 as functions of m_1, together with these upper bounds. Also, the best available experimental upper bounds on Υ_{3S} branching ratios are: (i) $\mathrm{Br}\left(\Upsilon_{3S} \to \gamma + \mu\mu\right) < 3 \times 10^{-6}$ for $1\text{GeV} < m_1 < 10\text{GeV}$; (ii) $\mathrm{Br}\left(\Upsilon_{3S} \to \gamma + \text{Invisible}\right) < 3 \times 10^{-6}$ for $1\text{GeV} < m_1 < 7.8\text{GeV}$ [39]. Typical corresponding branching ratios are shown in figure 2.

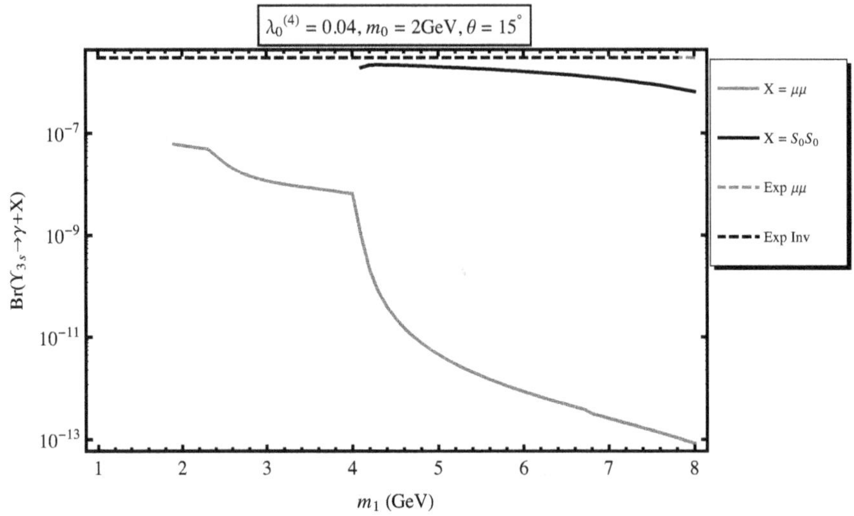

Figure 2. Typical branching ratios of Υ_{3S} decaying into muons and dark matter as functions of m_1. The corresponding experimental upper bounds are shown.

When scanning the parameter space, we see that the Higgs-dark-matter coupling constant $\lambda_0^{(4)}$ and the dark-matter mass m_0 have little effect on the shapes of the branching ratios, apart from excluding, via the relic density and perturbativity constraints, regions of applicability of the model. This shows in figures 1 and 2 where the region $m_1 \lesssim 1.9\text{GeV}$ is excluded. Also, the onset of the S_0S_0 channel for $m_1 \geq 2m_0$ abates sharply the other channels as this one becomes dominant by far. The effect of the mixing angle θ is to enhance all branching ratios as it increases, due to the factor $\sin^2\theta$. Furthermore, we notice that the dark matter decay channel reaches the invisible upper bound already for $\theta \simeq 15^\circ$, for fairly small m_0, say 0.5GeV, whereas the other channels find it hard to get to their respective experimental upper bounds, even for large values of θ.

4. B meson decays

Next we look at the flavor changing process in which the meson B^+ decays into a K^+ plus invisible. The corresponding Standard-Model mode is a decay into K^+ and a pair of neutrinos, with a branching ratio $\mathrm{Br}^{\mathrm{SM}}\left(B^+ \to K^+ + \nu\bar{\nu}\right) \simeq 4.7 \times 10^{-6}$ [40]. The experimental upper bound is $\mathrm{Br}^{\mathrm{Exp}}\left(B^+ \to K^+ + \mathrm{Inv}\right) \simeq 14 \times 10^{-6}$ [41]. Here too, the most prominent B

invisible decay is into S_0S_0 via S_1. The process $B^+ \to K^+ + S_1$ has the following branching ratio:

$$\mathrm{Br}\left(B^+ \to K^+ + S_1\right) = \frac{9\sqrt{2}\tau_B G_F^3 m_t^4 m_b^2 m_+^2 m_-^2}{1024\pi^5 m_B^3 \left(m_b - m_s\right)^2} \left|V_{tb}V_{ts}^*\right|^2 f_0^2\left(m_1^2\right) \times \sqrt{\left(m_+^2 - m_1^2\right)\left(m_-^2 - m_1^2\right)}\ \sin^2\theta\ \Theta\left(m_- - m_1\right). \quad (26)$$

In the above relation, $m_\pm = m_B \pm m_K$ where m_B is the B^+ mass, τ_B its lifetime, and V_{tb} and V_{ts} are flavor changing CKM coefficients. The function $f_0(s)$ is given by the relation:

$$f_0(s) = 0.33 \exp\left[\frac{0.63s}{m_B^2} - \frac{0.095s^2}{m_B^4} + \frac{0.591s^3}{m_B^6}\right]. \quad (27)$$

The different S_1 decay modes are given in (19) - (24) above. The branching ratio of B^+ decaying into $K^+ + S_0S_0$ via the production and propagation of an intermediary S_1 will be:

$$\mathrm{Br}^{(S_1)}\left(B^+ \to K^+ + S_0S_0\right) = \mathrm{Br}\left(B^+ \to K^+ + S_1\right) \times \mathrm{Br}\left(S_1 \to S_0S_0\right). \quad (28)$$

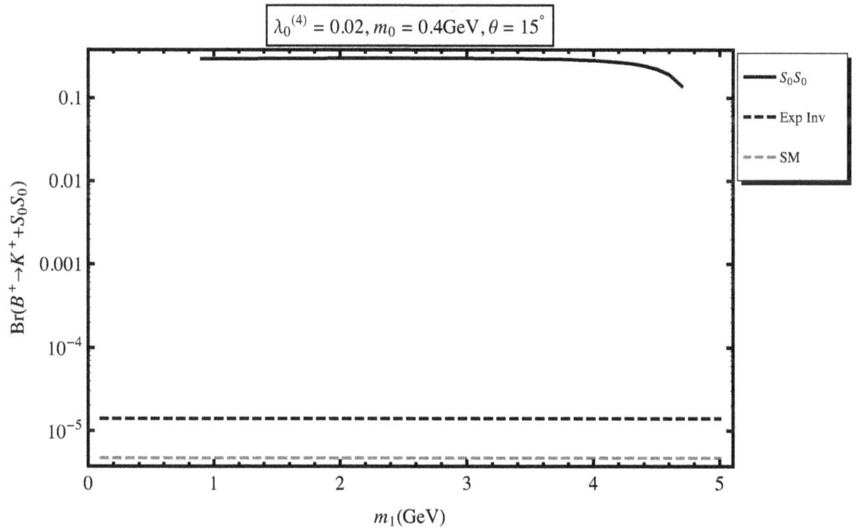

Figure 3. Typical branching ratio of B^+ decaying into dark matter via S_1 as a function of m_1. The SM and experimental bounds are shown.

Figure 3 displays a typical behavior of $\mathrm{Br}^{(S_1)}\left(B^+ \to K^+ + S_0S_0\right)$ as a function of m_1. As we see, the branching ratio is well above the experimental upper bound, and a mixing angle θ

as small as 1° will not help with this, no matter what the values for $\lambda_0^{(4)}$ and m_0 are. So, we conclude that for $m_1 \lesssim 4.8$GeV, this process excludes the two-singlet model for $m_0 < m_1/2$. For $m_1 \gtrsim 4.8$GeV or $m_0 \geq m_1/2$, the decay does not occur, so no constraints on the model from this process.

Another process involving B mesons is the decay of B_s into predominately a pair of muons. The Standard Model branching ratio for this process is $\mathrm{Br}^{\mathrm{SM}}\left(B_s \to \mu^+\mu^-\right) = (3.2 \pm 0.2) \times 10^{-9}$ [42], and the experimental upper bound is $\mathrm{Br}^{\mathrm{Exp}}\left(B_s \to \mu^+\mu^-\right) < 1.08 \times 10^{-8}$ [43]. In the present model, two additional decay diagrams occur, both via intermediary S_1, yielding together the branching ratio:

$$\mathrm{Br}^{(S_1)}(B_s \to \mu^+\mu^-) = \frac{9\tau_{B_s} G_F^4 f_{B_s}^2 m_{B_s}^5}{2048\pi^5} m_\mu^2 m_t^4 \left|V_{tb}V_{ts}^*\right|^2 \frac{\left(1 - 4m_\mu^2/m_{B_s}^2\right)^{3/2}}{\left(m_{B_s}^2 - m_1^2\right)^2 + m_1^2\Gamma_1^2} \sin^4\theta. \qquad (29)$$

In this relation, τ_{B_s} is the B_s life-time, $m_{B_s} = 5.37$GeV its mass, and f_{B_s} a form factor that we take equal to 0.21GeV. The quantity Γ_1 is the total width of the particle S_1 [17].

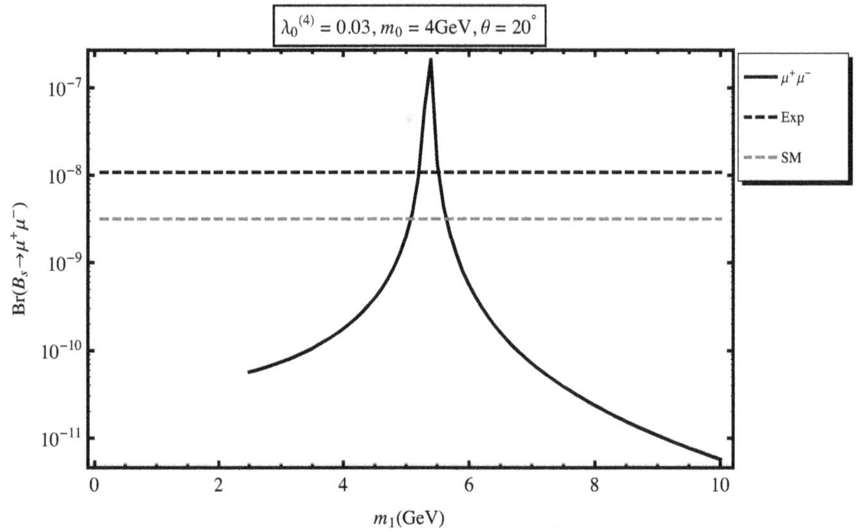

Figure 4. Typical behavior of $\mathrm{Br}^{(S_1)}(B_s \to \mu^+\mu^-)$ as a function of m_1, together with the SM and experimental bounds.

A typical behavior of $\mathrm{Br}^{(S_1)}(B_s \to \mu^+\mu^-)$ as a function of m_1 is displayed in figure 4. The peak is at m_{B_s}. All three parameters $\lambda_0^{(4)}$, m_0 and θ combine in the relic density constraint to exclude few regions of applicability of the model. For example, for the values of figure 4, the region $m_1 < 2.25$GeV is excluded. However, a systematic scan of the parameter space shows that outside the relic density constraint, $\lambda_0^{(4)}$ has no significant direct effect on the shape of $\mathrm{Br}^{(S_1)}(B_s \to \mu^+\mu^-)$. As m_0 increases, it sharpens the peak of the curve while pushing it up.

This works until about 2.7GeV, beyond which m_0 ceases to have any significant direct effect. Increasing θ enhances the values of the branching ratio without affecting the width. Also, for all the range of m_1, all of $\mathrm{Br}^{(S_1)} + \mathrm{Br}^{\mathrm{SM}}$ stays below $\mathrm{Br}^{\mathrm{Exp}}$ as long as $\theta < 10^o$. As θ increases beyond this value, the peak region pushes up increasingly above $\mathrm{Br}^{\mathrm{Exp}}$, like in figure 4, and hence gets excluded, but all the rest is allowed.

5. Higgs decays

We finally examine the implications of the model on the Higgs different decay modes. In this section, we allow the Higgs mass m_h to vary in the interval 100GeV – 200GeV. First, h can decay into a pair of leptons ℓ, predominantly τ's. The corresponding decay rate $\Gamma\left(h \to \ell^+\ell^-\right)$ is given by the relation (23) where we replace m_1 by m_h and $\sin\theta$ by $\cos\theta$. It can also decay into a pair of quarks q, mainly into b's and, to a lesser degree, into c's. Here too the decay rate $\Gamma(h \to q\bar{q})$ is given in (21) with similar replacements. Then the Higgs can decay into a pair of gluons. The corresponding decay rate that includes the next-to-next-to-leading QCD radiative corrections is given by:

$$\Gamma\left(h \to gg\right) = \frac{G_F m_h^3}{4\sqrt{2}\pi}\left|\sum_q \frac{m_q^2}{m_h^2}\int_0^1 dx \int_0^{1-x} dy \frac{1-4xy}{\frac{m_q^2}{m_h^2} - xy}\right|^2 \times \left(\frac{\bar{\alpha}_s}{\pi}\right)^2 \left[1 + \frac{215}{12}\frac{\bar{\alpha}_s}{\pi} + \frac{\bar{\alpha}_s^2}{\pi^2}\left(156.8 - 5.7\log\frac{m_t^2}{m_h^2}\right)\right]\cos^2\theta, \quad (30)$$

where the sum is over all quark flavors q. A systematic study of the double integral above shows that, with m_h in the range 100GeV – 200GeV, the t quark dominates in the sum over q, but with non-negligible contributions from the c and b quarks.

For m_h smaller than the W or Z pair-production thresholds, the Higgs can decay into a pair of one real and one virtual gauge bosons, with rates given by:

$$\Gamma\left(h \to VV^*\right) = \frac{3G_F^2 m_V^4 m_h}{16\pi^3}\cos^2\theta\, A_V\, R\left(\frac{m_V^2}{m_h^2}\right)\Theta\left[\left(m_h - m_V\right)\left(2m_V - m_h\right)\right]. \quad (31)$$

In this expression, m_V is the mass of the gauge boson V, the factor $A_V = 1$ for W and $\left(\frac{7}{12} - \frac{10}{9}\sin^2\theta_w + \frac{40}{9}\sin^4\theta_w\right)$ for Z with θ_w the Weinberg angle, and we have the definition:

$$R(x) = \frac{3(1-8x+20x^2)}{\sqrt{4x-1}}\arccos\left(\frac{3x-1}{2x^{3/2}}\right) - \frac{1-x}{2x}\left(2 - 13x + 47x^2\right) - \frac{3}{2}\left(1 - 6x + 4x^2\right)\log x. \quad (32)$$

For a heavier Higgs particle, the decay rates into a V pair is given by:

$$\Gamma(h \to VV) = \frac{G_F m_V^4 \cos^2\theta}{\sqrt{2}\pi m_h} B_V \left(1 - \frac{4m_V^2}{m_h^2}\right)^{\frac{1}{2}} \left[1 + \frac{\left(m_h^2 - 2m_V^2\right)^2}{8m_V^4}\right] \Theta(m_h - 2m_V), \quad (33)$$

with $B_V = 1$ for W and $\frac{1}{2}$ for Z.

While all these decay modes are already present within the Standard Model, the two-singlet extension introduces two additional (invisible) modes, namely a decay into a pair of S_0's and a pair of S_1's. The corresponding decay rates are:

$$\Gamma(h \to S_i S_i) = \frac{\lambda_i^2}{32\pi m_h} \left(1 - \frac{4m_i^2}{m_h^2}\right)^{\frac{1}{2}} \Theta(m_h - 2m_i), \quad (34)$$

where $\lambda_i = \lambda_{0(2)}^{(3)}$ for $S_{0(1)}$ given in (9). The total decay rate $\Gamma(h)$ of the Higgs particle is the sum of these partial rates. The branching ratio corresponding to a particular decay will be $\mathrm{Br}(h \to X) = \Gamma(h \to X)/\Gamma(h)$.

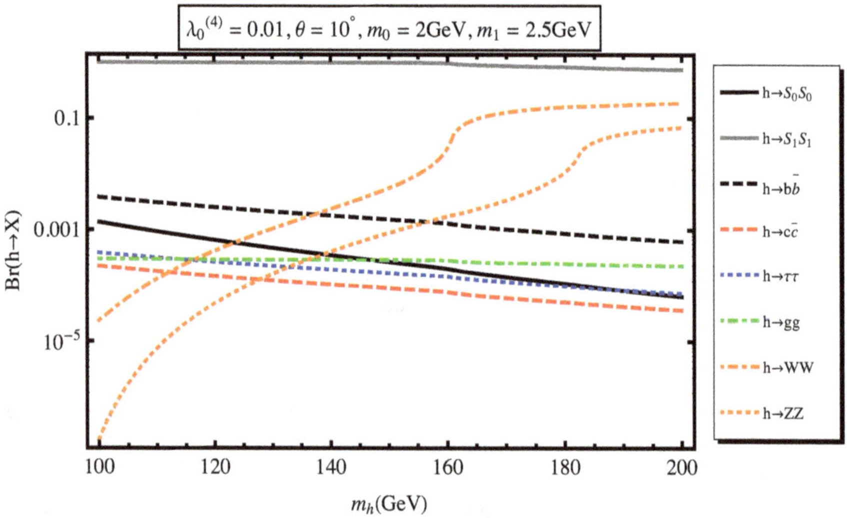

Figure 5. Branching ratios for Higgs decays. Very small dark-matter Higgs coupling.

Typical behaviors of the most prominent branching ratios are displayed in figure 5. A systematic study shows that for all ranges of the parameters, the Higgs decays dominantly into invisible. The production of fermions and gluons is comparatively marginal, whereas that of W and Z pairs takes relative importance towards and above the corresponding thresholds, and more significantly at larger values of the mixing angle θ.

However, the decay distribution between S_0 and S_1 is not even. The most dramatic effect comes from the coupling constant $\lambda_0^{(4)}$. When it is very small, the dominant production is that of a pair of S_1. This is exhibited in figure 5 for which $\lambda_0^{(4)} = 0.01$. As it increases, there is a gradual shift towards a more dominating dark-matter pair production, a shift competed against by the increase in θ. Figure 6 displays the branching ratios for $\lambda_0^{(4)} = 0.1$ and figure 7 for the larger value $\lambda_0^{(4)} = 0.7$. In general, increasing θ smoothens the crossings of the WW and ZZ thresholds, and lowers the production of everything except that of a pair of S_1, which is instead increased.

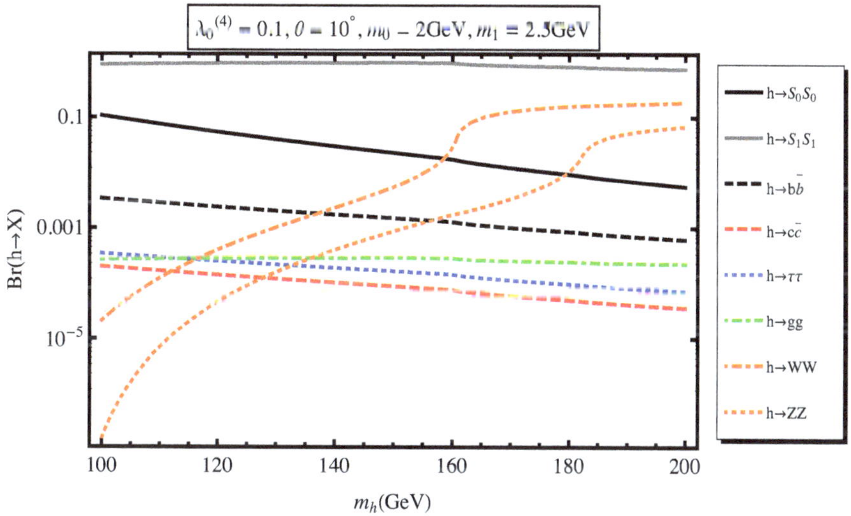

Figure 6. Branching ratios for Higgs decays. Small dark-matter Higgs coupling.

Like in the Standard Model, the production of a pair of b quarks dominates over the production of the other fermions, and all fermions are not favored by increasing $\lambda_0^{(4)}$. Changes in m_0 and m_1 have very little direct effects on all the branching ratios except that of S_0S_0 production where, at small θ, increasing m_1 (m_0) increases (decreases) the branching ratio, with reversed effects at larger θ. Note though that these masses have indirect impact through the relic density constraint by excluding certain regions [17].

6. Concluding remarks

Understanding light dark matter is one of the challenges facing popular extensions of the Standard Model. In this chapter, we have furthered the study of a two-singlet extension of the SM we proposed as a model for light dark matter by exploring some of its phenomenological aspects. We have looked into the rare decays of Υ and B mesons and studied the implications of the model on the decay channels of the Higgs particle.

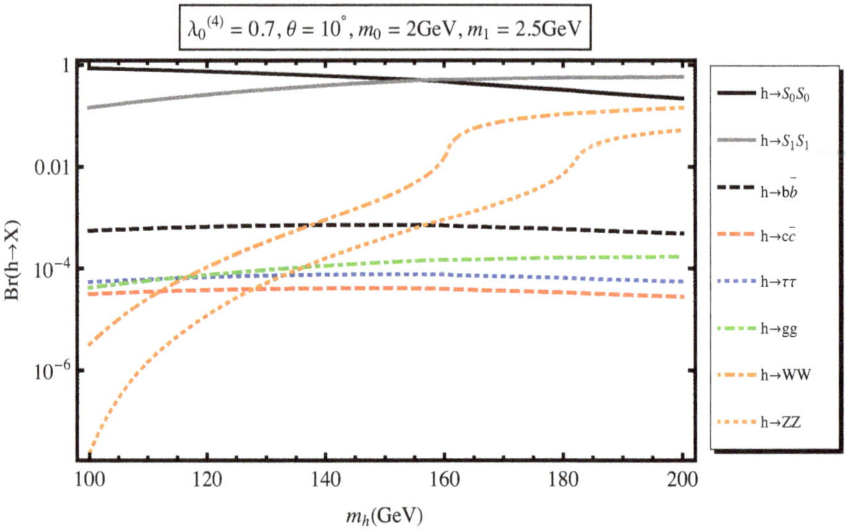

Figure 7. Branching ratios for Higgs decays. Larger dark-matter Higgs coupling.

In brief, for both Υ and B decays, the Higgs-DM coupling constant $\lambda_0^{(4)}$ and the DM mass m_0 have little effect on the shapes of the branching ratios, apart from combining with the other two parameters in the relic-density and perturbativity constraints to exclude regions of applicability of the model. Also, the effect of increasing the $h - S_1$ mixing angle θ is to enhance all branching ratios. For Υ decays, the DM channel dominates over the other decay modes in regions where kinematically allowed. It reaches the experimental invisible upper bound for already fairly small values of θ and m_0. From B^+ decays, we learn that our model is excluded for $m_1 < 4.8\text{GeV}$ $(= m_B - m_K)$ and $m_0 < m_1/2$. From B_s decay into muons, we learn that for the model to contribute a distinct signal to this process, it is best to restrict $4\text{GeV} \lesssim m_1 \lesssim 6.5\text{GeV}$ with no additional constraint on m_0 [20]. Also, in general, keeping $\lambda_0^{(4)} \lesssim 0.1$ to avoid systematic exclusion from direct detection for all these processes is safe.

Before closing the chapter, it is useful to comment briefly on how light dark matter in our model affects Higgs searches. Since $m_h \gg 2m_0$, the process $h \to S_0S_0$ is kinematically allowed and, for a large range of the parameter space, the ratio

$$\mathcal{R}_{\text{decay}}^{(b)} = \frac{\text{Br}\left(h \to S_0S_0\right)}{\text{Br}(h \to b\bar{b})} \tag{35}$$

can be larger than one for $m_h < 120\text{GeV}$ as can be seen in figure 6. In this situation, the LEP bound on the Higgs mass can be weaker. Also, in our model, the Higgs production at LEP via Higgstrahlung can be smaller than the one in the Standard Model, and so the Higgs can be as light as 100GeV. Such a light Higgs would be in good agreement with the electroweak precision tests. As to the Higgs searches at the LHC, the ATLAS and CMS collaborations have reported the exclusion of a Higgs mass in the interval 145 – 460 GeV [23, 24], which seems to suggest that we should have limited our analysis of the Higgs branching ratios to $m_h < 145$GeV. However, it is important to note that these experimental constraints apply to the SM Higgs and can not therefore be used as such if the Higgs interactions are modified. In our model, the mixing of h with S_1 will result in a reduction of the statistical significance of the Higgs discovery at the LHC. Indeed, the relevant quantity that allows one to use the experimental limits on Higgs searches to derive constraints on the parameters of the model is the ratio:

$$\mathcal{R}_{X_{\rm SM}} \equiv \frac{\sigma\left(gg \to h\right) {\rm Br}\left(h \to X_{\rm SM}\right)}{\sigma^{\rm (SM)}\left(gg \to h\right) {\rm Br}^{\rm SM}\left(h \to X_{\rm SM}\right)} = \frac{\cos^4\theta}{\cos^2\,\theta + \Gamma\left(h \to X_{\rm inv}\right)/\Gamma_h^{\rm SM}}. \qquad (36)$$

In this expression, $X_{\rm SM}$ corresponds to all the Standard Model particles, $X_{inv} = S_0 S_0$ and $S_1 S_1$, σ is a cross-section, ${\rm Br}^{\rm SM}\left(h \to X\right)$ the branching fraction of the SM Higgs decaying into any kinematically allowed mode X, and $\Gamma_h^{\rm SM}$ the total Higgs decay rate in the Standard Model. To open up the region $m_h > 140$GeV requires the ratio $\mathcal{R}_{X_{\rm SM}}$ to be smaller than 0.25 [23, 24], a constraint easily fulfilled in our model. By comparison, the minimal extensions of the Standard Model with just one singlet scalar or a Majorana fermion, even under a Z_2 symmetry, are highly constrained in this regard [44]. Finally, if the recent data from ATLAS and CMS turn out to be a signal for a SM-like Higgs with mass about 125GeV, then this will put a very strong constraint on the mixing angle θ. Indeed, only for $\theta \lesssim 0.5^0$ will the ratio $\Gamma(h \to b\bar{b})/\Gamma\left(h \to {\rm inv}\right) \gtrsim 1$. For larger values, the model is ruled out, independently of the dark matter mass, but as long as $m_1 \lesssim 50$GeV, which we are assuming in this work.

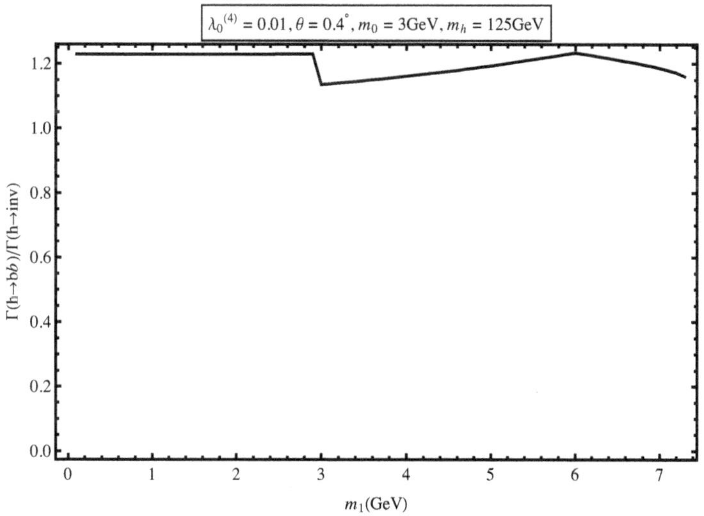

Figure 8. The ratio $\Gamma(h \to b\bar{b})/\Gamma(h \to {\rm inv})$ is larger than one for $\theta = 0.4^0$ and $m_h = 125$GeV.

Finally, it is important to find the bounds on the mass SM Higgs that can satisfy the triviality and perturbativity constraints on the coupling constants in the scalar sector of this model up to a scale higher then 1TeV. This requires studying the renormalization group equations of these coupling constants [45].

Author details

Abdessamad Abada[1,2] and Salah Nasri[2]

* Address all correspondence to: a.abada@uaeu.ac.ae; snasri@uaeu.ac.ae

1 Laboratoire de Physique des Particules et Physique Statistique, Ecole Normale Supérieure, BP 92 Vieux Kouba, Alger, Algeria
2 Physics Department, United Arab Emirates University, POB, Al Ain, United Arab Emirates

References

[1] Ellis, J., Hagelin, J.S., Nanopoulos. D.V., Olive, K.A., and Srednicki, M., Nucl. Phys. B238, 453 (1984).

[2] Jungman, G., Kamionkowski, M., and Griest, K., Phys. Rept. 267, 195 (1996) (arXiv:hep-ph/9506380).

[3] Aalseth, C.E. *et al.*, Phys. Rev. Lett. 107 , 141301 (2011); Phys. Rev. Lett. 106, 131301 (2011).

[4] Fornengo, N., Scopel, S., and Bottino, A., Phys. Rev. D 83, 015001 (2011); Niro, V., Bottino, A., Fornengo, N. and Scopel, S., Phys. Rev. D 80, 095019 (2009).

[5] Cumberbatch, D.T., Lopez-Fogliani, D.E., Roszkowski, L., de Austri, R.R., and Tsai, Y.L., arXiv:1107.1604 [astro-ph.CO]; Vasquez, D.A., Belanger, G., Boehm, C., Pukhov, A., and Silk, J., Phys. Rev. D 82, 115027 (2010); Kuflik, E., Pierce, A., and Zurek, K.M., Phys. Rev. D 81, 111701 (2010); Feldman, D., Liu, Z., and Nath, P., Phys. Rev. D 81, 117701 (2010).

[6] Gunion, J.F., Belikov, A.V., and Hooper, D., arXiv:1009.2555 [hep-ph]; Cumberbatch,, D.T., Lopez-Fogliani, D.E., Roszkowski, L., de Austri, R.R., and Tsai, Y.L., arXiv:1107.1604 [astro-ph.CO].

[7] Mambrini, Y., arXiv:1108.0671 [hep-ph]; Kajiyama, Y., Okada, H., and Toma, T., arXiv:1109.2722 [hep-ph]; Cline, J.M., and Frey, A.R., Phys. Rev. D 84, 075003 (2011); Kappl, R., Ratz, M., and Winkler, M.W., Phys. Lett. B 695, 169 (2011); Mambrini, Y., and Zaldivar, B., JCAP 1110, 023 (2011); Draper. P., Liu, T., Wagner, C.E.M, Wang, L.T.M., and Zhang, H., Phys. Rev. Lett. 106, 121805 (2011); Foot, R., Phys. Lett. B 703, 7 (2011); Gonderinger, M., Li, Y., Patel, H., and Ramsey-Musolf, M.J., JHEP 053, 1001 (2010); Andreas, S., Arina, C., Hambye, T., Ling, F.S., and Tytgat, M.H.G., Phys. Rev. D 82, 043522 (2010); Bae, K.J., Kim, H.D., and Shin, S., Phys. Rev. D 82, 115014 (2010); Farina, M., Pappadopulo, D., and Strumia, A., Phys. Lett. B 688, 329 (2010); Barger, V., Langacker, P., McCaskey, M., Ramsey-Musolf, M.J., and Shaughnessy, G., Phys. Rev. D 77, 035005 (2008).

[8] Silveira, V., and Zee, A.; Phys. Lett. B161, 136 (1985).

[9] McDonald, J., Phys. Rev. D 50, 3637 (1994).

[10] Burgess, C., Pospelov, M., and ter Veldhuis, T.; Nucl. Phys. B619, 709 (2001).

[11] Barger, V., Langacker, P., McCaskey, M., Ramsey-Musolf, M, and Shaughnessy, G.; Phys. Rev. D 77, 035005 (2008) (arXiv:0706.4311 [hep-ph]).

[12] Gonderinger, M., Li, Y., Patel, H., and Ramsey-Musolf, M.; JHEP 053, 1001 (2010) (arXiv:0910.3167 [hep-ph]).

[13] He, X., Li, T., Li, X., Tandean, J. and Tsai, H.; Phys. Rev. D 79, 023521 (2009) (arXiv:0811.0658 [hep-ph]).

[14] Angle, J., *et al.* [XENON Collaboration], Phys. Rev. Lett. 100, 021303 (2008) (arXiv:0706.0039 [astro-ph]).

[15] Ahmed, Z., *et al.* [CDMS Collaboration], Phys. Rev. Lett. 102, 011301 (2009) (arXiv:0802.3530 [astro-ph]); Ahmed, Z., *et al.* [CDMS Collaboration], Science 327, 1619 (2010) (arXiv:0912.3592 [astro-ph.CO]).

[16] Arina, C. and Tytgat, M.; JCAP 1101, 011 (2011).

[17] Abada, A., Ghaffor, D., and Nasri, S., Phys. Rev. D 83 095021 (2011) [arXiv:1101.0365[hep-ph]].

[18] Ahriche, A. and Nasri, S., Phys. Rev. D 85, 093007 (2012) (arXiv:1201.4614 [hep-ph]).

[19] Barger, V., Langacker, P., McCaskey, M., Ramsey-Musolf, M., and Shaughnessy, G., Phys. Rev. D 79, 015018 (2009) [arXiv:0811.0393 [hep-ph]].

[20] Abada, A., and Nasri, S., Phys. Rev. D 85, 075009 (2012) (arXiv:1201.1413 [hep-ph]).

[21] Kolb, E.W., and Turner, M.S., '*The Early Universe*', Addison-Wesley, (1998).

[22] Nakamura, K., *et al.* (Particle Data Group), JP G 37, 075021 (2010).

[23] [ATLAS Collaboration], arXiv:1112.2577 [hep-ex]; Aad, G., *et al.* [ATLAS Collaboration], Eur. Phys. J. C 71, 1728 (2011).

[24] [CMS Collaboration], CMS-PAS-HIG-11-025, CMS-PAS-HIG-11-029, CMS-PAS-HIG-11-030, CMS-PAS-HIG-11-031, CMS-PAS-HIG- 11-032; Chatrchyan, S., *et al.*, Phys. Lett. B 699, 25 (2011).

[25] Aprile, E., *et al.* [XENON100 Collaboration], Phys. Rev. Lett. 105, 131302 (2010) (arXiv:1005.0380 [astro-ph.CO]); arXiv:1005.2615 [astro-ph.CO].

[26] Bernabei, R., *et al.* [DAMA Collaboration], Eur. Phys. J. C 67, 39 (2010) [arXiv:1002.1028 [astro-ph.GA]]; Eur. Phys. J. C 56, 333 (2008) [arXiv:0804.2741 [astro-ph]].

[27] Cai, Y., He, X.G., and Ren, B., arXiv:1102.1522 [hep-ph].

[28] Asano, M., and Kitano, R., Phys. Rev. D 81, 054506 (2010).

[29] Eidelman, S., *et al.*, Phys. Lett. B592, 1 (2004).

[30] Nason, P., Phys. Lett. B 175, 223 (1986).

[31] Chivukula, R.S., Cohen, A.G., Georgi, H., Grinstein, B., and Manohar, A.V., Ann. Phys. 192, 93 (1989).

[32] Voloshin, M.B., Sov. J. Nucl. Phys. 44, 478 (1986) [Yad. Fiz. 44, 738 (1986)].

[33] Gunion, J.F., Haber, H.E., Kane, G., and Dawson, S., *'The Higgs Hunters Guide'*, Perseus Publishing, Cambridge, MA, (1990).

[34] McKeen, D., Phys. Rev. D 79, 015007 (2009).

[35] Djouadi, A., Phys. Rept. 459, 1 (2008).

[36] Love, W., *et al.*, Phys. Rev. Lett. 101, 151802 (2008).

[37] Anastassov, A., *et al.* [CLEO Collaboration], Phys. Rev. Lett. 82, 286 (1999).

[38] Athar, S.B., *et al.* [CLEO Collaboration], Phys. Rev. D 73, 032001 (2006).

[39] Auber, B., *et al* [BABAR Collaboration], Phys. Rev. Lett. 103, 081803 (2009).

[40] Bartsch, M., Beylich, M., Buchalla, G., and Gao, D.N., JHEP 0911, 011 (2009).

[41] Chen, K.F., *et al.* [BELLE Collaboration], Phys. Rev. Lett. 99, 221802 (2007).

[42] Buras, A.G., Acta Phys. Polon. B 41, 2487 (2010).

[43] The CMS and LHCb Collaborations, LHCb-CONF-2011- 047, CMS PAS BPH-11-019.

[44] Low, I., Schwaller, P., Shaughnessy, G., and Wagner, C.E.M., `arXiv:1110.4405 [hep-ph]`. Note that in deriving the constraint on the invisible width of the simplest singlet dark-matter model, the authors have used the ATLAS and CMS older analysis available at that time. Using the latest analysis [23, 24] will rule out the possibility of opening up the Higgs window above 140GeV in the one singlet model.

[45] Abada, A., and Nasri, S., work in progress.

Chapter 12

Where Is the PdV in the First Law of Black Hole Thermodynamics?

Brian P. Dolan

Additional information is available at the end of the chapter

1. Introduction

Ever since Hawking's discovery in 1974, [1–3], that black holes have a temperature associated to them, in the simplest case a temperature inversely proportional to their mass,

$$T = \frac{\hbar}{8\pi GM} \tag{1}$$

(we use units in which $c = 1$), the thermodynamics of black holes has been a fascinating area of research. Equation (1) immediately implies that a Schwarzschild black hole in isolation is unstable: it will radiate and in so doing loses energy hence the mass decreases, thus increasing the temperature causing it to radiate with more power leading to a runaway effect.

Hawking's result is fundamentally quantum mechanical in nature and came after a number of important developments in the classical thermodynamics of black holes. Penrose [4] realised that the mass of a rotating black hole can decrease, when rotational energy is extracted, and this was followed by the observation that the area never decreases in any classical process. Nevertheless there is still a minimum, irreducible, mass below which one cannot go classically [5, 6]. This lead Bekenstein's to propose that an entropy should be associated with a black hole that is proportional to the area, A, of the event horizon, [7, 8]. In natural units,

$$S = \alpha \frac{A}{\hbar G}, \tag{2}$$

where α is an undetermined constant, presumed of order one, and $\hbar G$ is the Planck length squared. In the classical limit the temperature vanishes and the entropy diverges.

The first law of black hole thermodynamics, in its simplest form, associates the internal energy of a black hole with the mass, $U(S) = M$, (more precisely the ADM mass, as defined with reference to the time-like Killing vector at infinity [9]) and reads

$$dU = TdS. \tag{3}$$

In particular, for a Schwarzschild black hole, the event horizon radius is $r_h = 2GM$ and the event horizon area is

$$A = 4\pi r_h^2 = 16\pi G^2 M^2 \qquad \Rightarrow \qquad M = \frac{1}{4G}\sqrt{\frac{A}{\pi}}. \tag{4}$$

Hence

$$U = \frac{1}{4}\sqrt{\frac{\hbar S}{\pi \alpha G}} \tag{5}$$

and

$$T = \frac{\partial U}{\partial S} = \frac{1}{8}\sqrt{\frac{\hbar}{\pi \alpha G S}} = \frac{\hbar}{32\pi \alpha G M}. \tag{6}$$

Hawking's result (1) then fixes the constant of proportionality in (2) to be one quarter.

The black hole instability referred to above is reflected in the thermodynamic potentials by the fact that the heat capacity of a Schwarzschild black hole,

$$C = T\frac{\partial S}{\partial T} = -\frac{\hbar G}{8\pi T^2} < 0, \tag{7}$$

is negative.

The first law generalises to electrically charged, rotating black holes as

$$dU = TdS + \Omega dJ + \Phi dQ \tag{8}$$

where J is the angular momentum of the black hole, Ω its angular velocity, and Q the electric charge and the electrostatic potential (see e.g. [9]).

In contrast to elementary treatments of the first law of black hole thermodynamics it is noteworthy that (8) lacks the familiar PdV term, but a little thought shows that it is by no means obvious how to define the volume of a black hole. For a Schwarzschild black hole the radial co-ordinate, r, is time-like inside the event horizon, where $r < r_h$, so it would seem non-sensical to associate a volume $V = 4\pi \int_0^{r_h} r^2 dr = \frac{4\pi}{3} r_h^3$ with the black hole. In fact identifying any function of r_h alone with a volume, $V(r_h)$, will lead to inconsistencies in a thermodynamic description since the area, and hence the entropy, is already a function of r_h, $S = \pi r_h^2$, so any volume $V(r_h)$ would be determined purely in terms of the entropy. The internal energy, $U(S,V)$, should be a function of two variables, so giving $V(S)$ uniquely as a specific function of S is liable to lead to inconsistencies. We shall see below how this potential problem is avoided.

2. Pressure and enthalpy

From the point of view of Eintein's equations a pressure is associated with a cosmological constant. There is now very strong evidence that the cosmological constant in our Universe is positive [10, 11]. This poses a problem for the study of black hole thermodynamics for two reasons: firstly there is no asymptotic regime in de Sitter space which allows the unambiguous identification of the ADM mass of a black hole embedded in a space with a positive Λ; secondly positive Λ corresponds to negative pressure, implying thermodynamic instability. The first problem is related to the fact that there are two event horizons for a de Sitter black hole, a black hole horizon and a cosmological horizon, and the radial co-ordinate is time-like for large enough values of r, outside the cosmological horizon. The second problem is not necessarily too serious as one can still glean some information from negative pressure systems which are thermodynamically unstable [12] (instability is not an insurmountable barrier to obtaining physical information from a thermodynamic system, after all, as described above, Hawking's formula (1) shows that black holes can have negative heat capacity but it is still a central formula in the understanding of black hole thermodynamics). In contrast for negative Λ there is no cosmological horizon and the pressure is positive, the thermodynamics is perfectly well defined, so we shall restrict our considerations here to negative Λ and identify the thermodynamic pressure $P = -\frac{\Lambda}{8\pi G}$ with the fluid dynamical pressure appearing in Einstein's equations.

The notion that the cosmological constant should be thought of as a thermodynamic variable is not new, and its thermodynamic conjugate is often denoted Θ in the literature, [13–21], but Θ was not given a physical interpretation in these works.

It may seem a little surprising to elevate Λ to the status of a thermodynamic variable. Λ is usually thought of as a coupling constant in the Einstein action, on the same footing as Newton's constant, and it would seem bizarre to think of Newton's constant as a thermodynamic variable. However the nature of Λ has long been mysterious [22] and we should keep an open mind as to its physical interpretation. Indeed in [23] it was argued that Λ must be included in the pantheon of thermodynamic variables for consistency with the Smarr relation [24], which is essentially dimensional analysis applied to thermodynamic functions. Furthermore [23] suggested that, for a black hole embedded in anti-de Sitter (AdS) space-time, the black hole mass is more correctly interpreted as the enthalpy, H beloved of chemists, rather than the more traditional internal energy,

$$M = H(S,P) = U(S,V) + PV. \tag{9}$$

The PV term in this equation can be though of as the contribution to the mass-energy of the black hole due the negative energy density of the vacuum, $\epsilon = -P$, associated with a positive cosmological constant. If the black hole has volume V then it contains energy $\epsilon V = -PV$ and so the total energy is $U = M - PV$.

This interpretation forces us to face up to the definition of the black hole volume. In [23] V is defined as the volume relative to that of empty AdS space-time: the black hole volume is the volume excluded from empty AdS when the black hole is introduced. We shall refer to this as the 'geometric volume' below. Other suggestions for the volume of a black hole have been made in [25, 26]

An alternative definition of the black-volume is that it is the thermodynamic conjugate of the pressure, under the Legendre transform (9),

$$V := \frac{\partial H}{\partial P}, \tag{10}$$

which we shall call the 'thermodynamic volume'.

With the definition of the thermodynamic volume (10) we are in a position to state the definitive version of the first law of black hole thermodynamics,

$$\boxed{dU = TdS + \Omega dJ + \Phi dQ - PdV} \tag{11}$$

which follows from the Legendre transform of

$$dM = dH = TdS + \Omega dJ + \Phi dQ + VdP. \tag{12}$$

Equation (12), in $\Theta d\Lambda$ notation, appeared in [27].

3. Thermodynamic volume

The suggested definition of the thermodynamic volume (10) must be tested for consistency. For example, for a non-rotating black hole in four-dimensional space-time, the line element is given, in Schwarzschild co-ordinates, by,

$$d^2s = -f(r)dt^2 + f^{-1}(r)dr^2 + r^2 d\Omega^2, \tag{13}$$

with

$$f(r) = 1 - \frac{2m}{r} - \frac{\Lambda}{3}r^2, \tag{14}$$

and $d\Omega^2 = d\theta^2 + \sin^2\theta d\phi^2$ the solid angle area element.[1] The event horizon is defined by $f(r_h) = 0$,

$$\frac{\Lambda}{3}r_h^3 - r_h + 2m = 0, \tag{15}$$

but we do not need to solve this equation explicitly in order to analyse (10). We already know that

$$S = \pi r_h^2, \qquad P = -\frac{\Lambda}{8\pi} \tag{16}$$

[1] From now on we set $G = \hbar = 1$ to avoid cluttering formulae.

and, for negative Λ, the ADM mass is $M = m$ [15], which, following the philosophy of [23], we identify with the enthalpy, $H(S,P)$. Solving (15) for m immediately yields

$$m = \frac{r_h}{2}\left(1 - \frac{\Lambda}{3}r_h^2\right), \tag{17}$$

from which $H(S,P) = M = m$, with (16), identifies the enthalpy as

$$H(S,P) = \frac{1}{2}\left(\frac{S}{\pi}\right)^{\frac{1}{2}}\left(1 + \frac{8SP}{3}\right). \tag{18}$$

The usual thermodynamic relations can now be used to determine the temperature and the volume,

$$T = \left(\frac{\partial H}{\partial S}\right)_P \quad \Rightarrow \quad T = \frac{1}{4}\left(\frac{1}{\pi S}\right)^{\frac{1}{2}}(1 + 8PS) = \frac{(1 - \Lambda r_h^2)}{4\pi r_h} \tag{19}$$

$$V = \left(\frac{\partial H}{\partial P}\right)_S \quad \Rightarrow \quad V = \frac{4}{3}\frac{S^{\frac{3}{2}}}{\sqrt{\pi}} = \frac{4\pi r_h^3}{3}. \tag{20}$$

That the resulting thermodynamic volume (for a non-rotating black hole) is identical to the geometric volume is quite remarkable, but appears co-incidental as this equality no longer holds for rotating (Kerr-AdS) black holes, as we shall see. It does however hold for non-rotating black holes in all dimensions [28].

As mentioned in the introduction, equation (20) has a potential problem associated with it, in that it implies that the volume and the entropy cannot be considered to be independent thermodynamic variables, S determines V uniquely – they cannot be varied independently and so V seems redundant. Indeed this may the reason why V was never considered in the early literature on black hole thermodynamics. But this is an artifact of the non-rotating approximation, V and S can, and should, be considered to be independent variables for a rotating black hole.

The line element for a charged rotating black hole in 4-dimensional AdS space is [29]

$$ds^2 = -\frac{\Delta}{\rho^2}\left(dt - \frac{a\sin^2\theta}{\Xi}d\phi\right)^2 + \frac{\rho^2}{\Delta}dr^2 + \frac{\rho^2}{\Delta_\theta}d\theta^2 + \frac{\Delta_\theta \sin^2\theta}{\rho^2}\left(adt - \frac{r^2 + a^2}{\Xi}d\phi\right)^2, \tag{21}$$

where

$$\Delta = \frac{(r^2 + a^2)(L^2 + r^2)}{L^2} - 2mr + q^2, \qquad \Delta_\theta = 1 - \frac{a^2}{L^2}\cos^2\theta,$$

$$\rho^2 = r^2 + a^2\cos^2\theta, \qquad \Xi = 1 - \frac{a^2}{L^2}, \tag{22}$$

and the cosmological constant is $\Lambda = -\frac{3}{L^2} = -8\pi P$.

The physical properties of this space-time are well known [15]. The metric parameters m and q are related to the ADM mass M and the electric charge Q by

$$M = \frac{m}{\Xi^2}, \qquad Q = \frac{q}{\Xi}. \tag{23}$$

The event horizon, r_+, lies at the largest root of $\Delta(r) = 0$, so, in terms of geometrical parameters,

$$M = \frac{(r_+^2 + a^2)(L^2 + r_+^2) + q^2 L^2}{2r_+ L^2 \Xi^2} \tag{24}$$

and the area of the event horizon is

$$A = 4\pi \frac{r_+^2 + a^2}{\Xi}, \tag{25}$$

giving

$$S = \pi \frac{r_+^2 + a^2}{\Xi}. \tag{26}$$

The angular momentum is $J = aM$ and the relevant thermodynamic angular velocity is

$$\Omega = \frac{a(L^2 + r_+^2)}{L^2(r_+^2 + a^2)}. \tag{27}$$

As explained in [27], Ω here is the difference between the asymptotic angular velocity and the angular velocity at the black hole outer horizon.

The electrostatic potential, again the difference between the potential at infinity and at the horizon, is

$$\Phi = \frac{q r_+}{r_+^2 + a^2}. \tag{28}$$

To determine the thermodynamic properties, M must be expressed in terms of S, J, Q and P (or, equivalently, L). This was done in [27] and the result is

$$H(S,P,J,Q) := \frac{1}{2}\sqrt{\frac{\left(S + \pi Q^2 + \frac{8PS^2}{3}\right)^2 + 4\pi^2\left(1 + \frac{8PS}{3}\right) J^2}{\pi S}}. \tag{29}$$

This generalises the Christodoulou-Ruffini formula [5, 6] for the mass of a rotating black hole in terms of its irreducible mass, M_{irr}. (The irreducible mass for a black hole with entropy S is the mass of a Schwarzschild black hole with the same entropy, $M_{irr}^2 = \frac{S}{4\pi}$).

The temperature follows from

$$T = \frac{\partial H}{\partial S}\bigg|_{J,Q,P} = \frac{1}{8\pi H}\left[\left(1+\frac{\pi Q^2}{S}+\frac{8PS}{3}\right)\left(1-\frac{\pi Q^2}{S}+8PS\right)-4\pi^2\left(\frac{J}{S}\right)^2\right], \tag{30}$$

from which we immediately see that $T \geq 0$ requires

$$J^2 \leq \frac{S^2}{4\pi^2}\left(1+\frac{\pi Q^2}{S}+\frac{8PS}{3}\right)\left(1-\frac{\pi Q^2}{S}+8PS\right). \tag{31}$$

The maximum angular momentum,

$$|J_{max}| = \frac{S}{2\pi}\sqrt{\left(1+\frac{\pi Q^2}{S}+\frac{8PS}{3}\right)\left(1-\frac{\pi Q^2}{S}+8PS\right)}, \tag{32}$$

is associated with an extremal black hole.

From (10) and (29) the thermodynamic volume is [30]

$$V = \frac{\partial H}{\partial P}\bigg|_{S,J,Q} = \frac{2}{3\pi H}\left[S\left(S+\pi Q^2+\frac{8PS^2}{3}\right)+2\pi^2 J^2\right], \tag{33}$$

which is manifestly positive.

The angular velocity and the electric potential also follow from (29) via

$$\Omega = \frac{\partial H}{\partial J}\bigg|_{S,Q,P} = \frac{4\pi^2 J\left(1+\frac{8PS}{3}\right)}{2H\sqrt{\pi S}} \tag{34}$$

and

$$\Phi = \frac{\partial H}{\partial Q}\bigg|_{S,J,P} = \frac{2\pi Q\left(S+\pi Q^2+\frac{8PS^2}{3}\right)}{2H\sqrt{\pi S}}. \tag{35}$$

The Smarr relation follows from (29), (30), (33), (34) and (35), namely

$$\frac{H}{2}+PV-ST-J\Omega-\frac{Q\Phi}{2}=0, \tag{36}$$

from which it is clear that the PV-term must be included for consistency, as pointed out in [23].

It is clear from (33) that, in general, V is a function of all the four independent thermodynamical variables, S, P, J and Q, but for the limiting case $J = 0$,

$$V = \frac{4}{3}\frac{S^{\frac{3}{2}}}{\sqrt{\pi}}, \tag{37}$$

is determined purely in terms of S alone, independent of both P and Q. Thus, as explained in the introduction, V and S cannot be viewed as thermodynamically independent variables as $J \to 0$, rendering the description in terms of the thermodynamic potential $U(S, J)$ impossible in this limit.

Expressing the thermodynamic volume (33) in terms of geometrical variables one gets [30]

$$V = \frac{2\pi}{3}\left\{\frac{(r_+^2 + a^2)(2r_+^2 L^2 + a^2 L^2 - r_+^2 a^2) + L^2 q^2 a^2}{L^2 \Xi^2 r_+}\right\}. \tag{38}$$

Given that the area of the event horizon is

$$A = 4\pi \frac{r_+^2 + a^2}{\Xi} \tag{39}$$

then, if we define a naïve volume

$$V_0 := \frac{r_+ A}{3} = \frac{4\pi}{3}\frac{r_+(r_+^2 + a^2)}{\Xi}, \tag{40}$$

equations (24) and (38) give

$$V = V_0 + \frac{4\pi a^2 M}{3} = V_0 + \frac{4\pi}{3}\frac{J^2}{M}, \tag{41}$$

a formula first derived in [31]. As pointed out in that reference, equation (40) implies that the surface to volume ratio of a black hole is always less than that of a sphere with radius r_+ in Euclidean geometry. This is the opposite of our usual intuition that a sphere has the smallest surface to volume ratio of any closed surface — the isoperimetric inequality of Euclidean geometry. Thus the surface to volume ratio of a black hole satisfies the reverse of the usual isoperimetric inequality (a similar result holds in higher dimensions [31]). At least this seems to be the case if quantum gravity effects are not taken into account. In one case where quantum gravity corrections can be calculated using the techniques in [32] , the three-dimensional Bañados–Zanelli–Teitelboim (BTZ) black hole [33], they tend to reduce the black hole volume [28] so it seems possible that quantum gravity effects may affect the reverse isoperimetric inequality.

4. The First Law

To examine the consequences of the PdV term in the first law we need to perform a Legendre transform on the enthalpy to obtain the internal energy $U(S,V,J,Q)$ from $U = H - PV$. We first write the enthalpy (29) in the form

$$H = \sqrt{a + bP + cP^2}, \tag{42}$$

where

$$\begin{aligned} a &:= \frac{\pi}{S}\left\{\frac{1}{4}\left(\frac{S}{\pi} + Q^2\right)^2 + J^2\right\} \\ b &:= \frac{4\pi}{3}\left\{\frac{S}{\pi}\left(\frac{S}{\pi} + Q^2\right) + 2J^2\right\} \\ c &:= \left(\frac{4\pi}{3}\right)^2\left(\frac{S}{\pi}\right)^3. \end{aligned} \tag{43}$$

Note that the discriminant,

$$b^2 - 4ac = \frac{64\pi^2}{9}J^2\left(J^2 + \frac{SQ^2}{\pi}\right), \tag{44}$$

is positive.

Now

$$V = \left.\frac{\partial H}{\partial P}\right|_{S,J,Q} = \frac{b + 2cP}{2H} \qquad \Rightarrow \qquad P = \frac{2HV - b}{2c}. \tag{45}$$

This allows us to re-express H as a function of V,

$$H = \frac{1}{2}\sqrt{\frac{b^2 - 4ac}{V^2 - c}}. \tag{46}$$

We can immediately conclude that

$$V^2 \geq c = \left(\frac{4\pi}{3}\right)^2\left(\frac{S}{\pi}\right)^3, \tag{47}$$

with equality only when

$$b^2 - 4ac = \frac{64\pi^2}{9}J^2\left(J^2 + \frac{SQ^2}{\pi}\right) = 0, \tag{48}$$

i.e. when $J = 0$.

It is now straightforward to determine

$$U = H - PV = H - \left(\frac{HV^2}{c} - \frac{bV}{2c}\right) = \frac{bV}{2c} - \frac{\sqrt{(V^2 - c)(b^2 - 4ac)}}{2c}, \tag{49}$$

which immediately gives

$$U(S,V,J,Q) = \left(\frac{\pi}{S}\right)^3 \left[\left(\frac{3V}{4\pi}\right)\left\{\left(\frac{S}{2\pi}\right)\left(\frac{S}{\pi} + Q^2\right) + J^2\right\}\right. \tag{50}$$

$$\left. - |J| \left\{\left(\frac{3V}{4\pi}\right)^2 - \left(\frac{S}{\pi}\right)^3\right\}^{\frac{1}{2}} \left(\frac{SQ^2}{\pi} + J^2\right)^{\frac{1}{2}}\right].$$

Note the subtlety in the $J \to 0$ limit, (50) is not differentiable at $J = 0$ unless

$$\left(\frac{3V}{4\pi}\right)^2 = \left(\frac{S}{\pi}\right)^3 \tag{51}$$

there.

Equation (50) can now be used to study the efficiency of a Penrose process. If a black hole has initial mass M_i, with internal energy U_i, and is taken through a quasi-static series of thermodynamic steps to a state with final internal energy U_f, then energy can be extracted if $U_f < U_i$. This is the thermodynamic description of a Penrose process [4] and the efficiency is

$$\eta = \frac{U_i - U_f}{M_i}. \tag{52}$$

We can determine the maximum efficiency for a process at constant P by first expressing U in (50) in terms of S, P, J and Q:

$$U = \frac{\left(S + \pi Q^2\right)\left(S + \pi Q^2 + \frac{8PS^2}{3}\right) + 4\pi^2\left(1 + \frac{4PS}{3}\right)J^2}{2\sqrt{\pi S\left[\left(S + \pi Q^2 + \frac{8PS^2}{3}\right)^2 + 4\pi^2\left(1 + \frac{8PS}{3}\right)J^2\right]}}, \tag{53}$$

which is manifestly positive.

For simplicity consider first the $Q = 0$ case, for which

$$dU = TdS + \Omega dJ - PdV. \tag{54}$$

The work extracted at any infinitesimal step is

$$dW = -dU = -TdS - \Omega dJ + PdV \tag{55}$$

and, since $dS \geq 0$, this is maximised in an isentropic process $dS = 0$. Now with $Q = 0$ and S and P held constant, the internal energy in equation (53) can be thought of as a function of J only, $U(J)$. The greatest efficiency is then obtained by starting with an extremal black hole and reducing the angular momentum from J_{max} to zero, it is given by

$$\eta_{ext} = \frac{U(J_{max}) - U(0)}{H(J_{max})} \tag{56}$$

where $H(J_{max}) = M_{ext}$ is the initial extremal mass. One finds

$$\eta_{ext} = \frac{1+2PS}{1+4PS} - \frac{1}{\sqrt{2+8PS}}\frac{3}{(3+8PS)}. \tag{57}$$

In asymptotically flat space, $\Lambda = 0$, we set $P = 0$ in η_{ext} and obtain the famous result [9]

$$\eta_{ext} = 1 - \frac{1}{\sqrt{2}} \approx 0.2929. \tag{58}$$

More generally, η_{ext} is a maximum for $SP = 1.837\ldots$ (obtained by solving a quartic equation) and attains there the value $0.5184\ldots$. Thus turning on a negative cosmological constant increases the efficiency of a Penrose process, as first observed in [30].

What is happening here is that, as $|J|$ decreases (giving a positive contribution to dW) the volume decreases, which actually tends to *decrease* the work done because of the PdV term in (55). But when $P > 0$, the extremal value $|J_{max}|$ in (32) is increased, which more than compensates, and overall η_{ext} is increased.

For a charged black hole the internal energy is a function of J and Q for an isobaric isentropic process, $U(J,Q)$. The requirement $J^2_{max} \geq 0$ in (32) imposes the constraint

$$Q^2 \leq Q^2_{max} = \left(\frac{S}{\pi}\right)(1+8PS) \tag{59}$$

on the charge. The greatest efficiency is achieved starting from an extremal black hole with $Q^2 = Q^2_{max}$ and reducing both J and Q to zero in the final state,

$$\eta_{ext} = \frac{U(J_{max}, Q_{max}) - J(0,0)}{H(J_{max}, Q_{max})} = \frac{3}{2}\left(\frac{1+8PS}{3+16PS}\right), \tag{60}$$

with $H(J_{max}, Q_{max})$ the initial extremal mass, M_{ext}. For large S efficiencies of up to 75% are possible [30], which should be compared to 50% in the $\Lambda = 0$ case, [9].

5. Critical behaviour

With knowledge of both H and U general questions concerning the heat capacity of black holes can be addressed. The heat capacity at constant volume, $C_V = T/\left(\frac{\partial T}{\partial S}\right)_{V,J,Q}$, tends to zero when $J = 0$, though C_V can be non-zero for $J \neq 0$ it does not diverge. For comparison the heat capacity at constant pressure, $C_P = T/\left(\frac{\partial T}{\partial S}\right)_{P,J,Q}$, C_P vanishes when $T = 0$ and diverges when $\frac{\partial T}{\partial S}$ =0.

A full stability analysis was given in [27] and there are both local and global phase transitions. Local stability can be explored visually, by plotting thermodynamic functions, or analytically, examining the curvature of the derivatives of thermodynamics functions.

5.1. $Q = 0$

Let us first focus on the $Q = 0$ case. The blue (lower) curve in the figure below shows the locus of points where C_P diverges in the $J - S$ plane, it is given by setting the denominator of C_P,

$$144\,(\pi JP)^4(9+32\,SP)+24\,(\pi PJ)^2(PS)^2(3+16\,SP)(3+8\,SP)^2-(PS)^4(1-8\,SP)(3+8\,SP)^3 \qquad (61)$$

to zero. The red (upper) curve is the $T = 0$ locus, all points above and left of this curve are unphysical as $T < 0$ in this region.

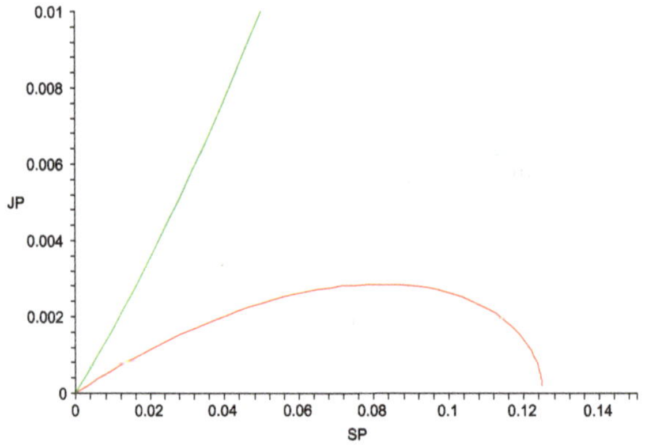

Figure 1. $T = 0$ and $C_P \to \infty$ curves in $J - S$ plane.

There is also a global phase transition, not shown in the figure, when the free energy of pure AdS is lower than that of a black hole in asymptotically AdS space-time, the famous Hawking-Page phase transition [34]. We shall focus on the second order local phase transition here and examine its critical properties.

In general, at fixed P and J, there are two values of S at which C_P diverges, and there is a critical point where these two values coalesce into one, the maximum of the lower curve in figure 1. This critical point was first identified in [27]. On purely dimensional grounds PC_P

can be expressed as a function of PS and PJ and the critical point can be found analytically, by solving a cubic equation, but the explicit form is not very illuminating. Numerically it lies at

$$(PS)_{crit} \approx 0.08204, \qquad (PJ)_{crit} \approx 0.002857. \tag{62}$$

The critical temperature is obtained from (30), with $Q = 0$,

$$\left(\frac{T}{\sqrt{P}}\right)_{crit} \approx 0.7811 \tag{63}$$

and the critical volume likewise from (33)

$$\left(VP^{3/2}\right)_{crit} \approx 0.01768 \tag{64}$$

(the authors of [27] fix $P = \frac{3}{8\pi} \approx 0.1194$, corresponding to $L = 1$, and find a critical value of J at $J_c \approx 0.0236$).

The equation of state cannot be obtained analytically, but its properties near the critical point can be explored by a series expansion and critical exponents extracted. Define the reduced temperature and volume as

$$t = \frac{T - T_c}{T_c} \qquad v = \frac{V - V_c}{V_c}. \tag{65}$$

It is convenient to expand around the critical point using

$$p := 16\pi\left(PJ - (PJ)_{crit}\right) \tag{66}$$

and

$$q := 8\left(PS - (PS)_{crit}\right). \tag{67}$$

Expanding the temperature (30) around the critical point, with Q set to zero, gives

$$t = 2.881\, p + 2.201\, pq + 0.3436\, q^3 + o(p^2, pq^2, q^4). \tag{68}$$

while similar expansion of the thermodynamic volume (33) yields

$$v = -10.44\, p + 2.284\, q + o(p^2, pq, q^2). \tag{69}$$

For a given fixed $J > 0$, p is the deviation from critical pressure in units of $1/(16\pi J)$, but one must be aware that this interpretation precludes taking the $J \to 0$ limit in this formulation. Bearing this in mind, (68) and (69) give the $J > 0$, $Q = 0$ equation of state parametrically in terms of q. Eliminating q one arrives at

$$p = 0.3472\, t - 0.1161\, tv - 0.02883\, v^3 + o(t^2, tv^2, v^4). \tag{70}$$

The critical exponent α is defined by

$$C_V \propto t^{-\alpha} \tag{71}$$

and, since as already stated, C_V does not diverge at $t = 0$, $\alpha = 0$. To see this explicitly note that $C_V = T / \frac{\partial T}{\partial S}\big|_V$ and

$$\left.\frac{\partial T}{\partial S}\right|_V = \left.\frac{\partial T}{\partial S}\right|_P + \left.\frac{\partial T}{\partial P}\right|_S \left.\frac{\partial P}{\partial S}\right|_V = T_c \left(\left.\frac{\partial t}{\partial S}\right|_P + \left.\frac{\partial t}{\partial P}\right|_S \left.\frac{\partial P}{\partial S}\right|_V \right). \tag{72}$$

Now, near the critical point, (68) gives

$$\left.\frac{\partial t}{\partial S}\right|_P = 8P \left.\frac{\partial t}{\partial q}\right|_p = o(p, q^2), \tag{73}$$

$$\left.\frac{\partial t}{\partial P}\right|_S = 8S \left.\frac{\partial t}{\partial q}\right|_p + 16\pi J \left.\frac{\partial t}{\partial p}\right|_q = 2.881(16\pi J) + o(p, q), \tag{74}$$

while (69) implies $dp = 0.2188\, d\, q$ for constant v, from which is follows that $\frac{\partial P}{\partial S}\big|_V$ is non-zero at the critical point, hence $\frac{\partial T}{\partial S}\big|_V$ does not vanish at the critical point and so $\alpha = 0$.

The exponent β is defined by

$$v_> - v_< = |t|^\beta \tag{75}$$

where $v_>$ is the greater volume and $v_<$ the lesser volume across the phase transition, at constant pressure, when $t < 0$ ($v_<$ is negative, since $v = 0$ at the critical point).

Keeping p and t constant in (70) implies that

$$p \int_{v_<}^{v_>} dv = 0.3742\, t \int_{v_<}^{v_>} dv - \int_{v_<}^{v_>} \left(0.1161\, tv + 0.02883\, v^3\right) dv. \tag{76}$$

Allowing for the area of the rectangle in figure 3, namely $0.3742\,|t|(v_> - v_<)$, Maxwell's equal area law then requires

$$\int_{v_<}^{v_>} \left(0.1161\, tv + 0.02883\, v^3\right) dv = 0 \qquad \Rightarrow \qquad |t| \propto (v_>^2 + v_<^2). \tag{77}$$

It is clear from the figure that $v_> - v_< \gg v_> + v_<$ so

$$(v_>^2 + v_<^2) = \frac{1}{2}\left((v_> - v_<)^2 + (v_> + v_<)^2\right) \approx \frac{1}{2}(v_> - v_<)^2 \tag{78}$$

giving

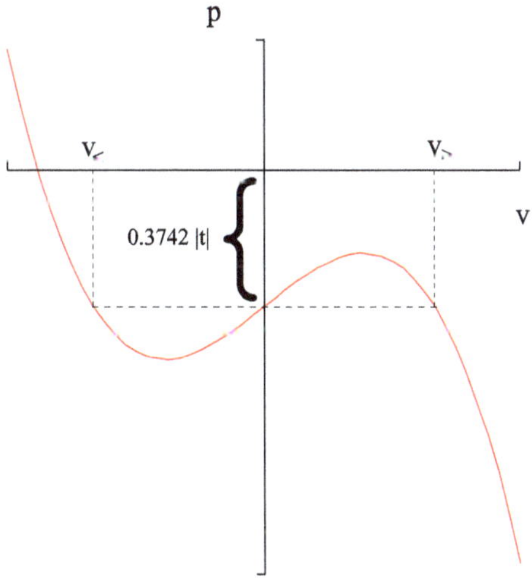

Figure 2. Construction associated with Maxwell's equal area law

$$|t| \propto (v_> - v_<)^2 \tag{79}$$

and $\beta = \frac{1}{2}$.

The critical exponent γ is related to the isothermal compressibility,

$$\kappa_T = -\frac{1}{V}\left(\frac{\partial V}{\partial P}\right)_{T,J} = -\frac{1}{V}\left(\frac{\partial V}{\partial P}\right)_{S,J} - \left(\frac{\partial V}{\partial S}\right)_{P,J} \frac{\left(\frac{\partial T}{\partial P}\right)_{S,J}}{\left(\frac{\partial T}{\partial S}\right)_{P,J}} \tag{80}$$

which diverges along the same curve as C_P does (the adiabatic compression, $\kappa_S = -\frac{1}{V}\left(\frac{\partial V}{\partial P}\right)_{S,J}$, is everywhere finite — see equation (92)). γ gives the divergence of the isothermal compressibility near the critical point,

$$\kappa_T \propto t^{-\gamma}. \tag{81}$$

γ can be found by expanding the denominator of C_P in (61) around the critical point, but a quicker method, since we have the equation of state, is to differentiate (70) with respect to v, keeping t constant, giving

$$\left.\frac{\partial p}{\partial v}\right|_t \propto -t, \tag{82}$$

hence

$$\kappa_T \propto -\left.\frac{\partial v}{\partial p}\right|_t \propto \frac{1}{t} \tag{83}$$

and $\gamma = 1$.

Lastly setting $t = 0$ in (70) we see that

$$|p| \propto |v|^{\delta} \tag{84}$$

with $\delta = 3$, again the mean field result.

To summarise, the critical exponents are

$$\boxed{\alpha = 0, \quad \beta = \frac{1}{2}, \quad \gamma = 1, \quad \delta = 3.} \tag{85}$$

These are the same critical exponents as the Van der Waals fluid and, more importantly, are mean field exponents. The same critical exponents have been found using a virial expansion in [50].

It was first pointed in [35, 36] that a non-rotating, charged black hole has a critical point of the same nature as that of of a Van der Waals fluid, and the critical exponents for the black hole phase transition in this case were calculated in [37] and verified to be mean field exponents, which are indeed the those of a Van der Waals fluid. A similarity between the neutral rotating black hole and the Van der Waals phase transition was first pointed out in [27] and further explored in [30].

The critical point can be visualised by plotting the Gibbs free energy

$$G(T, P, J) = H(S, P, J) - TS, \tag{86}$$

for $J = 1$ and $Q = 0$, as a function of P and T as in figure 3. We see the "swallow-tail catastrophe" that is typical of the Van der Waals phase transition [38].

This structure is a straightforward consequence of Landau theory, [39]. Near the critical point the Landau free energy is

$$L(T, P, v) = G(T, P) + A\left\{(p - Bt)v + Ctv^2 + Dv^4\right\} + \dots, \tag{87}$$

where $G(T, P)$ is the Gibbs free energy and A, B, C and D are positive constants (for simplicity the constant J is not made explicit). As stressed in [40] L is not strictly speaking a thermodynamic function as it depends on three variables, p, t and v instead of two: v is to be determined in terms of p and t by extremising L to obtain the equation of state.

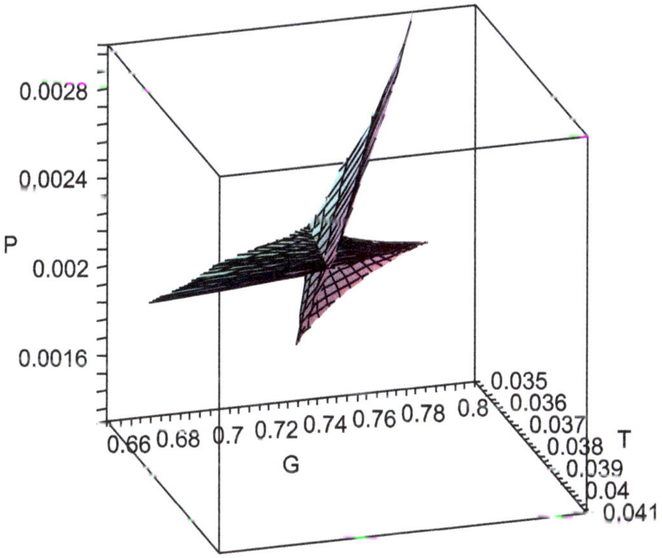

Figure 3. Gibbs free energy as a function of pressure and temperature, at fixed angular momentum.

For notational convenience equation (87) can be written, for fixed p and t, as

$$L = a + bv + cv^2 + v^4 \tag{88}$$

where a, b and c need not be positive and $L \to \frac{1}{AD} L$ has been rescaled by a trivial positive constant. We are to think of b and c are control parameters that can be varied by varying p and t.

Extremising (88) with respect to v determines the value of v in terms of b and c through

$$b = -2cv - 4v^3. \tag{89}$$

Using this in L leads to

$$L = a - cv^2 - 3v^4. \tag{90}$$

Equations (89) and (90) together give $L(a, b, c)$ implicitly: a parametric plot of $L(b, c)$, for any fixed a, reveals a characteristic "swallow tail catastrophe" structure. With hindsight the swallow-tail structure is clear: in the $A - D - E$ classification of critical points of functions, [41], (88) has three control parameters and is derived from type A_4 in Arnold's classification.

5.2. $Q \neq 0$

The above structure was first found in AdS black hole thermodynamics in the charged $J = 0$ case [35, 36], where the equation of state can be found exactly and the critical exponents can be determined [37]. When both J and Q are non-zero an analytic analysis is much more difficult, for example finding the zero locus of the denominator of C_P requires solving a quintic equation. However numerical studies show that for a charged rotating black hole, as long as the charge is below the extremal value, the picture is qualitatively the same: the critical exponents are the same, the Landau free energy is still related to type A_4 and the Gibbs free energy still takes on a characteristic swallow-tail shape. For fixed values of J and Q, not both zero, all that changes is the numerical value of the co-efficients in equations (68), (69) and (70) or, equivalently the numerical values of the constants A, B, C and D in (87). As long as none of these constants actually changes sign the nature of the critical point does not change and the critical exponents are the same.

6. Compressibility and the speed of sound

In the previous section, the nature of the singularity in the isothermal compressibility near the critical point was discussed, but the adiabatic compressibility

$$\kappa_S = -\frac{1}{V}\left(\frac{\partial V}{\partial P}\right)_{T,J,Q} \tag{91}$$

is also of interest, and this was studied in [42] on which most of this section is based. From (10) one finds, setting $Q = 0$ for simplicity, that

$$\kappa_S = \frac{36(2\pi J)^4 S}{(3+8PS)\{(3+8PS)S^2 + (2\pi J)^2\}\{2(3+8PS)S^2 + 3(2\pi J)^2\}}. \tag{92}$$

This is finite at the critical point, indeed it never diverges for any finite values of S, P and J, and it vanishes as $J \to 0$: non-rotating black holes are completely incompressible. Black holes are maximally compressible in the extremal case $T = 0$, when $J = J_{max}$ in (32),

$$\kappa_S|_{T=0} = \frac{2S(1+8PS)^2}{(3+8PS)^2(1+4PS)}. \tag{93}$$

A speed of sound, c_S, can also be associated with the black hole, in the usual thermodynamic sense that

$$c_S^{-2} = \left.\frac{\partial \rho}{\partial P}\right|_{S,J} = 1 + \rho\kappa_S = 1 + \frac{9(2\pi J)^4}{\{2(3+8PS)S^2 + 3(2\pi J)^2\}^2}, \tag{94}$$

where $\rho = \frac{M}{V}$ is the density. c_S is unity for incompressible non-rotating black holes and is lowest for extremal black holes in which case

$$c_S^{-2}\Big|_{T=0} = 1 + \left(\frac{1+8PS}{3+8PS}\right)^2. \tag{95}$$

giving $c_S^2 = 0.9$ (in units with $c = 1$) when $P = 0$. In the limiting case $PS \to \infty$, c_S^2 achieves a minimum value of $1/2$.

These results show that the equation of state is very stiff for adiabatic variations of non-rotating black holes and gets softer as J increases. For comparison, the adiabatic compressibility of a degenerate gas of N relativistic neutrons in a volume V at zero temperature follows from the degeneracy pressure

$$P_{deg} = (3\pi^2)^{\frac{1}{3}} \frac{c\hbar}{4} \left(\frac{V}{N}\right)^{-\frac{4}{3}} \quad \Rightarrow \quad \kappa_S = \frac{3}{4P_{deg}}. \tag{96}$$

For a neutron star $\frac{N}{V} \approx 10^{45}\, m^{-3}$ and $\kappa_S \approx 10^{-34}\, kg^{-1}\, m\, s^2$. With zero cosmological constant the black hole adiabatic compressibility at zero temperature is given by (93) with $P = 0$,

$$\kappa_S|_{T=P=0} = \frac{2S}{9} = \frac{4\pi M^2 G^3}{9\, c^8}, \tag{97}$$

where the relevant factors of c and G are included, and the entropy has been set to the extremum value of $2\pi M$. Putting in the numbers

$$\kappa_S|_{T=0} = 2.6 \times 10^{-38} \left(\frac{M}{M_\odot}\right)^2 kg^{-1}\, m\, s^2, \tag{98}$$

which is four orders of magnitude less than that of a solar-mass neutron star. We conclude that the zero temperature black hole equation of state, although "softer" than that of a non-rotating black hole, is still very much stiffer than that of a neutron star.

The "softest" compressibility for a neutral black hole however is the isothermal compressibility: for an extremal black hole

$$\kappa_T\Big|_{T=0} = \frac{2S(11+80PS+128(PS)^2)}{(1+4PS)(3+48PS+128(PS)^2)} \underset{P\to 0}{\longrightarrow} \frac{22\,S}{3}, \tag{99}$$

some 33 times larger than $\kappa_S\big|_{T=P=0}$ in (97), but still much larger than degenerate matter in a solar-mass neutron star.

7. Open questions

The obvious open question arising from the ideas presented here is: what about $\Lambda > 0$?

The analysis of critical behaviour in §5 is only valid for $\Lambda < 0$, this critical point lies deep in the region $P > 0$ and does not appear to be of any relevance to astrophysical situations. It is certainly of interest in the AdS-CFT correspondence [43] but the particular analysis of §5, being in $1+3$-dimensions could only be relevant to $2+1$-dimensional conformal field theory, which is of course of interest in its own right [44]. Of course one could perform a similar analysis for $4+1$-dimensional, or yet higher dimensional black holes, to try and gain insight into higher dimensional conformal field theory, and indeed this seems to have been the motivation in [31, 35, 36], but these ideas are not the focus of this volume and will not be pursued here.

The thermodynamics of black holes in de Sitter space-time is a notoriously difficult problem [17, 20, 45–48] as there are two event horizons and no "asymptotically de Sitter" region inside the cosmological horizon. Even with no black hole, a naïve interpretation of the cosmological horizon implies that the transition from $\Lambda = 0$ to any infinitesimally small $\Lambda > 0$ appears to involve a discontinuous jump from zero to infinite entropy, at least if one associates the usual Hawking-Bekenstein entropy with the cosmological horizon when $\Lambda > 0$.

Nevertheless it is argued in [45] that a consistent strategy is to fix the relevant components of the metric at the cosmological horizon, rather than at spacial infinity as would be done in asymptotically flat or AdS space-time. When that is done the same expression for the ADM mass (24) is obtained, but with $L^2 \to -L^2$, so $\Xi > 1$ while the angular momentum is still given by $J = aM$. In this picture, all of the formulae in §3 are applicable for positive Λ and negative P, provided P is not too negative. If Λ is too large the black hole horizon and the cosmological horizon coincide and demanding that this does not happen puts a lower bound on P, for any fixed S, J and Q: with $Q = 0$, for example, this requirement constrains P to

$$P > \frac{\sqrt{S^2 + 12\pi J^2} - 2S}{8S}. \tag{100}$$

Provided P lies above this lower bound we can analytically continue (29) to negative P, with the understanding that S is the entropy of the black hole event horizon only and does not include any contribution from the cosmological horizon.

Of course $P < 0$ is thermodynamically unstable, but it can be argued, in some circumstances at least, that positive pressures can be analytically continued to negative pressures [12, 39], and in a cosmological context there can now be little doubt that $P < 0$. Adopting the strategy of [45] the maximal efficiency of a rotating black hole in de Sitter space will be less than in the $\Lambda = 0$ case, based on simply changing the sign of Λ in §4, and the zero charge efficiency vanishes when the black hole horizon and the cosmological horizon coincide at $PS = -\frac{1}{8}$. Any such deviation from the $\Lambda = 0$ case will however be completely negligible for astrophysical black holes around one solar mass and the observed value of Λ, but it could be more significant during periods of inflation when Λ was larger.

It has been suggested that primordial black-holes may have formed in the early Universe [49] and, if this is the case and if they formed in sufficient numbers at any stage, then one should model the primordial gas as containing a distribution of highly incompressible black holes, like beads in a gas. These would certainly be expected to affect the overall compressibility of the gas as well as the speed of sound through the gas. In a radiation dominated Universe, ignoring the matter density, the speed of sound in the photon gas would be given by

$$c_\gamma^{-2} = \left.\frac{\partial \epsilon}{\partial P}\right|_S = 3c^{-2}, \tag{101}$$

where ϵ is the energy density (essentially since the equation of state is $P = \frac{1}{3}\epsilon$) so $c_\gamma = 0.577\,c$. Since the speed of sound associated with the embedded black hole "beads" is $c_S \geq \sqrt{0.9}\,c = 0.9487\,c$ the presence of a significant density of primordial black holes would expected to affect speed of sound in the photon gas and thus affect the dynamics.

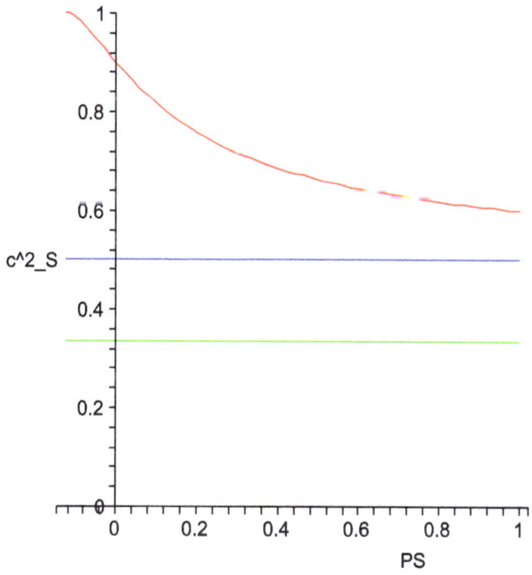

Figure 4. The speed of sound for an electrically neutral, extremal, black hole (with $c = 1$).

The square of the speed of sound for an extremal electrically neutral black hole is plotted in figure 4, for $PS > -1/8$. For comparison the asymptotic value ($c_S^2 = 1/2$ for $PS \to \infty$) and the speed of sound in a thermal gas of photons ($c_\gamma^2 = 1/3$) are also shown.

8. Conclusions

In conclusion there are strong reasons to believe that the cosmological constant should be included in the laws of black hole thermodynamics as a thermodynamic variable, proportional to the pressure of ordinary thermodynamics. The conjugate variable is a thermodynamic volume (10) and the complete first law of black hole thermodynamics is now (8),

$$dU = TdS + \Omega dJ + \Phi dQ - PdV. \tag{102}$$

With this interpretation the ADM mass of the black hole is identified with the enthalpy

$$M = H(S,P,J,Q) = U(S,V,J,Q) + PV \tag{103}$$

rather than the internal energy, U, of the system.

The inclusion of this extra term increases the maximal efficiency of a Penrose process: for a neutral black hole in asymptotically anti-de Sitter space the maximal efficiency is increased from 0.2929 in asymptotically flat space to 0.5184 in the asymptotically AdS case. For a charged black hole the efficiency can be as high as 75%. A positive cosmological constant is expected to reduce the efficiency of a Penrose process below the asymptotically flat space value.

This point of view makes the relation between asymptotically AdS black holes and the Van der Waals gas, first found in [35, 36], even closer as there is now a critical volume associated with the critical point. The thermodynamic volume then plays the rôle of an order parameter for this phase transition and the critical exponents take the mean field values,

$$\alpha = 0, \quad \beta = \frac{1}{2}, \quad \gamma = 1, \quad \delta = 3. \tag{104}$$

While there is no second order phase transition for a black hole in de Sitter space, there are other possible physical effects of including the PdV term in the first law. The adiabatic compressibility can be calculated (93) and the speed of sound for such a black hole (95) is greater even than that of a photon gas and approaches c when $PS = -1/8$.

Despite much progress the thermodynamics of black holes in de Sitter space-time is still very poorly understood and no doubt much still remains to be discovered.

Author details

Brian P. Dolan[1,2]

1 Department of Mathematical Physics, National University of Ireland Maynooth, Maynooth, Co. Kildare, Ireland
2 School of Theoretical Physics, Dublin Institute for Advanced Studies, Dublin, Ireland

References

[1] S W Hawking. Black hole explosions? *Nature*, 248:30, 1974.

[2] S W Hawking. Particle creation by black holes. *Comm. Math. Phys.*, 43:199, 1975. Erratum: ibid. vol. 46, p. 206 (1976).

[3] S W Hawking. Black holes and thermodynamics. *Phys. Rev.*, D13:191, 1976.

[4] R. Penrose. Gravitational collapse: the role of general relativity. *Riv. Nuovo Cimento*, 1:252, 1969.

[5] D. Christodoulou. Reversible and irreversible transformations in black-hole physics. *Phys. Rev. Lett.*, 25:1596, 1970.

[6] D. Christodoulou and R. Ruffini. Reversible transformations of a charged black hole. *Phys. Rev.*, D4:3552, 1971

[7] J D Bekenstein. Title. *Lett. Nuovo. Cimento*, 4:737, 1972.

[8] J D Bekenstein. Black holes and entropy. *Phys. Rev.*, D7:2333, 1973.

[9] R M Wald. *General Relativity*. University of Chicago Press, 1984.

[10] A G Riess et al. Observational evidence from supernovae for an accelerating universe and a cosmological constant. *Astronomical Journal*, 116:1009, 1998.

[11] S. Perlmutter et al. Measurements of Ω and Λ from 42 high-redshift supernovae. *Astrophysical Journal*, 517:565, 1999. [arXiv:astro ph/9812133]

[12] B. Lukács and K. Martinás. Thermodynamics of negative absolute pressures. *Acta. Phys. Pol.*, B21:177, 1990.

[13] M. Henneaux and C. Teitelboim. The cosmological constant as a canonical variable. *Phys. Lett.*, 143B:415, 1984.

[14] M. Henneaux and C. Teitelboim. The cosmological constant and general covariance. *Phys. Lett.*, 222B:195, 1989.

[15] M. Henneaux and C. Teitelboim. Asymptotically anti-de sitter spaces. *Commun. Math. Phys.*, 98:391, 1985.

[16] C. Teitelboim. The cosmological constant as a thermodynamic black hole parameter. *Phys. Lett.*, 158B:293, 198.

[17] Y. Sekiwa. Thermodynamics of de Sitter Black Holes: thermal cosmological constant. *Phys. Rev.*, D73:084009, 2006. [arXiv:hep-th/0602269].

[18] E A Larrañaga Rubio. Stringy generalization of the first law of thermodynamics for rotating BTZ black hole with a cosmological constant as state parameter. [arXiv:0711.0012 [gr-qc]], 2007.

[19] S. Wang, S-Q. Wu, F. Xie, and L. Dan. The first laws of thermodynamics of the (2+1)-dimensional BTZ black holes and Kerr-de Sitter spacetimes. *Chin. Phys. Lett.*, 23:1096, 2006. [arXiv:hep-th/0601147].

[20] S. Wang. Thermodynamics of high dimensional Schwarzschild de Sitter spacetimes: variable cosmological constant. [arXiv:gr-qc/0606109], 2006.

[21] M. Urano, A. Tomimatsu, and H. Saida. Mechanical first law of black hole space- times with cosmological constant and its application to Schwarzschild-de Sitter spacetime. *Class. Quantum Grav.*, 26:105010, 2009. arXiv:0903.4230 [gr-qc].

[22] S. Weinberg. The cosmological constant problem. *Rev. Mod. Phys.*, 61:1, 1989.

[23] D. Kastor, S. Ray, and J. Traschen. Enthalpy and the mechanics of AdS black holes. *Class. Quantum Grav.*, 26:195011, 2009. [arXiv:0904.2765 [hep-th]].

[24] L Smarr. Mass formula for Kerr black holes. *Phys. Rev. Lett.*, 30:71, 1973. Erratum: ibid. vol. 30, p. 521 (1973).

[25] M.K. Parikh. Volume of black holes. *Phys. Rev.*, D73:124021, 2006.

[26] W. Ballik and K. Lake. The volume of stationary black holes and the meaning of the surface gravity. 2012. [arXiv:1005.1116 [gr-qc]].

[27] M M Caldarelli, G. Cognola, and D. Klemm. Thermodynamics of Kerr-Newman-AdS black holes and conformal field theories. *Class. Quantum Grav.*, 17:399, 2000. [arXiv:hep-th/9908022].

[28] B P Dolan. The cosmological constant and black hole thermodynamic potentials. *Class. Quantum Grav.*, 28:125020, 2011. [arXiv:1008.5023].

[29] B. Carter. Hamiltonian-Jacobi and Schrödinger separable solutions of Einstein's equations. *Comm. Math. Phys.*, 10:280, 1968.

[30] B P Dolan. Pressure and volume in the first law of black hole thermodynamics. *Class. Quantum Grav.*, 28:235017, 2011. [arXiv:1106.6260].

[31] M. Cvetic, G W Gibbons, D. Kubizňák, and C N Pope. Black hole enthalpy and an entropy inequality for the thermodynamic volume. *Phys. Rev.*, D84:024037, 2011. [arXiv:1012.2888[hep-th].

[32] A. Maloney and E. Witten. Quantum gravity partition functions in three dimensions. *JHEP*, 1002:029, 2010. [arXiv:0712.0155 [hep-th]].

[33] M Bañadis, C Teitelboim, and J Zanelli. The black hole in three dimensional space time. *Phys. Rev. Lett.*, 69:1849, 1992.

[34] S W Hawking and D N Page. Thermodynamics of black holes in anti-de Sitter space. *Comm. Math. Phys.*, 87:577, 1983.

[35] A Chamblin, R Emparan, C V Johnson, and R C Myers. Charged AdS black holes and catastrophic holography. *Phys. Rev.*, D60:064018, 1999. [arXiv:hep-th/9902170v2].

[36] A Chamblin, R Emparan, C V Johnson, and R C Myers. Holography, thermodynamics and fluctuations of charged AdS black holes. *Phys. Rev.*, D60:104026, 1999. [arXiv:hep-th/9904197].

[37] D. Kubiznak and R B Mann. P-V criticality of charged AdS black holes. arXiv:[1205.0559], 2012.

[38] M Sewell. On legendre transformations and elementary catastrophes. *Math. Proc. Camb. Phil. Soc.*, 82:147, 1977.

[39] L D Landau and E M Lifshitz. *Statistical Physics*, volume 5 of *A Course of Theoretical Physics*. Pergamon, Oxford, Part I, 3rd edition, 1980.

[40] N. Goldenfeld. *Lectures On Phase Transitions And The Renormalization Group (Frontiers in Physics;85)*. Addison-Wesley, 1992.

[41] V I Arnold. *Catastrophe Theory*. Springer-Verlag, 1984.

[42] B P Dolan. Compressibility of rotating black holes. *Phys. Rev.*, D84:127503, 2011. [arXiv:1109.0198].

[43] O Aharony, S S Gubser, J Maldacena, H Ooguri, and Y Oz. Large N field theories, string theory and gravity. *Phys. Rep.*, 323:183, 2000. [arXiv:hep-th/9905111].

[44] S A Hartnoll. Lectures on holographic methods for condensed matter physics. *Class. Quant. Grav.*, 26:224002, 2009. [arXiv:0903.3246v3 [hep-th]].

[45] A Gomberoff and C Teitelboim. de Sitter black holes with either of the two horizons as a boundary. *Phys. Rev.*, D67:104024, 2003. [arXiv:hep-th/0302204].

[46] T Roy Choudhury and T Padmanabhan. Concept of temperature in multi-horizon spacetimes: Analysis of Schwarzschild-de Sitter metric. *Gen. Rel. Grav.*, 39:1789, 2007. [arXiv:gr-qc/0404091].

[47] A Corichi and A Gomberoff. Black holes in de Sitter space: Masses, energies and entropy bounds. *Phys. Rev.*, D69:064016, 2004. [arXiv:hep-th/0311030].

[48] R Aros. de Sitter thermodynamics: A glimpse into non equilibrium. *Phys. Rev.*, D77:104013, 2008. [arXiv:0801.4591].

[49] B J Carr. Primordial black holes as a probe of cosmology and high energy physics. *Lect. Notes Phys.*, 631:301, 2003. [arXiv:astro-ph/0310838].

[50] S. Gunasekaran, D. Kubizňák and R.B. Mann, "Extended phase space thermodynamics for charged and rotating black holes and Born-Infeld vacuum polarization, [arXiv:1208.6251]

Chapter 13

Plasma Vortices in Planetary Wakes

H. Pérez-de-Tejada, Rickard Lundin and
D. S. Intriligator

Additional information is available at the end of the chapter

1. Introduction

Measurements conducted in interplanetary space and in the vicinity of planets of the solar system have shown plasma structures produced by the solar wind that resemble fluid dynamic features; namely, vortex rotations within the earth´s magnetosphere and also along the Venus wake. In both planets the solar wind encounters different obstacles since the earth is protected by its intrinsic magnetic field that is compressed by the dynamic pressure of the solar wind to form a large size cavity (the magnetosphere) that bounds its direct approach to the earth´s vicinity. At Venus the conditions are different since the planet does not have an internal magnetization that would produce an earth-type magnetic obstacle to the solar wind. Instead, the latter reaches directly upon the upper layers of the planet´s atmosphere and interacts with its ionized components (the ionosphere). The outcome of this interaction is a plasma wake of large extent whose geometry is similar to that of the earth´s magnetospheric tail but that arises from conditions that are different in both planets. While there is evidence for the observation of fluid-like vortices as the solar wind streams along the wake of the earth and Venus there is a major issue as to the manner in which they are produced. In fact, since the solar wind is a collisionless plasma; namely its charged particles barely execute collisions among them (their mean free path is comparable to one astronomical unit) it should not be expected that it behaves as a continuum when it interacts with planetary obstacles. The opposite has been verified by a variety of observations with indications that the physical properties of both the solar wind fluxes and the planetary particles that are being eroded through their interaction can be described in terms of fluid dynamic processes (a review of this issue was presented by Pérez-de-Tejada, 2012).

The motion of the solar wind particles as they interact with the earth´s magnetic field is described in terms of gyromagnetic trajectories as they move across the magnetic field lines.

Their circular (Larmor) motion is influenced by local drift displacements that are produced by space gradients of the magnetic field intensity and that carry them along the boundary of the magnetosphere. Such process also occurs while there are local electric currents produced by the different drift of the positive (mostly proton) and negative (electron) components of the solar wind. Throughout the front part (dayside) of the magnetosphere the motion of those particles is directed by magnetic (Lorentz) forces which guide them along the magnetic field lines in an environment where the local magnetic energy density is larger than their kinetic energy density (Under such conditions the transport speed of magnetic signals, i.e. the Alfven speed, is larger than the particles' speed and the flow is labeled subalfvenic). The gyrotropic motion of the solar wind particles as they encounter the earth's magnetic field provides a mechanism that leads to a continuum transport of their properties despite the fact that they do not collide with each other. Alternate conditions are encountered along the magnetospheric tail where the magnetic field intensity has decreased significantly with the downstream distance from the earth and the speed of the solar wind particles has increased to nearly its freestream values (the local flow is expected to achieve superalfvenic conditions). Data that will be addressed below show the presence of vortical plasma structures within the magnetospheric tail and that reveal effects that can be related to Kelvin-Helmholtz instabilities (Wolfe et al., 1980: Terada et al:, 2002) associated to a fluid dynamic response in their motion rather than to trajectories directed by the magnetic field.

The overall manner in which the solar wind responds as it interacts with the Venus ionosphere is different since in the absence of an intrinsic planetary magnetization its particles reach the ionospheric plasma and, at the same time, also encounter ions that are located in the outer exosphere of the planet. Thus, the solar wind particles do not enter a region dominated by an intrinsic planetary magnetic field where they would be guided to carry out gyromagnetic trajectories as it is the case in the earth's magnetosphere. Instead, the processes produced as a result of their interaction with the planetary ions are more complex since the latter ions are accelerated by a (convective) electric field that derives from the relative velocity difference that exists between them and the solar wind (the planetary oxygen ions of the Venus exosphere are subject to the effects of that electric field). The result of that process is labeled mass loading and leads the planetary ions to execute gyromagnetic trajectories as they travel through the solar wind and its convected solar magnetic field. The magnetic field intensity of the solar wind becomes enhanced around the upper boundary of the dayside ionosphere (ionopause) in a layer labeled °magnetic barrier° that is produced by the solar wind – ionosphere interaction. Despite the acceleration of the exospheric planetary ions by the solar wind measurements show that the bulk of the solar wind momentum is mostly transferred to the upper Venus ionosphere where it strongly contributes to produce the nightward directed trans-terminator flow inferred from the Pioneer Venus Orbiter (PVO) data that was reported by Knudsen et al., (1980).

Processes associated with this effect are different from those expected in gyrotropic trajectories since they lead to the observation of vortical plasma structures in the near Venus wake associated with superalfvenic flow conditions (the kinetic energy density of the plasma being larger than the local magnetic energy density). The origin of the continuum response of the solar wind as it streams around and behind Venus does not seem to be only related to gyromagnetic trajectories but to a fluid-like behavior generated by other particle motions. Strong magnetic and plasma turbulence that has been inferred from different spacecraft measurements (Bridge et al., 1967; Vörös et al., 2008) suggest that the motion of the particles may be dictated by local turbulence which should lead to a stochastic distribution of their trajectories. It is possible that wave-particle interactions may be ultimately responsible for the fluid-like character of the solar wind interaction with planetary ionospheres and the generation of plasma and magnetic vortical structures as those reported by Pope et al., (2009) from the Venus Express measurements.

2. Plasma vortices in the earth's magnetosphere

Measurements conducted with various spacecraft that have probed the earth's magnetospheric tail have shown that the plasma flow that streams within that region of space exhibits changes in its direction of motion that are suggestive of vortical structures. From the analysis of plasma data first obtained with the International Sun Earth Explorer (ISEE) satellites and more recently with the Cluster spacecrafts it has been possible to derive that there are regions along the tail direction of the magnetosphere where the plasma exhibits a sense of rotation (clockwise in the morning sector of the magnetosphere and counter clockwise in the evening sector both viewed from above the ecliptic plane) which suggests a wave motion that in some instances leads to vortical structures that move tailward with a speed of 300-400 km/s. The inferred features have a scale size of several earth radii nearly comparable to the width of the magnetospheric tail and it has been inferred that their rotation period is of the order of 5-20 min (Hones et al. 1978; 1981; 1983). The vortex structures form part of a wave which has a several earth radii wavelength and that increases downstream along the tail direction. From observations carried out with the Cluster spacecrafts it has also been possible to examine the plasma composition and the vortex flow dynamics at the time when there are enhanced values of the solar wind dynamic pressure (Tian et al. 2010). As a whole it is believed that vortices may derive from Kelvin-Helmholtz instabilities at the boundary of the magnetosphere (magnetopause) or at the inner edge of a plasma boundary layer within the magnetosphere.

A suitable example of the plasma and magnetic field data obtained with a Cluster spacecraft by the morning flank of the magnetosphere and far downstream from the earth (at $x = -11\ R_e$; $y = -15\ R_e$; $z = 3\ R_e$; R_e being the earth radius) was presented by Tian et al., (2010) and is reproduced in Figure 1. Those measurements lead to the observation of a series of vortices that can be inferred from the ion velocity components marked in the lower panels of that figure. The vortex regions are indicated by the shaded bands when the v_x component becomes large, the v_y component changes sign (red profile), and the vortex rotation is derived from changes in the velocity vector orientation in the frame at the left side of the bottom panel. Comparable changes are also seen in the value of the B_x and B_y magnetic field components (second panel) which are derived by tracing the z-axis along the mean magnetic field direction.

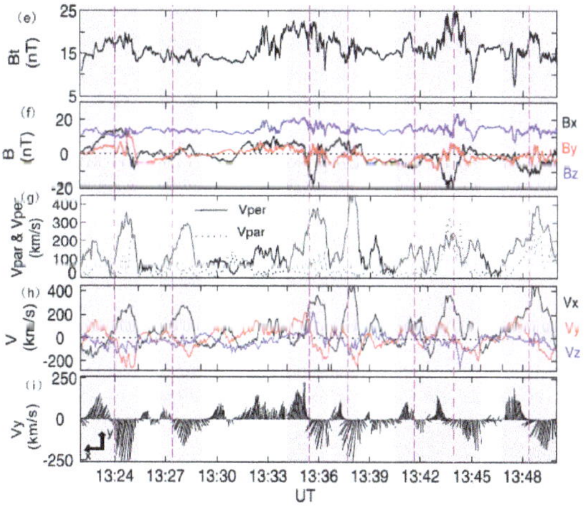

Figure 1. – A series of flow vortices in the earth´s magnetosphere observed by a Cluster spacecraft on July 6, 2003. The magnetic field and its components are indicated in the top two panels and the ion velocity and its components in the two lower panels (the velocity components in a reference frame in which the z and the x axes are parallel and perpendicular to the average magnetic field are in the middle panel). The profiles were selected from those presented in Figure 7 of Tian et al., (2010).

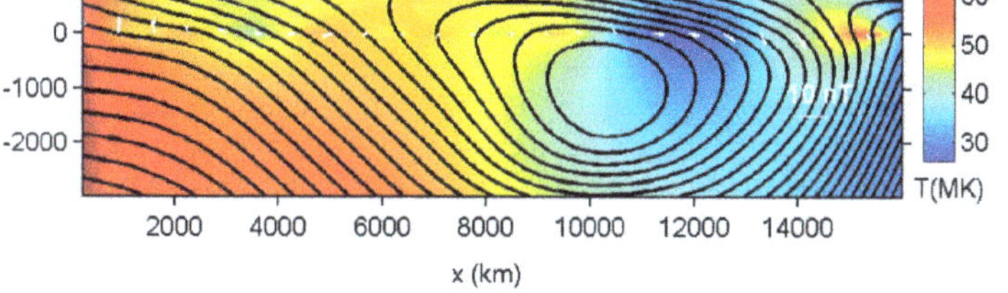

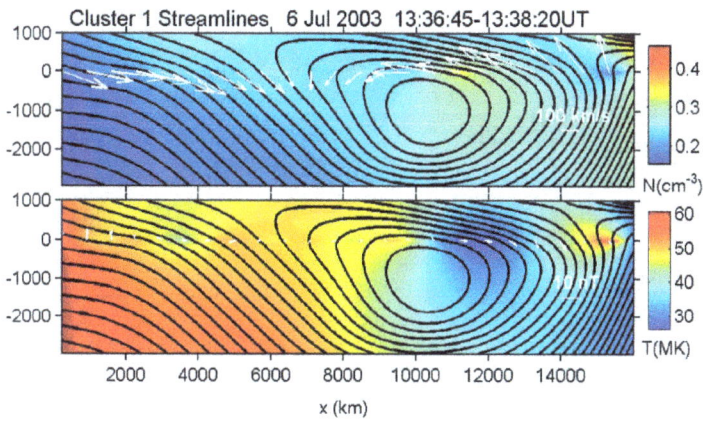

Figure 2. – Map of flow streamlines projected on the xy plane derived from the velocity rotation measured in the 13:36:45 UT – 13:38:20 UT time interval of July 6, 2003 in Figure 1. The streamlines are traced on the ion density (top panel) and on the ion temperature (lower panel) distributions (the white arrows in the top panel represent the measured velocity vector direction, and those in the lower panel the direction of the magnetic field). The streamlines describe conditions from the inner plasma sheet across the dawn side of the magnetosphere and were selected from those presented in Figure 11 of Tian et al., (2010).

The rotation of the velocity vector implied from the lower panels in Figure 1 serves to construct streamline maps of the flow using a numerical code applied to the data points along the spacecraft trajectory. The results are reproduced in Figure 2 where the observed direction of the velocity vector projected on the xy plane is marked by the white arrows in the upper panel and the direction of the magnetic field is indicated by white arrows in the lower panel (after Tian et al., 2010). The gradual and persistent change in the direction of the velocity vector with distance in the upper panel is indicative of an anticlockwise rotating vortex structure with an approximated two dimensional 1-2 R_e scale size. The reported observations occurred at the time when there is an enhanced solar wind dynamic pressure thus implying an intensified compression of the magnetosphere which could in this case have led to earthward moving directed vortices.

Measurements conducted with the ISEE vehicles have shown evidence of tailward directed vortical structures along the sides of the magnetosphere. However, there are indications that in the central plasma sheet of the magnetosphere the motion of the particles is observed to be earthward from the tail. An example of such measurements is reproduced in Figure 3 with evidence in the lower panel of a repeated rotation in the direction of the flow which in most cases leads to large (> 100 km/s) speed values (upper panel) at the time when the solar ecliptic longitude ϕ_{SE} becomes small; i.e. when the flow is sunward directed (Hones et al., 1981).

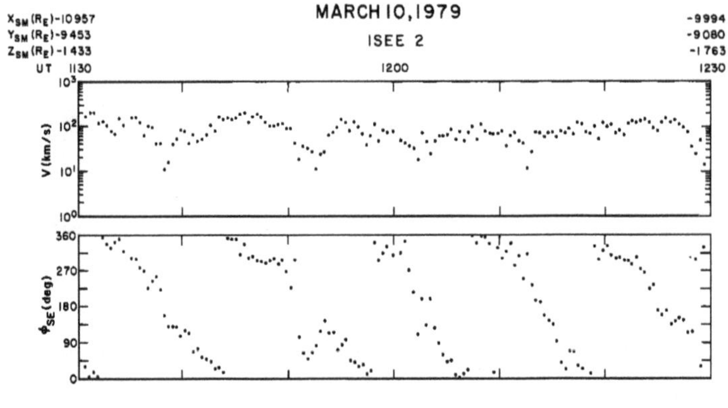

Figure 3. – Speed values (upper panel) and flow direction (lower panel) measured with the ISEE 2 spacecraft on March 10, 1979 within the earth´s magnetosphere. A repeated counterclockwise rotation of the solar ecliptic longitude angle ϕ is detected in the evening sector of the plasma sheet (after Hones et al., 1981).

A general schematic view of the flow distribution along the sides of the magnetosphere that has been derived from the ISEE measurements is reproduced in Figure 4. Tailward directed vortex structures occur by those regions and the diagram illustrates plasma patterns that may arise from Kelvin Helmholtz instabilities at the magnetopause and that lead to sudden changes in the flow direction when observed in the earth´s rest frame.

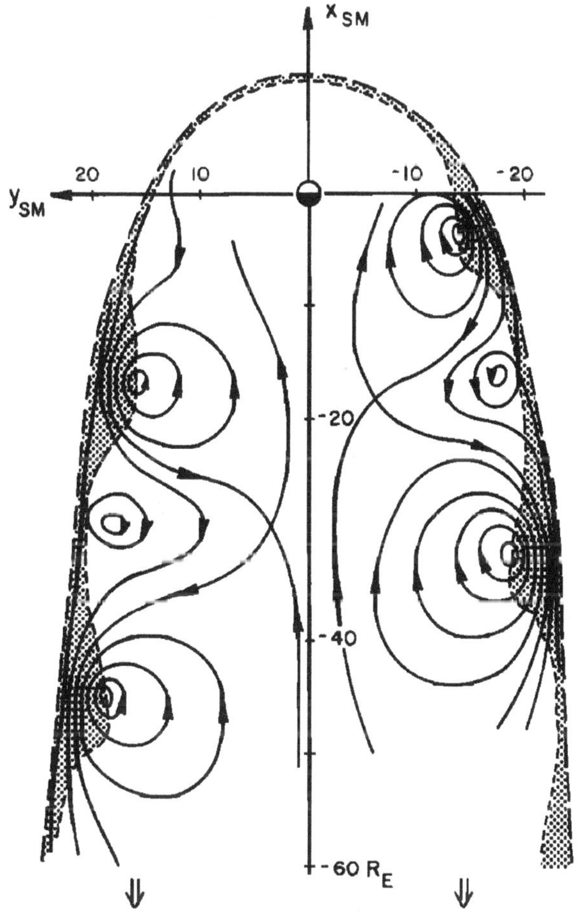

Figure 4. – Schematic description of plasma vortices within the earth's magnetosphere inferred from the ISEE measurements. The flow pattern represented by the solid lines is mostly tailward through the magnetosphere as it is indicated by the white arrows at the bottom (Hones et al., 1981).

The vortex structures are superimposed on a general circulation flow pattern within the magnetosphere in which the plasma is driven tailward along the sides by the kinetic pressure of the solar wind and then is forced back up in the sunward direction through the central plasma sheet region (Axford and Hines, 1961). The detail variation due to flow motion within and around those vortices is described in Figure 5a as it would be expected at four different distances from the magnetopause and that are represented by the dash lines A, B, C, and D. For a tailward moving vortex pattern (indicated by changes in the arrows of the flow direction) a waving motion should be apparent along the path line A which is located closer to the center of the tail. While along the path line B there should be a complete clockwise rotation (measured from the bottom to the top) a wavy periodic reversal will occur again along the path line C, and a counterclockwise rotation will prevail along the path line D. These variations serve to describe the structure of the vortices as seen in the earth´s rest frame. On the other hand, the flow pattern in the rest frame of the tailward moving wave is sketched in Figure 5b to show how the shape of the vortex structure is maintained within that wave (Hones et al., 1981).

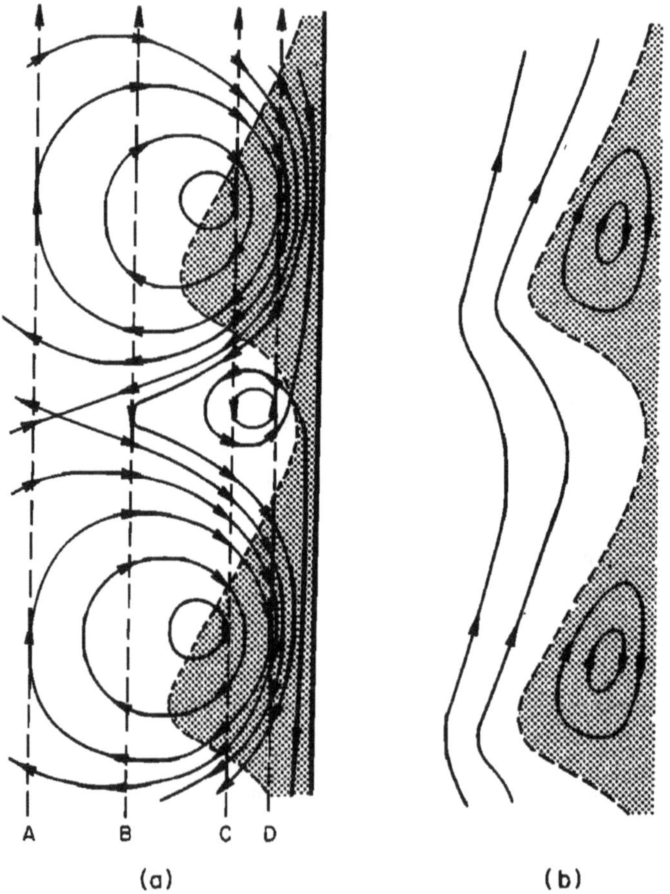

Figure 5. – (left panel) Dawn sector of the flow pattern in Figure 4 depicting the sequence of flow vectors observed at four different distances from the magnetopause. (right panel) Flow in the rest frame of the tailward moving wave (Hones et al., 1981).

The wavy pattern in the velocity direction that is implied from the measurements reproduced in Figures 2 and 4 could result from effects associated with Kelvin-Helmholtz instabilities produced along the boundary of the magnetosphere. Such structures require, however, fluid dynamic processes produced under conditions in which the flow is superalfvenic; namely, that the local value of the kinetic energy density of the plasma is larger than the local magnetic energy density. As a whole this peculiarity is not applicable to the plasma that pervades within the inner magnetosphere where the magnetic energy density is far larger than the local kinetic energy density. However, different conditions occur along the tail where the magnetic field intensity gradually decreases with the downstream distance from the earth thus reducing the ratio of the magnetic energy density to the kinetic energy density. A suitable comparative calculation of the conditions that were present at the time when the plasma vortices of the Cluster data of Figure 1 were identified can be conducted by using the corresponding pressure profiles of those data and that are reproduced in Figure 6. Notable is that the plasma vortices in Figure 1 were detected after the dynamic pressure of the solar wind exhibited a sudden increase indicated by the dashed vertical line at 12:55 UT in Figure 6. The dynamic pressure $P_D \sim 3$ nPa of the solar wind after that event shown in the upper panel is over 10 times larger than the ~ 0.16 nPa magnetic pressure that can be derived from the ~ 20 nT magnetic field intensity values measured at the same time and that are shown in the third panel. The implication here is that the local flow is superalfvenic thus suggesting conditions that could have led to Kelvin-Helmholtz waves along the magnetopause.

3. Plasma vortices in the Venus wake (PVO measurements)

Measurements conducted within and along the flanks of the Venus wake have revealed conditions that are also suitable for the onset of fluid dynamic processes that lead to the generation of vortex structures. From observations made with the early Mariner 5 spacecraft it was noted that the solar wind that streams around Venus is subject to a large decrease of its momentum flux as its speed U and density n in the wake become significantly smaller than the values measured under freestream conditions (Bridge et al., 1967; Shefer et al., 1979). A summary of those measurements is presented in Figure 7 to show the presence of a velocity boundary layer that extends along the flanks of the Venus wake (indicated by the black shaded region) where both plasma properties exhibit strong deficient values. Such changes are bounded by a plasma transition labeled with the items 2 and 4 along the trajectory of the spacecraft in the lower panel and also at the top of the upper panel, being different from the Venus bow shock encountered at the items labeled 1 and 5 (in the latter case the plasma density is larger in the downstream side as a result of the compressed density values that the solar wind acquires across the bow shock). Much of the missing amount of momentum flux (nU^2) of the solar wind in the wake that is implied from the profiles in Figure 7 between items 2 and 4 has been found

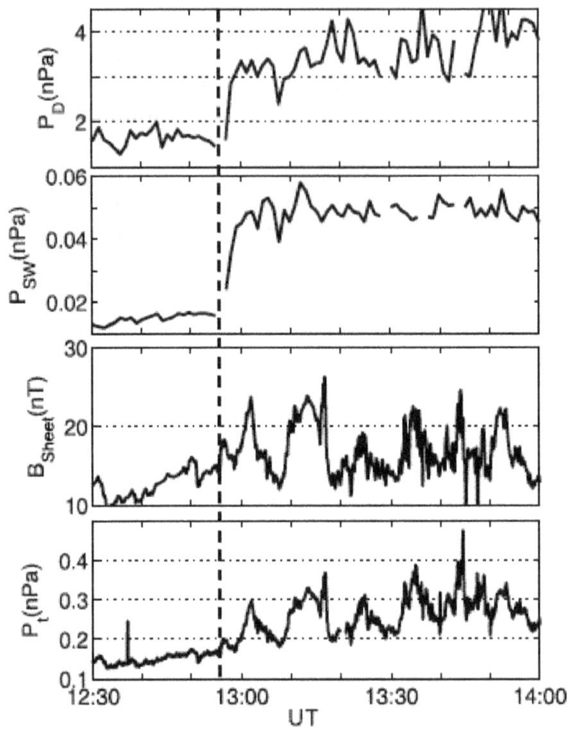

Figure 6. – Pressure variations of the solar wind dynamic pressure (top panel) and its static pressure (second panel), together with the magnetic field intensity (third panel) and the plasma pressure (bottom panel) of the plasma sheet that were measured on July 6, 2003 with a Cluster spacecraft (Figure 12 in Tian et al., 2010).

to be comparable to the momentum flux of the plasma that streams in the trans-terminator flow within the Venus upper ionosphere (Perez-de-Tejada, 1986) and that was measured with the Pioneer Venus Orbiter (PVO) spacecraft (Knudsen et al: 1980). The implication of this agreement is that there is an efficient transport of solar wind momentum to the Venus upper ionosphere that could be accounted for through the onset of viscous forces in a continuum fluid interpretation. As it was indicated before such processes require small scale interactions among the particles of both populations in view that the solar wind is a collisionless plasma. Wave-particle interactions resulting from the turbulence associated with strong fluctuations in the magnetic field that are measured along the flanks of the wake (Bridge et al., 1967; Vőrős et al., 2008) should be ultimately responsible for the erosion of the ionospheric particles produced by the solar wind and, at the same time, for the the Kelvin-Helmholtz instabilities that develop under the measured superalfvenic flow conditions (Pèrez-de-Tejada et al., 2011).

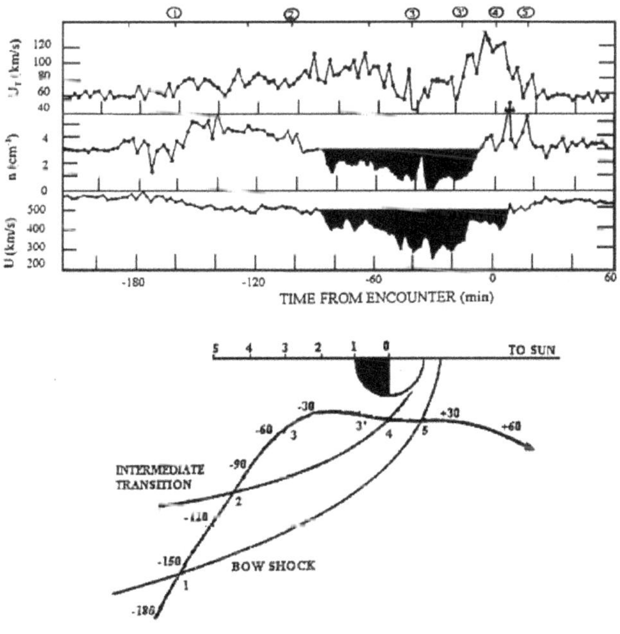

Figure 7. – Thermal speed, density, and bulk speed of the solar wind measured with the Mariner 5 spacecraft (its trajectory projected in cylindrical coordinates is shown in the lower panel). The labels 1 through 5 along the trajectory and at the top of the upper panel mark important events in the plasma properties (bow shock, intermediate plasma transition) (after Bridge et al..1967).

The presence of a velocity boundary layer along the flanks of the Venus wake has been confirmed with observations conducted with the PVO and more recently with the Venus Express (VEX) spacecraft (Pérez-de-Tejada et al., 2011). In the data obtained with both vehicles it has been possible to identify features within that layer that can be interpreted in terms of fluid dynamic processes and, in particular, vortical structures with properties similar to those detected in the earth´s magnetosphere. The first indication on the existence of plasma vortices in the Venus wake was inferred from the observation of changes in the velocity direction of plasma particles along the PVO trajectory across the Venus wake (Pérez-de-Tejada et al., 1982). In some instances the ion fluxes are not directed away from the sun but their motion has a sunward directed component. A description of such change serves to identify the manner in which the flow is arranged to form vortical structures and their position within the wake.

From the collection of plasma ion fluxes measured during the first seasons of observation of the PVO we will first refer to a set of plasma data in which the velocity vectors of the ion fluxes exhibit sudden changes in the direction of the particle motion. Those data are presented in Table I and apply to PVO measurements conducted in orbits 80, 68, 65 and 66, which were conducted in the vicinity of the midnight plane during the first season of observation of that spacecraft along its nearly polar oriented trajectory. The measurements conducted in the energy cycle for each orbit correspond to those obtained near the outbound (southern hemisphere) crossing of the upper boundary of the ionosphere (ionopause) and represent the most intense ion fluxes that were observed during that cycle. Each reading in the first and second columns describes the time (in UT) when it was made together with the number of the corresponding energy step and the volt per charge value (in parenthesis). The count number of the flux intensity (which leads to the intensity values given in the right side column) and the latitudinal collector of the plasma instrument where it was received (marked by the number 3 or 4) are indicated in the third column. These values together with the calculated particle speed if they represent either H+ or O+ ions, and the azimuthal sector (and also the azimuthal angle α), are presented in the fourth through the seventh columns. The most notable example in the data for orbit 80 shown at the top are measurements in which there are fluxes with a sunward directed component (negative α values). These are observed between the energy steps number 11 and 16 (first column), and there are also ion fluxes measured along the solar wind direction (positive α values) between the energy steps number 17 and 21. Comparable variations are also present in the data set of the other orbits with a distribution of the azimuthal α angle that is persistent in orbit 68 (as in orbit 80), and then fluctuating orientations in orbits 65 and 66.

A schematic representation of the direction of the particle fluxes in orbit 80 is described in the upper panel of Figure 8 for 3 different energy cycles of measurements together with the PVO trajectory projected on one quadrant using cylindrical coordinates (the outbound pass with cycles II and III occurs in the southern hemisphere but has been projected to the same quadrant of the inbound pass of cycle I that occurs in the northern hemisphere). The position of the spacecraft during the energy cycle I initiated at 19:34.27 UT in the inbound pass, and the energy cycles II and III initiated at 19:59.16 UT and at 20:08.43 UT in the outbound pass are shown in that figure (each cycle is marked by a rectangular shape). The arrows indicate the latitudinal direction of arrival of the particle fluxes corresponding to the orientation of the collector of the plasma instrument where they were observed (measurements made in collector labeled 4 in Table I during the outbound pass correspond to particle fluxes reaching the most northbound direction detected by the instrument ($22.5^{\circ} < \theta < 69^{\circ}$) and is opposite to those made in collector 1 which would be the most southbound directed collector of the instrument [Intriligator et al., 1980]. During the energy cycle I in the inbound pass and also in the energy cycle III in the outbound pass there is a tendency for the particle fluxes to be directed away from the wake, implying the observation of northbound fluxes in the northern hemisphere (collectors 3 and 4) and also the observation of southbound fluxes in the southern hemisphere (collectors 1 and 2). Different conditions can be identified in the energy cycle II measured in the vicinity of the outbound crossing of the ionopause that is reported in Table I. In this cycle the (dominant) particle fluxes now converge toward the wake (all are detected in collectors 3 or 4) and exhibit directions either with a sunward component (negative α values) or with an anti-sunward component (positive α values). This later variation is indicated by large differences in the azimuthal sector number of the measurements which implies angles that may differ by up to 180º between both cases ($\alpha = 0$ corresponding to the antisolar direction).

Orbit 80 (19:59.16 UT)

energy step (V/q)		count	speed (km/s)		sector	α(°)	ions/cm² s sr
11 (257)	20:01.32 UT	107-4	236	59	119	-161	$3.7\ 10^7$
15 (495)	20:02.22 UT	101-4	308	77	109	-168	$3.2\ 10^7$
16 (583)	20:02.34 UT	101-3	336	84	101	-173	$3.2\ 10^7$
17 (686)	20:02.59 UT	95-4	360	90	367	14	$2.8\ 10^7$
18 (808)	20:03.11 UT	106-4	392	98	369	15	$3.6\ 10^7$
21 (1321)	20:03.48 UT	101-4	504	126	375	19	$3.5\ 10^7$

Orbit 68 (20:16.59 UT)

energy step (V/q)		count	speed (km/s)		sector	α(°)	ions/cm² s sr
1(50)	20:16.59 UT	102-3	96	24	438	65	$3.3\ 10^7$
2(59)	20:17.11 UT	104-4	104	26	430	58	$3.5\ 10^7$
4(82)	20:17.28 UT	107-4	124	31	192	-109	$3.7\ 10^7$
14(420)	20:19.53 UT	96-4	284	71	195	-107	$2.9\ 10^7$
15(495)	20:20.06 UT	97-4	308	77	190	-111	$2.9\ 10^7$
16(583)	20:20.18 UT	98-3	336	84	200	-104	$3.0\ 10^7$

Orbit 65 (20:16.39 UT)

energy step no. (V/q)		count	speed (km/s)		sector	α(°)	ions/cm² s sr
2(59)	20:16.51 UT	105-4	104	26	435	6	$3.5\ 10^7$
3(69)	20:17.04 UT	99-4	116	29	437	63	$3.1\ 10^7$
4(82)	20:17.16 UT	107-4	124	31	176	-120	$3.7\ 10^7$
10(218)	20:19.03 UT	101-4	208	52	185	111	$3.2\ 10^7$
15(495)	20:19.45 UT	113-4	308	77	176	-120	$4.2\ 10^7$
18(808)	20:20.35 UT	98-4	392	98	439	65	$3.0\ 10^7$
21(1122)	20:21.12 UT	100-4	504	126	189	-111	$3.2\ 10^7$
25(2543)	20:22.14 UT	100-4	700	175	449	71	$3.2\ 10^7$
26(2996)	20:22.26 UT	100-4	732	183	195	- 107	$3.2\ 10^7$

Orbit 66 (20:19.25 UT)

energy step (V/q)		count	speed (km/s)		sector	α(°)	ions/cm² s sr
1(50)	20:19.25 UT	107-4	96	24	182	-11	$3.1\ 10^7$
9(185)	20:21.16 UT	116-4	188	47	449	71	$4.6\ 10^7$
11(257)	20:21.41 UT	102-4	220	55	200	-104	$3.3\ 10^7$
14(420)	20:22.18 UT	109-4	284	71	127	-155	$3.9\ 10^7$
17(656)	20:23.08 UT	117-4	364	91	456	76	$4.7\ 10^7$
24(2159)	20:24.35 UT	100-4	644	161	350	2	$3.2\ 10^7$

Table 1. Table I: Values of the ion flux intensity (right side column) measured during an energy cycle in the outbound pass of the VO in orbits 80, 68, 65, and 66, through the Venus near wake. The time when the ion fluxes are measured are given in the second column, and the corresponding energy step number with its volts per charge value (in parenthesis) are presented in the first column. Their count number and the latitudinal collector of the plasma instrument where the fluxes were measured is shown in the third column (fluxes in collectors 3 and 4 are northbound-directed implying that fluxes converge toward the plasma wake in the southern (outbound measured) hemisphere). Their speed for H+ and O+ ions is indicated in the fourth and fifth columns. The azimuthal sector and the corresponding azimuthal angle α where the fluxes were detected are indicated in the sixth and seventh columns (sectors above and below ~347.5 correspond to positive and negative α values where in the later case it implies sunward directed fluxes).

A similar distribution of velocity vectors in the Venus wake is observed in the data of orbit 68 of Table I in energy cycles whose position along the PVO trajectory is reproduced in the lower panel of Figure 8. As it was the case for orbit 80 in the upper panel of that figure the northbound directed diverging particle fluxes are measured in the inbound pass (energy cycle I initiated at 19:53.22 UT), and southbound directed fluxes that also diverge from the wake are observed in the energy cycle III of the outbound pass initiated at 20:26.12 UT.

In addition, particle fluxes measured in the inner ionosheath during cycle II in the outbound pass (initiated at 20:16.59 UT) and that were included in Table I, converge toward the wake and like those in orbit 80 also include velocity vectors with a sunward directed component (most notable flux intensities in the 4 – 16 energy steps) but there are also sporadic weak fluxes with an anti-sunward directed component in the lowest energies that were measured. The tendency in the data set of the outbound pass of orbits 80 and 68 is that the solar wind streaming near the ionopause (cycle labeled II in Figure 8) can acquire an orientation with a sunward directed component leading to a vortex structure (~180° change of the azimuthal α angle in that region).

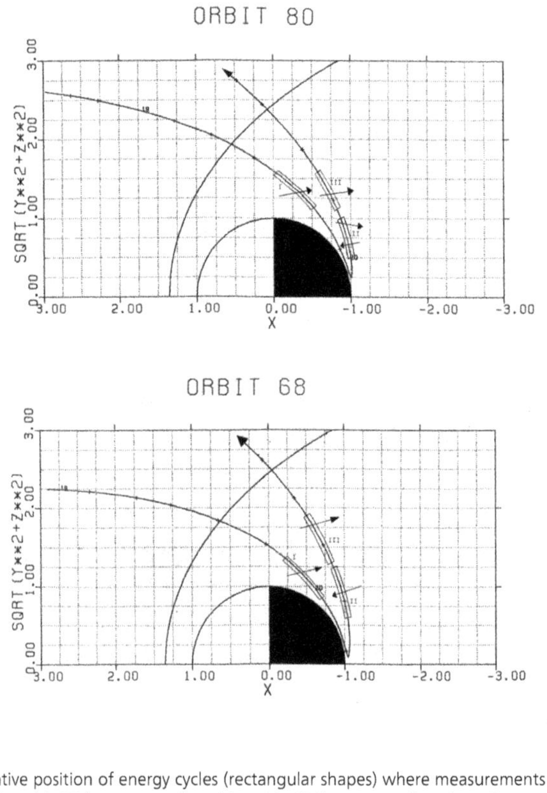

Figure 8. - Representative position of energy cycles (rectangular shapes) where measurements were made along the trajectory of the PVO in orbit 80 (upper panel) and in orbit 68 (lower panel) projected on a quadrant in cylindrical coordinates. The arrows show schematically the (latitudinal) velocity direction of ion fluxes detected at different energy steps within each cycle

4. Plasma vortices in the Venus wake (VEX measurements)

More conclusive evidence in the observation of sunward directed plasma fluxes in the Venus wake has been recently reported from measurements conducted with the ASPERA-4 instrument of the Venus Express spacecraft [Lundin et al., 2011]. From a collection of velocity vectors derived from measurements conducted in 380 orbits it was possible to produce the average pattern of O+ and solar wind (SW) H+ ion velocity vectors presented in Figure 9 which are projected in cylindrical coordinates. Notable is that most of the velocity vectors of the O+ ions are deviated towards the inner wake. On the other hand, for the majority of the solar wind H+ ions the velocity vectors by the central tail at $x < -2R_V$ have a sunward directed component. Such variation mostly occurs in the central wake but in the solar wind H+ ion panel there is also evidence of that behavior in the velocity vectors located downstream from the polar region. The overall pattern of the velocity vectors is consistent with plasma fluxes that reverse direction and agrees with a similar structure that was inferred from the PVO observations; that is, velocity vectors with a sunward component and some directed away from the wake can be separately identified in Figure 9, thus indicating the presence of divergent and sunward directed ion fluxes as it was inferred from the PVO data.

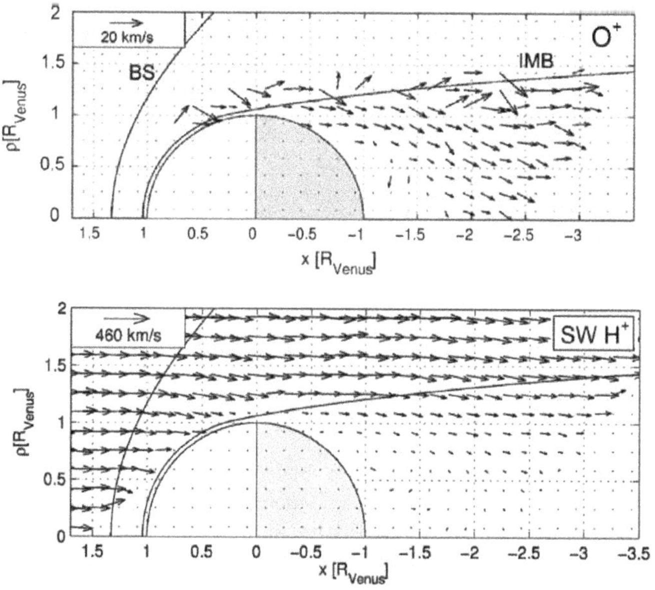

Figure 9. – Velocity vectors of out-flowing O+ (upper panel) and solar wind H+ (lower panel) ions measured in the Venus wake with the ASPERA-4 instrument of the Venus Express spacecraft projected in cylindrical coordinates. The velocity scales for the O+ and the H+ ions are noted in the upper left corner in each diagram. Notice that the H+ flow velocities in the wake are low and barely discernible. However, applying unit vectors we find that most solar wind H+ velocity vectors are sunward directed near the tail central plane at $x < 2\ R_V$.

A more detail description of the distribution of velocity vectors measured with the ASPERA-4 instrument of the VEX spacecraft is presented in Figure 10 with an account of separate observations of the solar wind H+ and the O+ ion fluxes with values now averaged along the z-direction and subsequently projected to the xy plane. Most notable in the left panel of Figure 10 is a pattern located in the central wake within which the velocity vectors of the H+ ions in the dusk side (negative y-values) show a common tendency to be deflected towards the dawn side (positive y-values), and also vectors located at and in the vicinity of the midnight meridian that exhibit a strong solar oriented component. This pattern is consistent with the onset of reverse flow conditions that seem to occur slightly shifted towards the dawn side. As a whole the velocity vectors in that region contain cases in which they are directed anti-sunward, some are deflected toward positive values in the y-axis, and in various observations they are mostly sunward oriented. A scheme with different properties is observed in the distribution of velocity vectors of the planetary O+ ion population shown in the right panel of Figure 10. In this case the velocity vectors across the wake show a more thorough deflection toward the dawn side (positive y-values) but there is little or no indication in the xy-projection (values averaged along the z-axis) that they have a sunward directed component.

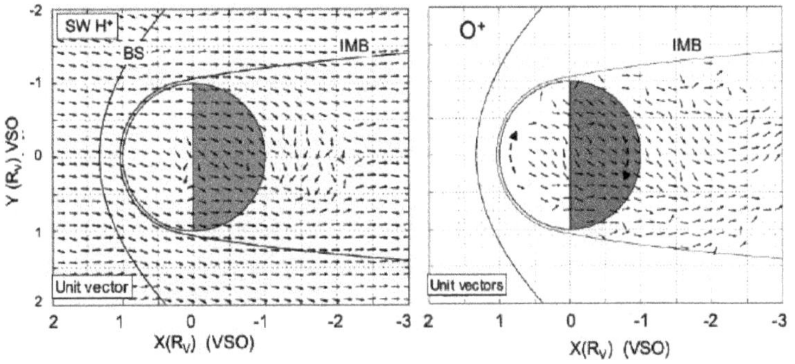

Figure 10. – (left panel) Distribution of velocity vectors of the solar wind H+ ion population in the Venus wake obtained with the ASPERA-4 instrument of the Venus Express projected on the xy plane (the data show evidence of a region in the central wake where the velocity vectors acquire a sunward directed component). (right panel) Distribution of velocity vectors of the O+ ion population indicating that ions in the dusk (upper) side of the retrograde rotating Venus ionosphere are deviated toward the positive (dawn side) y-direction (Lundin et al., 2011).

From the sunward directed fluxes reported in the left panel of Figure 10 the reversed flow direction should mostly occur for the H+ population and, as it is the case for obstacles immersed in directional flows, the solar wind particles in the wake will be forced in the upstream direction leading to a vortex structure. A schematic representation of the distribution of flow streamlines that is compatible with this view is indicated in the left panel of Figure 11 with a circulation pattern in the wake that is consistent with the plasma data. The rotation of the flow direction in the wake is supported by the diagram presented in the right panel of Figure 11 where a wide velocity boundary layer with a thickness that is larger downstream from the terminator stresses the low local flow speeds that should occur near the ionosphere. Such velocity layer derives from the observations shown in Figure 7 and is produced by viscous transport of solar wind momentum to the upper ionospheric plasma. Since low speeds are expected close to the ionosphere the solar wind will reach a position within the wake where the large local flow pressure values will produce a change in its direction thus leading to a vortex structure as that depicted for a flow past an obstacle indicated in the right panel of Figure 11. It is to be noted that the fluid dynamic response of the solar wind to the large pressure values present in the wake is different from processes associated with Kelvin-Helmholtz instabilities along the boundary of the ionosphere and that, as it is the case in the earth's magnetosphere, also contribute to produce vortex structures.

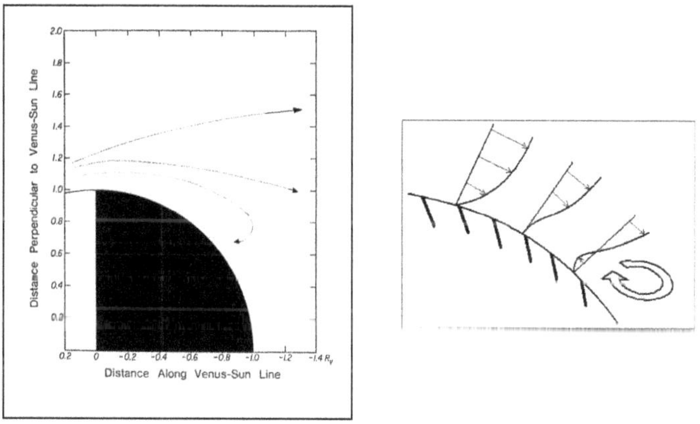

Figure 11. – (left panel) Schematic representation of a vortex structure generated downstream from a polar region of the Venus ionosphere (the streamlines indicate the direction of motion of the solar wind). (right panel) Flow pattern within a velocity boundary layer that extends past an obstacle (higher pressures in the wake reverse the flow direction near the obstacle leading to a vortex structure, Schlichting, 1968).

The fluid dynamic interpretation of both phenomena serves to account for the processes that produce the vortex structures in the earth's magnetosphere and in the Venus plasma wake, and should also be applicable to the conditions that are expected in other planets; namely, those with an appreciable or strong intrinsic magnetic field (Jupiter, Saturn, Uranus, Neptune, and Mercury), or in other un-magnetized planets in which there may only be fossil remnants of an early magnetic field as it is the case in Mars (Acuña et al., 1999).

Author details

H. Pérez-de-Tejada[1*], Rickard Lundin[2] and D. S. Intriligator[3]

1 Institute of Geophysics, UNAM, México, México

2 Institute of Space Physics, Umea, Sweden

3 Carmel Research Center, Santa Mónica, California, USA

References

[1] Acuña, M., et al. (1999). Global distribution of crustal magnetization discovered by the Mars Global Surveyor MAG/ER experiment. *Science*, 790 EOF.

[2] Axford, I., & Hines, C. (1961). A unifying theory of high latitude geophysical phenomena and geomagnetic storms in Canada. *Journal of Physics*,., 39, 1433 EOF-1464 EOF.

[3] Bridge, H., , A., Lazarus, C., Snyder, E., Smith, I., Davies, P., Coleman, , & Jones, D. (1967). Plasma and magnetic fields observed near Venus,. *Science*,, 158, 1669.

[4] Hones Jr, E., et al. (1978). Vortices in magnetospheric plasma flow. *Geophys. Res. Letters*,, 5, 1059 EOF-1062 EOF.

[5] Hones Jr, E., et al. (1981). Further determination of the characteristics of magnetospheric plasma vortices with Isee 1 and 2. *J. Geophys. Res.*,, 86, 814 EOF-820 EOF.

[6] Hones Jr, E., et al. (1983). New observations of plasma vortices and insights into their interpretation. *Geophys. Res. Letters*,, 10, 674 EOF-677 EOF.

[7] Intriligator, D. J., Wolfe, , & Mihalov, J. (1980). The Pioneer Venus Orbiter plasma analyzer experiment. *IEEE Trans, Geosci. Remote Sensing*,, GE-18, 39 EOF-43 EOF.

[8] Knudsen, W., et al. (1980). Transport of ionospheric O^+ ions across the Venus terminator and implications. *J. Geophys. Res.*,, 85, 7803 EOF-7810 EOF.

[9] Lundin, R. S., Barabash, Y., Futaana, J. A., Sauvaud, A., Fedorov, , & Perez-de-Tejada, H. (2011). Ion flow and momentum transfer in the Venus plasma environment. *ICARUS*, 751 EOF-758 EOF.

[10] Pérez-de-Tejada, H., Gay-García, C., Intriligator, D., & Dryer, M. (1982). Plasma vortex in the Venus near wake,. *EOS*, 63(18), 368.

[11] Pérez-de-Tejada, H. (1986). Fluid dynamic constraints of the Venus ionospheric flow. *J. Geophys. Res.*,, 91, 6765 EOF-6770 EOF.

[12] Pérez-de-Tejada, H., Lundin, R., Barabash, S., Zhang, T. L., Coates, A., Sauvaud, J. A., Durand-Manterolla, H., & Reyes-Ruiz, M. (2011). Plasma transition along the flanks of the Venus ionosheath: Evidence from the Venus Express plasma data,. *Journal of Geophys. Res.*,, 116, A01103, doi: JA015216.

[13] Pérez-de-Tejada, H. (2012). Fluid dynamics in space sciences,. chapter 28, 611, *Fluid dynamics, computational modeling and applications*,, Hector Juarez, Ed., INTECH,, 978-9-53510-052-2, open access.

[14] Pope, S., et al. (2009). Giant vortices lead to ion escape from Venus and redistribution of plasma in the ionosphere,. *Geophys. Res. Lett.*, 36, L07202.

[15] Schlichting, H. (1968). Boundary Layer Theory,. McGraw Hill,, 40

[16] Shefer, R., Lazarus, A., & Bridge, H. (1979). A re-examination of plasma measurements from the Mariner 5 Venus encounter,. *J. Geophys. Res.,*, 84, 2109.

[17] Tian, A., et al. (2010). A series of plasma flow vortices in the tail plasma sheet associated with solar wind pressure enhancement. *J. Geophys. Res.*, 115, A09204, doi: JA014989.

[18] Terada, N., Machida, S., & Shinagawa, H. (2002). Global hybrid simulation of the Kelvin-Helmholtz instability at the Venus ionopause,. *J. Geophys. Res.,*, 107, 14.

[19] Wolfe, R., et al. (1980). The onset and development of Kelvin-Helmholtz instability at the Venus ionopause,. *Journal of Geophys. Res.,*, 85, 7697-7707.

[20] Vőrős, Z., et al. (2008). Intermittent turbulence, noisy fluctuations, and wavy structures in the Venusian magnetosheath and wake,. *Journal of. Geophys. Res.-Planets,*, 113, E00B21, doi:10.1029/2008.

Permissions

The contributors of this book come from diverse backgrounds, making this book a truly international effort. This book will bring forth new frontiers with its revolutionizing research information and detailed analysis of the nascent developments around the world.

We would like to thank Gonzalo J. Olmo, for lending his expertise to make the book truly unique. He has played a crucial role in the development of this book. Without his invaluable contribution this book wouldn't have been possible. He has made vital efforts to compile up to date information on the varied aspects of this subject to make this book a valuable addition to the collection of many professionals and students.

This book was conceptualized with the vision of imparting up-to-date information and advanced data in this field. To ensure the same, a matchless editorial board was set up. Every individual on the board went through rigorous rounds of assessment to prove their worth. After which they invested a large part of their time researching and compiling the most relevant data for our readers. Conferences and sessions were held from time to time between the editorial board and the contributing authors to present the data in the most comprehensible form. The editorial team has worked tirelessly to provide valuable and valid information to help people across the globe.

Every chapter published in this book has been scrutinized by our experts. Their significance has been extensively debated. The topics covered herein carry significant findings which will fuel the growth of the discipline. They may even be implemented as practical applications or may be referred to as a beginning point for another development. Chapters in this book were first published by InTech; hereby published with permission under the Creative Commons Attribution License or equivalent.

The editorial board has been involved in producing this book since its inception. They have spent rigorous hours researching and exploring the diverse topics which have resulted in the successful publishing of this book. They have passed on their knowledge of decades through this book. To expedite this challenging task, the publisher supported the team at every step. A small team of assistant editors was also appointed to further simplify the editing procedure and attain best results for the readers.

Our editorial team has been hand-picked from every corner of the world. Their multi-ethnicity adds dynamic inputs to the discussions which result in innovative outcomes. These outcomes are then further discussed with the researchers and contributors who give their valuable feedback and opinion regarding the same. The feedback is then collaborated with the researches and they are edited in a comprehensive manner to aid the understanding of the subject.

Apart from the editorial board, the designing team has also invested a significant amount of their time in understanding the subject and creating the most relevant covers. They scrutinized every image to scout for the most suitable representation of the subject and create an appropriate cover for the book.

The publishing team has been involved in this book since its early stages. They were actively engaged in every process, be it collecting the data, connecting with the contributors or procuring relevant information. The team has been an ardent support to the editorial, designing and production team. Their endless efforts to recruit the best for this project, has resulted in the accomplishment of this book. They are a veteran in the field of academics and their pool of knowledge is as vast as their experience in printing. Their expertise and guidance has proved useful at every step. Their uncompromising quality standards have made this book an exceptional effort. Their encouragement from time to time has been an inspiration for everyone.

The publisher and the editorial board hope that this book will prove to be a valuable piece of knowledge for researchers, students, practitioners and scholars across the globe.

List of Contributors

Emilio Elizalde
Consejo Superior de Investigaciones Científicas, ICE/CSIC and IEEC Campus UAB, Facultat de Ciències, Spain

Luigi Toffolatti
Department of Physics, University of Oviedo, Oviedo, Spain
IFCA-CSIC, University of Cantabria, Santander, Spain

Carlo Burigana
National Institute of Astrophysics - Institute of Space Astrophysics and Cosmic Physics (INAF-IASF), Bologna, Italy
Department of Physics, University of Ferrara, Ferrara, Italy

Francisco Argüeso
IFCA-CSIC, University of Cantabria, Santander, Spain
Department of Mathematics, University of Oviedo, Oviedo, Spain

José M. Diego
IFCA-CSIC, University of Cantabria, Santander, Spain

Adrienne Leonard, Jean-Luc Starck and Sandrine Pires
Laboratoire AIM (UMR 7158), CEA/DSM-CNRS-Université Paris Diderot, IRFU, SEDI-SAP, Service d'Astrophysique, Centre de Saclay, France

François-Xavier Dupé
Laboratoire d'Informatique Fondamentale, Centre de Mathématique et d'Informatique, UMR CNRS, Aix-Marseille Université, Marseille, France

L. Fatibene and M. Francaviglia
Department of Mathematics, University of Torino, Italy
INFN- Sezione Torino, Iniziativa Specifica Na12, Italy

Katie Auchettl and Csaba Balázs
Monash University, Australia

Sergio Mendoza
Instituto de Astronomia, Universidad Nacional Autónoma de Mexico, Ciudad Universitaria, Distrito Federal CP 04510, Mexico

Salvador J. Robles Pérez
Grupo de Relatividad y Cosmología, Instituto de Física Fundamental - Consejo Superior de Investigaciones Científicas (IFF-CSIC), Madrid, Spain, and Estación Ecológica de Biocosmología (EEBM), Medellín, Spain

Gonzalo J. Olmo
Departamento de Física Teórica & IFIC, Centro Mixto Universidad de Valencia & CSIC, Facultad de Física, Universidad de Valencia, Burjassot, Valencia, Spain

C. Pallis and N. Toumbas
Department of Physics, University of Cyprus, Nicosia, Cyprus

Paulo A. Rodríguez, O. Núñez-Soltero, Rafael Hernández and Abraham Espinoza-García
Departamento de Física de la DCeI de la Universidad de Guanajuato-Campus León, Guanajuato, México

J. Socorro
Departamento de Física de la DCeI de la Universidad de Guanajuato-Campus León, Guanajuato, México
Departamento de Física, Universidad Autónoma Metropolitana, Apartado Postal 55 534, C.P. 09340 México, DF, México

Abdessamad Abada
Laboratoire de Physique des Particules et Physique Statistique, Ecole Normale Supérieure, BP 92 Vieux Kouba, Alger, Algeria

Salah Nasri
Physics Department, United Arab Emirates University, POB, Al Ain, United Arab Emirates

Brian P. Dolan
Department of Mathematical Physics, National University of Ireland Maynooth, Maynooth, Co. Kildare, Ireland
School of Theoretical Physics, Dublin Institute for Advanced Studies, Dublin, Ireland

H. Pérez-de-Tejada
Institute of Geophysics, UNAM, México, México

Rickard Lundin
Institute of Space Physics, Umea, Sweden

D. S. Intriligator
Carmel Research Center, Santa Mónica, California, USA

www.ingramcontent.com/pod-product-compliance
Lightning Source LLC
Chambersburg PA
CBHW070727190326
41458CB00004B/1062
* 9 7 8 1 6 3 2 3 9 3 1 8 0 *